STATISTICS:
A Conceptual Approach

STATISTICS:
A Conceptual Approach

K. LAURENCE WELDON

Simon Fraser University

Prentice-Hall, Inc., Englewood Cliffs, New Jersey 07632

Library of Congress Cataloging in Publication Data

Weldon, K. L. (Kenneth Laurence), (date)
   Statistics, a conceptual approach.

   Bibliography: p.
   Includes index.
   1. Statistics. I. Title.
QA276.12.W456 1986   519.5   85-9549
ISBN 0-13-845819-7

Editorial/production supervision: *Kathleen Lafferty*
Interior design: *Karen J. Clemments*
Editorial assistant: *Susan Pintner*
Cover design: *Photo Plus Art*
Manufacturing buyer: *John B. Hall*
Cover illustration: This diagram shows that sample averages vary less than the population sampled, and that they tend to be close to the population average. The top four levels show four random samples of size five from a particular population. The fulcrum (the triangle) under each level shows the position of the sample average. The sample averages are shown by themselves on the bottom level, and the fulcrum under this bottom level is the average of all 20 values. The vertical strip indicates the position of the population average.

© 1986 by Prentice-Hall, Inc., Englewood Cliffs, New Jersey 07632

*All rights reserved. No part of this book
may be reproduced, in any form or by any means,
without permission in writing from the publisher.*

Printed in the United States of America

10 9 8 7 6 5 4 3 2 1

ISBN 0-13-845819-7 01

Prentice-Hall International (UK) Limited, *London*
Prentice-Hall of Australia Pty. Limited, *Sydney*
Prentice-Hall Canada Inc., *Toronto*
Prentice-Hall Hispanoamericana, S.A., *Mexico*
Prentice-Hall of India Private Limited, *New Delhi*
Prentice-Hall of Japan, Inc., *Tokyo*
Prentice-Hall of Southeast Asia Pte. Ltd., *Singapore*
Editora Prentice-Hall do Brasil, Ltda., *Rio de Janeiro*
Whitehall Books Limited, *Wellington, New Zealand*

To the memory of my grandfather,
Claris Edwin Silcox

Contents

Preface   xv

1 Introduction to Statistics   1

- **1.1** Some Examples of Statistical Problems   1
  - *1.1.1*   Survey of higher-education intentions   1
  - *1.1.2*   Weather forecasting and public lotteries   3
  - *1.1.3*   Public opinion polls   5
  - *1.1.4*   Evaluation of vitamin C therapy for colds   6
  - *1.1.5*   The baby boom and the echo boom   6
  - *1.1.6*   Lecturer effectiveness and ratings of instruction   8
  - *1.1.7*   Computers, statistics, and management of airlines   8
- **1.2** The Training Required to Solve Statistical Problems   10
- **1.3** Statistics: Past, Present, and Future   12
- **1.4** Answers to Exercises   14
  - Problems   15

2 Study Design I: Logical Keys to Statistical Thinking — 18

- **2.1** Search and Research in Everyday Living — 19
- **2.2** Learning the Logic of Numerical Assessment — 20
- **2.3** A First Look at Sampling Strategies — 21
  - *2.3.1 Simple random sampling* — 22
  - *2.3.2 More complex sampling strategies: clusters and strata* — 25
- **2.4** Inferences from Comparisons — 29
- **2.5** Experiments and Surveys — 31
- **2.6** Special Problems of Studies with Human Subjects — 38
- **2.7** Sources of Variation in Measurements — 41
- **2.8** Answers to Exercises — 44
- **2.9** Summary and Glossary — 48
- Problems and Projects — 50

3 Study Design II: Principles of Experimentation — 54

- **3.1** The Terminology of Experimentation — 54
- **3.2** Use of Averaging to Reduce Confusion — 57
- **3.3** Blocking, for More Precise Comparisons — 59
- **3.4** Factorial Designs — 63
- **3.5** Answers to Exercises — 65
- **3.6** Summary and Glossary — 69
- Problems and Projects — 70

4 Descriptive Methods for One Variable — 73

- **4.1** Introduction — 74
- **4.2** Measurement — 74
  - *4.2.1 Measurements and instruments* — 74
  - *4.2.2 Measurements of quantity and quality* — 75
  - *4.2.3 Precision, bias, and accuracy* — 76

4.3	Frequency Distributions and Histograms 80	
4.4	Summary Measures: Average and Standard Deviation 91	
	4.4.1 Measures of center 93	
	4.4.2 Measures of spread 96	
	4.4.3 Some invariance properties of descriptive summaries 100	
4.5	Normal Approximation to Frequency Distributions 103	
	4.5.1 Introduction to the normal curve 103	
	4.5.2 Table of normal percents and percentiles 107	
	4.5.3 Description of nonnormal frequency distributions 111	
4.6	Answers to Exercises 114	
4.7	Notation and Formulas 121	
	4.7.1 About mathematical notation in statistics 121	
	4.7.2 Notation for this chapter 122	
4.8	Summary and Glossary 123	
	Problems and Projects 125	

5 Descriptive Methods for Two Variables   132

5.1	Some Preliminaries 133
	5.1.1 Terminology 133
	5.1.2 Scatter diagrams 134
	5.1.3 Prediction, association, and causation 137
5.2	Correlation 140
5.3	Regression 150
	5.3.1 Regression as prediction 150
	5.3.2 Use of averages for prediction 151
	5.3.3 The SD line and the regression line 155
	5.3.4 Prediction error with the regression method 157
	5.3.5 Assumptions of the regression method 159
	5.3.6 Causality in regression 162
	5.3.7 Regression pitfalls 164
5.4	Other Methods for Quantitative Variables 168
	5.4.1 Graphical display of three or more variables 168
	5.4.2 Multivariable strategies 172
	5.4.3 Adjustment of averages 173
5.5	Methods for Qualitative Variables 175
	5.5.1 Contingency tables 175
	5.5.2 The histogram 178

x   Contents

5.6	Answers to Exercises	180
5.7	Notation and Formulas	185
5.8	Summary and Glossary	186
	Problems and Projects	189

6 Descriptive Methods for Time Series — 199

6.1	Extraction of Trends and Seasonal Effects	200
6.2	Patterns of Residuals	209
6.3	Answers to Exercises	215
6.4	Notation and Formulas	216
6.5	Summary and Glossary	217
	Problems and Projects	218

7 Sampling and Probability — 221

- 7.1 Introduction to Inferential Statistics   222
- 7.2 Probability Modeling   222
  - 7.2.1 Statistical populations   222
  - 7.2.2 Variation in samples   225
  - 7.2.3 Sampling models   227
- 7.3 Determination of Probabilities   228
  - 7.3.1 The notion of probability   228
  - 7.3.2 Counting the ways   230
  - 7.3.3 Combination of probabilities   233
  - 7.3.4 Probability models   237
  - 7.3.5 Probabilities for binomial experiments   239
- 7.4 Estimation of Parameters   243
  - 7.4.1 Parameters and statistics   243
  - 7.4.2 The expected value and the standard error   245
  - 7.4.3 The standard error in sampling without replacement   254
  - 7.4.4 Stratified sampling and cluster sampling   256
- 7.5 Probabilities from a Normal Approximation   259
  - 7.5.1 Normal probabilities for averages   259
  - 7.5.2 Other normal approximations   267

7.6	Nonparametric Probabilities	270
7.7	Answers to Exercises	271
7.8	Notation and Formulas	277
7.9	Summary and Glossary	279
	Problems and Projects	282

8 Estimation 290

8.1	Introduction	290
8.2	Estimation of Population Averages	291
8.3	Estimation of Population Percentages	299
8.4	Determination of Sample Size	303
	8.4.1 *Sample size for estimating population percentages*	*303*
	8.4.2 *Sample size for estimating population averages*	*305*
8.5	Small-Sample Estimation of Population Averages	307
8.6	Estimation of Probability Distributions	310
8.7	Answers to Exercises	312
8.8	Notation and Formulas	315
8.9	Summary and Glossary	316
	Problems and Projects	317

9 Testing Hypotheses with Data 324

9.1	The Logic of Hypothesis Testing	324
9.2	Tests Concerning Averages	333
	9.2.1 *Normal tests for an average*	*333*
	9.2.2 *The t test for an average*	*335*
9.3	Tests Concerning Percentages	338
	9.3.1 *The normal test for a percentage*	*338*
	9.3.2 *The binomial test for a percentage*	*339*
9.4	Two-Sample Tests	340
	9.4.1 *Normal and t tests for averages*	*340*
	9.4.2 *2 by 2 contingency table tests*	*344*
	9.4.3 *Dependent samples*	*349*

xii Contents

9.5	Many-Sample Tests 352	
	9.5.1 Analysis of Variance 352	
	9.5.2 Chi-square tests 358	
9.6	Nonparametric Tests: An Introduction 362	
	9.6.1 One-sample median test 362	
	9.6.2 Two-sample tests: the median test and the sign test 363	
9.7	The Logic of Hypothesis Testing, Revisited 366	
9.8	Answers to Exercises 370	
9.9	Notation and Formulas 378	
9.10	Summary and Glossary 380	
	Problems and Projects 383	

Appendix 1 Use of Computers for Statistical Tasks 391

A1.1 Introduction to the Use of Packages of Statistical Computer Programs 392

A1.2 Statistical Experimentation: Monte Carlo Simulation 395

Appendix 2 List of Applications 399

A2.1 Sports and Entertainment 399
A2.2 Social Sciences and Business 400
A2.3 Life Sciences and Medicine 401
A2.4 Natural Sciences 402
A2.5 General 402

Appendix 3 Table of Random Digits 404

Appendix 4 Probability Tables 406

A4.1 Probabilities for the Normal Distribution 406
A4.2 Probabilities for the Binomial Distribution 408
A4.3 Probabilities for the t Distribution 412
A4.4 Probabilities for the Chi-Square Distribution 414
A4.5 Probabilities for the F Distributions 416

Sources and Notes **418**

Annotated Bibliography **423**

Solutions to Selected Problems **425**

Index **436**

Preface

Give a man a fish, and he will eat for a day. But teach a man to fish, and he will eat for the rest of his days.

Anonymous

Common sense is not so common.

Voltaire (1694–1778)

THE APPROACH: CONCEPTS WITH WORDS AND TECHNIQUES BY EXAMPLE

Statistics: A Conceptual Approach provides an introduction to statistics for those who prefer words to mathematical symbols. There are some subtle advantages to using words. One advantage is that students who can express verbally what they learn about statistics will be able to relate this knowledge to the material they learn in language-based disciplines, that is, in almost all disciplines except the "hard" sciences and mathematics itself. Another advantage of the verbal approach is that it relates directly to applied contexts: practical problems requiring elementary statistical techniques usually appear as verbal descriptions of data-based studies or study plans. Finally, an obvious advantage of the verbal approach is that it makes the widely applicable principles of statistics accessible to students who do not have any prior college-level mathematical training.

The approach is to emphasize the absorption of mental tools for thinking about statistical problems, rather than the memorization of mathematical formulas and calculation rituals. The text even encourages some skepticism of statistical dogma; uncritical acceptance of statistical procedures is as dangerous as wholesale acceptance of them. For different reasons the approach of this book is inappropriate for statistical clerks and for mathematicians. It is intellectually more demanding than a how-to-calculate manual, and it works hard to introduce the ideas clearly without resorting to mathematical symbolism.

For many years elementary statistics for nonmathematicians was taught as a "how-to-calculate" subject. It was widely believed that mathematical training was absolutely necessary to understand even elementary concepts of statistics. The landmark development that changed this was the appearance of the textbook *Statistics* by Freedman, Pisani, and Purves in 1978. This book showed that a rigorous development without mathematical symbolism was possible. They replaced the space usually devoted to the complexities of hand calculations with discussions of the rationale of the methods and their applicability. This text is a continuation of their initiative.

Today, there are computers to do the arithmetic of statistics, and relatively few people who know what procedures should be done, or what conclusions are justified, in statistical applications. Learning the strategies of statistical modeling, developing good judgment in the summarization of data, and becoming aware of the subtle hazards of numerical comparisons seem more important activities than learning how to perform the calculations. The goal in this text is to expose the big ideas of statistics, and to demonstrate some practical tools in the process.

ABOUT "EXERCISES" AND "PROBLEMS"

The problems play a key role in the text. Problems often require the student to recognize a concept they have learned from an earlier chapter, or require a combination of common sense and technical knowledge. In many of the problems, the aim is to simulate the conditions a student might find in the real world. Other problems explore ideas not covered explicitly in the text. All are intended to enrich the student's mental models of statistical phenomena. Problems appear at the end of each chapter and depend on the material of all the chapters covered up to that point. Solutions to selected problems are included in the text to demonstrate the style of acceptable solutions. However, a majority of problems are unanswered; the mental processes used to sort out novel statistical situations must be learned by firsthand experience.

On the other hand, answering questions is useful for the learning of particular definitions and procedures. This kind of question is termed an "exercise," and exercise sets appear at the end of each section in each chapter. Answers to all exercises are included.

Some specially grouped problems require the use of a statistical computer package (such as BMDP, MINITAB, SAS, or SPSS). Such problems appear under the heading "Computer Projects" following the regular problems. Although only a few students will have this computer-package familiarity at the time they are first exposed to statistics, the inclusion of such problems may motivate students to gain or improve their computing skills. The charm of viewing the operation of chance in building up a frequency distribution has great pedagogic value.

In courses where very few students have computing skills, a lecturer can use the computing exercises for lecture demonstrations. In courses with a mixture of students, the computer problems can be looked upon as enrichment materials.

There is another option for computer use with this text: it is for the student to cover the material in Appendix A1 early in the course, perhaps after Chapter 1, or alternatively just after Chapter 6. This appendix instructs students on how to use a package of statistical programs to do their statistical calculations, and also discusses the mechanics of Monte Carlo simulation. Of course, the student will need some instruction on the details of the package that is locally available, and will need to obtain access to the computer system. This information is too varied to include in the Appendix, and will depend on the guidance of the instructor. With a good interactive statistical package, students should be able to learn to use the elementary parts of a statistical package, such as constructing histograms or calculating averages and standard deviations, in a very few hours of lecture time. Whether students are able to use the computer for both the exercises and the problems, or for the designated computer problems only, or not at all, will depend on local facilities and preferences.

THE STYLE OF THE EXPOSITION

It is not easy for students to learn how to apply statistical ideas and avoid misuses of statistics in one or two elementary courses. A text must exhort the student to invest quite a bit of intellectual effort. In conventional lecture courses, this motivation task is usually undertaken by the instructor. However, in the self-instruction mode, the text must be relied on to provide encouragement and feedback. This is one reason for the inclusion of learning exercises and for the conversational style of the text. The text is intended as a teaching aid rather than as a reference text.

The pattern for the introduction of new material is to introduce each technique as an example with a loosely worded query; the query is then developed and formulated into an unambiguous question and a numerical procedure. It is felt that this approach parallels both the historical development of statistics and the usual scenario of real-life applications of statistical methods.

The book has been organized to assist recognition of application categories, as much as possible, while still keeping the order logical from a pedagogic standpoint.

NOTES ON THE COVERAGE OF THE TEXT

The conflict between the mathematical basis of statistical methods and the non-mathematical complications of many applications is recognized in the text. This conflict is evident particularly in the discussions of sampling models, study design, and hypothesis testing.

Mathematical notation is kept to a bare minimum. However, since a few students find the mathematical notation clarifying, separate sections on notation are included *after* the concept has been introduced. The concepts are often better learned without the notation, even for those who are comfortable with mathematical symbolism. The instructor can choose the degree of emphasis to give to the notation; it is possible to avoid it almost entirely. Students who expect to take subsequent courses in statistics, in techniques such as multiple regression and analysis of variance, have the option of coming to grips with the notation in this first course.

Descriptive statistics has been given more emphasis than in some other texts. In industrial and business applications, by far the greatest use of statistics is for purely descriptive purposes. Certainly, most exploratory studies begin with descriptive summaries. Consider, for example, the collective personnel time involved in the construction and interpretation of time charts of stock prices, interest rates, and regional unemployment rates. There appears to be a great need for methods to describe time-series data. Although analysis of time series can be a very complex subject, the descriptive techniques that are commonly used can be simply outlined. This descriptive approach to time series is included in a chapter that is optional in that subsequent material does not depend on it. A different area of descriptive statistics that is covered in this text is the adjustment of data for a fair comparison between groups. Again, the descriptive aspects of this complex subject are simple enough for a first course.

On the other hand, the plethora of avant-garde descriptive methods have been avoided. It is felt that wise use of such techniques (for example, distribution summary by hinges and midhinges, the use of families of transformations, the details of "hanning" and "splitting" in smoothing time series) requires more mathematical sophistication than may be expected of students in their first course in statistics.

An attempt has been made at a serious presentation of the relationship between probability models and the applied settings of statistics. Probability techniques are motivated by sampling and inference problems. The meaning of "population" in the measurement context is explained. The notion of variation among population items is distinguished from measurement variation and variation due to sample selection. It is felt that although these issues make the material more complex, they also make it more applicable.

The text includes a few topics that are unusual in an elementary text. Students in the social sciences will likely meet, in the required readings in their field, references to factor analysis, analysis of covariance, multiple regression, Monte Carlo simulation, and star plots. Although a full coverage of these terms is impossible within the scope of this text, a hint is provided of their meaning and utility in applications.

An optional chapter on the structure of designed experiments is included. Only the most basic ideas are covered, and these in the most elementary way possible: blocking, randomization, replication, factorial designs, interaction, and confounding. This chapter is most appropriate for students in fields such as biology or psychology, where data collection by designed experiments is feasible.

The text covers enough material to accompany between 45 and 65 hours of instruction. The depth of coverage is dependent on the emphasis given to problems. Some selection would be required for the usual one-semester course, and some amplification, possibly the inclusion of instruction with statistical computer packages, for a two-semester course. The logical relationships between the chapters are displayed schematically below.

Logical dependence of chapters.

ACKNOWLEDGMENTS

I would like to thank Bob Sickles and Shelley Duke of Prentice-Hall for their initial faith in this project; the following reviewers, Robert Berk, Rutgers University; William I. Notz, Purdue University; Galen R. Shorack, University of Washington; Bette Warren, State University of New York, Binghamton; and Douglas A. Wolf, Ohio State University; my friends John Weymouth and John MacKay, for their detailed and constructive critiques of early drafts of the manuscript; my colleagues Jonathan Berkowitz, David Eaves, Richard Lockhart, Richard Routledge, Michael Stephens, and Cesareo Villegas for their criticism and encouragement; students Gordon Brown, Mark Donnelly, Oonagh Enright and Jarek Kulikowski for their criticisms of the exercises and problems; editors Karen Clemments and Kathleen Lafferty for their cheerful communications and professional expertise; Eloise Starkweather for persisting with my exacting specifications for the cover design; Adam Osborne, whose ingenuity advanced the production of the initial manuscript by several years; and especially my wife, Jill, for tolerating my single-minded schedule and for editing all the manuscript several times.

K. Laurence Weldon

Introduction to Statistics

Statistical thinking will one day be as necessary for efficient citizenship as the ability to read and write.

H. G. Wells (1866–1946)

Example is always more efficacious than precept.

Samuel Johnson (1709–1784)

This is an orientation chapter. You will become aware of the very broad range of problems to which statistical techniques apply. You will be encouraged to work hard at understanding the underlying principles of statistics, as this understanding is necessary for the subject to have practical value to you. The reasons for the expanding role of statistics in modern careers will be outlined. Examples introduced here are referred to in subsequent chapters.

1.1 SOME EXAMPLES OF STATISTICAL PROBLEMS

This section has a dual purpose: first, to introduce some typical examples of problems that have been addressed by the techniques of statistics; and second, to begin the fairly complex task of providing a classification of statistical methods. We hope to demolish the myth that to study statistics is to study facts: the subject is more concerned with methods to detect facts rather than the facts themselves. This study of methods requires a mental filing system for the many categories of statistical problems. This system is not only a useful aid to memory but is essential for intelligent application of statistical methods.

1.1.1 Survey of Higher-Education Intentions

Institutions of higher education have had to adapt to major shifts in enrollments in recent decades. Numbers of applicants have surged and plunged, and technological programs have expanded more rapidly than the more general programs. The effects of World War II, the postwar baby boom, the Sputnik-induced increased interest in science, the computer revolution, and economic recessions have all caused major shifts in the size and type of enrollment in colleges and universities. Of course, major shifts are more clearly discerned a year or two after they have begun, and our institutions of higher education have had difficulty adapting quickly enough to these changes.

To anticipate changes in higher-education enrollments senior high school students have been surveyed regarding their education intentions.[1] Reproduced in Figure 1.1 is a very small portion of the questionnaire in one such study. Question 28 attempts to determine one aspect of the socioeconomic background of high school students. Question 29 seeks information on the motivating factors likely to guide a student's higher-education decisions. Such questions are obviously of interest to administrators of higher-education institutions.

28. Below is a list of different types of locations. Please indicate with the appropriate code how much of your life has been spent in these areas.

   (a) Major metropolitan centres .... [ ] 112
   (b) Major regional towns .... [ ] 113
   (c) Small towns .... [ ] 114
   (d) Unorganized rural areas and/or small remote communities (farming areas, isolated logging/mining centres, fishing villages, etc.) .... [ ] 115

   Code: 1. Never   4. 4-10 years
          2. Less than one year 5. More than 10 years
          3. 1-3 years

29. Below is a list of possible functions that the post-secondary education system may serve. Please indicate the relative importance to you of each by writing the appropriate code in each box.

   (a) To increase my general level of education .... [ ] 116
   (b) To allow me to meet more people .... [ ] 117
   (c) To prepare me for a job or career .... [ ] 118
   (d) To increase my level of income .... [ ] 119
   (e) To make me a more well-informed citizen .... [ ] 120
   (f) To satisfy my personal curiosity about a particular subject .... [ ] 121
   (g) To provide me with more opportunities for recreation and/or social activities .... [ ] 122
   (h) To make me a more knowledgeable and mature adult .... [ ] 123
   (i) To give me a more personal independence and a wider choice of occupations .... [ ] 124

   Code: 1. Very important   3. Somewhat important
          2. Important   4. Not important

Figure 1.1 Excerpt from a questionnaire used in a survey of senior high school students.[1]

If you were the person responsible for this survey you would want to determine answers to the following questions:

1. Which students should I ask to answer this questionnaire? and how many students?

2. When should I ask the students to fill in the questionnaire?
3. What is the most efficient way to distribute and collect the questionnaires?
4. How will I aggregate the information from the questionnaires (both physically and arithmetically)?
5. Will I be able to make any valid claim about the accuracy of the results?

To be able to answer the fifth question affirmatively, the other questions will have to be given some very careful thought. In fact, these questions are all statistical questions, and you will learn how to approach such questions in this first statistics course. Many of these issues are discussed in Chapter 2. The details of the calculations are part of the subject matter of Chapter 4. Yet descriptive methods are not merely calculations. For example, how would you display the relationship between high school grade-point average and the intention to enroll at a university? Graphical displays are an important part of descriptive methods and are discussed in Chapters 4, 5, and 6. Again, how would you combine the results of two surveys done in different years to predict a student's degree of success at college, based on data obtained from the student while in high school? One technique for this is regression analysis, discussed in Chapter 5.

1.1.2 Weather Forecasting and Public Lotteries

Weather forecasting and public lotteries are two areas where "chance" mechanisms appear to be operating. In the case of the weather, it is not clear who is rolling the dice. In lotteries, the chance mechanisms can be more completely understood. In each example the concept of probability is the key to a useful study of the phenomenon.

If you are planning a day of rest and relaxation at the beach, you will be very interested in the weather forecast. You know that the forecast is not always correct, but a statement such as "Today there is only a 10 percent chance of rain" would certainly be comforting. It is not even necessary to have a clear idea of what exactly is meant by "10 percent chance" to make use of this forecast. On the other hand, if you were thinking of building a resort on the coast of the Gulf of Mexico, and you were informed that the resort had a 10 percent chance of being destroyed by a hurricane during the next five years, you would treat this information as a serious matter: you would want to know how reliable the estimate is and just exactly what it means. See Figure 1.2. What would hurricane insurance cost? Could you work out the chance that a hurricane would destroy your investment within one year?

You may be able to hire an expert to advise you on these matters of chance, but will you know what questions to ask? Will the "expert" really understand *your* concerns? It helps to have a clear understanding of the ways "chances" or "probabilities" are derived, in order to assess properly potential applications.

Many countries use public lotteries as a way of producing revenues for government-funded operations. One such is called Lotto 6/49.[2] When a ticket is purchased, the buyer chooses six different numbers, each of them in the range 1, 2, ..., 49. A ticket costs $1, and it is usual for about 5 million tickets to be sold for

Figure 1.2 Sometimes the weather is important.

the weekly draw. Any holders of a particular combination of six numbers chosen by computer will split a grand prize of about $1 million. It is quite likely that there is no winner: about 14 million different combinations are possible. When there is no grand prize winner, this prize money is added to the next round.

In January 1984 the prize money accumulated to $14 million. Seventy million $1 tickets were sold for this particular issue of the lottery. Why did so many people want tickets for this issue?

Presumably, each ticket was worth $1 to the purchaser at the time it was purchased. The *average* value of a ticket after the lottery is usually 45 cents. (Most tickets are worth nothing, but a few are worth a great deal.) Each ticket that is purchased contributes only 45 cents to the prize pool. (The rest goes to various charities and to lottery administration expenses.) If one person had bought all the tickets, he or she would receive prizes worth 45 percent of the cost of purchasing the

NESTLINGS

Figure 1.3 Beating the system.[3]

tickets. In a sense the average return for a single ticket is 45 cents, and the average loss is 55 cents.

In the lottery with the $14 million grand prize, this average return crept up to about 57 cents. So the average loss dropped to 43 cents. (This is because about $8 million from previous issues of the lottery was transferred to the prize pool, which otherwise would have contained about $32 million; and 32 + 8 = 40 million is about 57 percent of 70 million.) The prospect of losing only 43 cents per ticket, on average, instead of 55 cents, was apparently a powerful inducement to millions of participants! See Figure 1.3.

Some of the mysteries of lotteries and weather forecasting can be analyzed using the ideas of probability, and these are introduced in Chapter 7. The connection between probabilities and descriptive statistical methods is discussed in Chapters 7 and 8.

1.1.3 Public Opinion Polls

Polls of public opinion such as those conducted by the Gallup and Harris organizations are regularly reported by the media. What is so special about these polls? Why do we not see more attention paid to polls conducted by television networks and radio stations? The answer is that for a poll to produce useful results, it must be designed according to certain statistical principles; polls that do not use these methods are notoriously misleading and are not given much weight by knowledgeable news reporters.

A famous example of a badly designed poll was the one the *Literary Digest* magazine organized in 1936, prior to the American presidential election in that year.[4] The *Digest*'s huge poll of 2.4 million people indicated a decisive victory for Alfred Landon, with only about 43 percent of the vote to go to Franklin Roosevelt. The actual result was 62 percent of the vote for Roosevelt to 38 percent for Landon. The *Literary Digest* went bankrupt after this fiasco. (Its misfortunes were compounded by the rise of *Time* magazine.)

One of George Gallup's first polls was done about the same time as the notorious *Literary Digest* poll.[4] It was based on the opinions of about 50,000 people, and predicted 56 percent for Roosevelt, which was closer to the truth than the poll based on 2.4 million people. It used, to the extent that was possible at the time, proper statistical methods in the design of the survey. The success of this poll was not a matter of good luck but rather of good method. In Chapters 2 and 7 we discuss the key elements of survey sampling and discuss their relationship to other statistical procedures.

1.1.4 Evaluation of Vitamin C Therapy for Colds

In 1970, Linus Pauling published a book summarizing his views on the effectiveness of very large doses of vitamin C for the prevention and amelioration of colds.[5] The experimental evidence in support of Pauling's claims was scanty; it was decided to do a large-scale clinical trial to determine whether or not these large doses of vitamin C really had any beneficial effect. One thousand volunteers were given a supply of pills with instructions for taking the pills regularly, and increasing the dose at the first sign of cold symptoms, and also were given a form for recording the presence and severity of cold symptoms. Five hundred of the volunteers were given a placebo which was known to have no pharmaceutical effect on the cold symptoms. The volunteers could not tell whether they were taking the vitamin C or the placebo.

The result was that the vitamin group had 30 percent fewer days off work because of colds than did the placebo group. In evaluating this result we should ask the following questions: Would the result be reproduced if the experiment were repeated? Is the difference between the groups of any practical importance? Is the difference between the groups attributable to the difference in pills only? Such questions are discussed in Chapter 9.

1.1.5 The Baby Boom and the Echo Boom

In the wake of World War II, disrupted families were reunited and many new families were formed. The prevailing social milieu was apparently just right for a remarkable increase in the birthrate. Figure 1.4 presents the trend in the number of live births per 1000 population, the "birthrate," for the period 1920–1983. During the period 1945–1965, higher-than-usual birthrates persisted; this era is often called the "baby boom." The female baby-boomers entered their childbearing age during 1965–1985, as shown in Figure 1.5, and one might expect an increase in the birthrate during these years. This increase appears not to have occurred. The "echo boom," which began about 1974, is an increase in the number of births, but not in the birth rate. The baby boom itself was caused by an increase in the birth rate.

Business people take demographic trends very seriously, as they can be important determinants of changing markets. But changes in the age mix of the population affect everyone. Certainly, students who are born at the end of a baby boom can be expected to face stiff competition for jobs when they finish their schooling.

Sec. 1.1   Some Examples of Statistical Problems   7

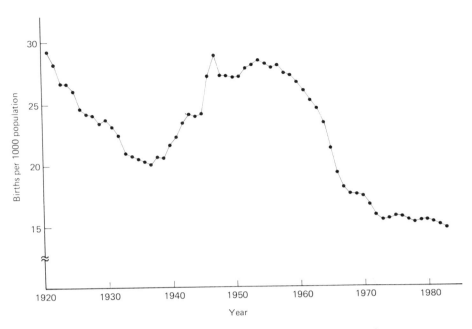

Figure 1.4 Live births per 1000 population, 1920–1983.[6]

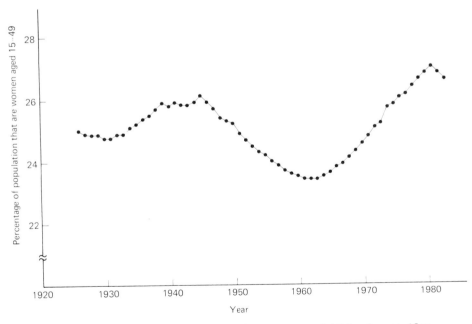

Figure 1.5 Percentage of the population that are women of childbearing age, 15 to 49 years, 1926–1983.[7]

Although it is certainly possible to predict the number of 21-year-olds next year if we know the number of 20-year-olds this year, the prediction of births next year is not quite so simple. Nevertheless, as you examine Figure 1.4 or 1.5 you may well feel that this sequence of numbers is, to some extent, predictable into the future. The issues of the description and predictability of such time series are discussed in Chapter 6.

1.1.6 Lecturer Effectiveness and Ratings of Instruction

At Southern Illinois University, a study was done to measure the relative importance of content and "seduction" in lectures.[8] The evaluation of satisfaction with the lecture was provided by the student audience. Earlier studies had suggested that lecture content was ignored in the evaluations, whereas the entertainment value was a key factor. However, the earlier studies had some design defects that made the interpretation of the results ambiguous. The new study was more carefully designed to eliminate these ambiguities.

A class of 207 students was divided into six groups. Each group was presented with a different version of the same lecture. The lectures were different in seductiveness (high and low) and the amount of content (low, medium, and high). Student satisfaction with the lectures was measured by means of a detailed questionnaire; a numerical index of overall satisfaction was constructed. The conclusion of the study was that lectures that were highly seductive received a higher rating regardless of their content. Lectures with low seductiveness were differentiated on the basis of content, the lectures with the high content receiving the higher rating. Another interesting finding was that the students apparently learned more from the seductive lectures, given a fixed lecture content.

Now, in assessment of this study, we must know certain aspects of the study design before we jump to conclusions. How were the six groups of students selected? Were students effectively prevented from changing to the instruction mode of their choice after they had been assigned to a particular instruction mode? Were the students' responses biased by a knowledge of the purpose of the study? Were the groups exposed to the same lecture conditions except for the intended differences? Such questions concern study design and are the subject of Chapters 3 and 4.

1.1.7 Computers, Statistics, and Management of Airlines

Most college and university students have some appreciation for the potential of computers to simplify the arithmetic side of the analysis of data. However, the impact of computers is not merely to speed up the old methods: computations previously unthinkable because of the time required are now performed in a few seconds by technicians with little knowledge of the complexities involved. Calculations and data manipulation that were of little interest in precomputer days because they were complex and time consuming can now be accomplished quickly and accurately using computers. The impact of computers on statistics is due to the many new kinds of computations that can now be done rather than to the speed with which a computer can do the old computations. Consequently, the impact that computer developments are having on the theory and practice of statistics is profound.

A simple example will illustrate this point. Suppose that an airline sells 20 million tickets in a particular year. If the airline wishes to study the number of tickets purchased per person, and perhaps to make a list of high-usage passengers, it is necessary to check each name against all other names. There are ways to improve the efficiency of this task, but any manual method would take a clerk at least 20 years. A computerized scheme for this task might be complex enough to require two weeks of a programmer's time, but the computer program itself would run in a few hours at most. Thus the impact of the computer in this example is that it changes the tasks that the airline can usefully undertake to assist its marketing efforts.

In Appendix 1 information about the use of computers for statistical analysis is presented. Some of the problems in the text are designed to make use of computer programs, and only if you do these problems will you begin to appreciate the relationship between computers and statistics. In the meantime we recommend that you grasp every opportunity to familiarize yourself with the capabilities of computers.

The exciting thing about modern applied statistics is that the availability of computers for the arithmetic of statistics has opened up the field to busy executives, journalists, lawyers, and many others. Their use of statistical methods is bound to motivate many novel developments of the methods themselves. To take advantage of these new opportunities, there will be a demand for professionals of all types who can understand the basic concepts of statistics.

EXERCISES

1. For the reader who wishes to make use of elementary statistical methods, it is useful to have a mental table of contents of the methods. Then, when the potential for statistical applications is recognized, the type of technique needed may be extracted from this table. The following exercise is intended to encourage you to begin to form such a mental organization chart.

    The subsequent chapters of this text are:

    | Chapter 2 | Study design I: Logical keys to statistical thinking |
    | Chapter 3 | Study design II: Principles of experimentation |
    | Chapter 4 | Descriptive methods for one variable |
    | Chapter 5 | Descriptive methods for two variables |
    | Chapter 6 | Descriptive methods for time series |
    | Chapter 7 | Sampling and probability |
    | Chapter 8 | Estimation |
    | Chapter 9 | Testing hypotheses with data |

In which chapter of this book would you expect to find a discussion relevant to the following questions? (Do not attempt to answer questions (a) to (g). These questions relate to a survey of textbooks currently for sale at a college bookstore; the study is aimed at comparing the costs of textbooks in different disciplines. Just consider what kind of statistical question is involved and in what chapter it may be discussed.)

(a) How many textbooks should be examined for cost, discipline, and need status?
(b) What summary numbers are likely to be necessary for the report of the survey?
(c) Are texts in the science disciplines more expensive, on average, than texts in the humanities?
(d) In view of the fact that the survey uses information on only some of the textbooks in the bookstore, how much difference in the summary numbers might we expect to find between our survey and a more exhaustive one that was based on all the textbooks?
(e) Since there are more titles in the social science disciplines than in the natural sciences, should the survey include more titles in the social sciences than in the natural sciences?
(f) How could it be determined whether doubling the number of salesclerks in the bookstore would increase revenues from textbooks enough to justify the extra salary expense?
(g) Did the initiation of a secondhand bookstore on campus appreciably reduce revenues at the store selling new books?

2. In Lotto 6/49, described in this section, a person who buys one ticket is very likely to lose $1. If someone buys all the tickets, he or she will certainly lose 55 cents for each dollar spent.
(a) Which is it more rational to do: to buy one ticket or to buy all the tickets? Explain.
(b) Why do rational people buy any tickets? (It does seem that rational people buy tickets, since participation in lotteries is so widespread.)
(c) A 1-in-14 million chance of winning is like the selection of one particular book from a tall stack of books. How many meters high do you think the stack would have to be to make the choice of the correct book as likely as winning a lottery with a 1-in-14 million chance? (Assume that the average book width is 1 in. or 2.5 cm.)

3. What relevance do the questions mentioned in Section 1.1.6 (concerning the comparison of teaching methods) have on the interpretation of the study of lecturer effectiveness?

1.2 THE TRAINING REQUIRED TO SOLVE STATISTICAL PROBLEMS

It is often said that the process of formal education is as important as the factual knowledge gained. This is certainly the case in statistical education. Applications of statistical techniques are seldom straightforward, and it is absolutely essential that you learn how to blend your common sense with the rather sophisticated logic of formal statistical methods. The British statesman Disraeli has been credited with the comment: "There are three kinds of lies: lies, damned lies, and statistics." Many modern politicians share this view. Perhaps this mistrust of statistics has resulted from the politicians' reliance on superficial analyses of numerical information. There is a need for people who can communicate with politicians (and others) and for those who can sort out sense from nonsense in numerical information.

Sec. 1.2 The Training Required to Solve Statistical Problems 11

It has repeatedly been demonstrated in the public media, academic journals, and in industrial research that a "doctor–patient" relationship between a statistician and a user of statistics does not work—the user of statistics has to understand quite a bit about the prescription recommended by the statistician to ensure that the prescription is appropriate (see Figure 1.6). All those beginning a study of statistics should realize that they will never be able to make wise use of this subject unless they are willing to learn the process of combining formal methods with practical requirements. The closest that one can come to this training in the context of a college course is to take problem solving seriously: that is, to realize that the act of solving a statistical problem has at least as much educational value as learning the answer to the problem. Good examinations in statistics present problems unlike any the student has seen before, because this is typical of the problems one meets in professional work. Applied statistics is neither entirely mathematics nor entirely common sense but rather a

Figure 1.6 Modern statistics cannot cure all data ills.

complex blend of these two components. To learn how to use the techniques of applied statistics, one has to learn how to bring abstract ideas into real-world contexts.

Here is an illustration of the difference between learning a statistical principle and learning to use the principle. One important principle is that "correlation does not imply causation." For example, just because fat people often eat yogurt, it should not be inferred that it is the yogurt that makes them fat. A study that recorded the health status of the population of the town of Framingham, Massachusetts, for 18 years showed that people with high-cholesterol diets tended to experience more heart disease.[9] Should we accept this as evidence that the high-cholesterol diet is a causal factor contributing to the increased frequency of heart disease? Having been warned by the preamble, you should doubt that the causal implication is justified on the basis of the facts presented. The big question is this: If you had not been forewarned, would you have responded to the Framingham news with the query to yourself: "Is this an example of correlation erroneously suggesting causation?"

The best way to arm yourself with the ability to detect an issue in new contexts is to have faced the same issue in many different settings. The abstraction about correlation and causation should become a part of your "intuition." In other words, you will have to solve many similar problems to really absorb the idea. Merely memorizing "correlation does not imply causation" will be of little use to you. You are unlikely to meet the question "Does correlation imply causation?" explicitly either on an examination or in your career. However, you will almost certainly benefit from recognizing this issue in both exam questions and in your working life.

EXERCISE

1. Explain the relationship of the yogurt example to the inference suggested by the Framingham study.

1.3 STATISTICS: PAST, PRESENT, AND FUTURE

The word *statistics* is derived from the Latin word for "the state." The first important accumulations of data were for the purposes of the state, and were simply lists of names and addresses of a state's residents. The data were used for tax collection (or extraction) and military impressment. More humane uses of similar data include the assessment of birth and mortality rates which appeared in England in the seventeenth century.

The mathematical development of the subject was greatly stimulated by the interest in gambling problems. The French mathematicians Blaise Pascal (1623–1662) and Pierre de Fermat (1601–1665) discovered certain useful principles for combining probabilities associated with games of chance.

The application of probability ideas to astronomy was extensively developed by

Pierre Simon Laplace (1749–1827) and Karl Friedrich Gauss (1777–1855). Several other European thinkers of the nineteenth century contributed their wisdom to the many paradoxical issues surrounding the meaning of probabilities.

One of the most influential applications of these ideas was the work of Sir Francis Galton (1822–1911) in his studies of heredity. Karl Pearson (1857–1936) and Sir Ronald Fisher (1890–1962) greatly extended both the methods themselves and their range of application, and this work formed the basis of modern statistics. Statistical studies of agricultural experiments and psychological measurements were important in the development of current analytical techniques.

Passport photograph of R. A. Fisher at age 34. At this age, Fisher had published about 40 of his approximately 300 research papers. [Courtesy of Joan Fisher Box, *R. A. Fisher: The Life of a Scientist* (New York: John Wiley and Sons, Inc.), 1978. Reproduced by permission.]

With this diversity of roots, it is perhaps not surprising that the word "statistics" means different things to different people. To many people the word is synonymous with "recorded numbers," for example, baseball batting averages. Even those who are also familiar with its alternative meaning as a subject area may well have a narrow view of what the subject area comprises. Some typical perceptions of statistics by workers in various fields are listed below.

Medicine: methods of detecting causal links between exposure and disease, and between treatment and cure

Business: methods of presenting numerical facts

Agriculture: methods for the design and analysis of planned experiments

Engineering, physics, chemistry: methods for coping with instrument or technician error

Biology: methods for describing natural variation

Sociology: methods for extracting trends from data which exhibit a large component of unexplained variation

Geography: methods for describing spatial relationships between human settlements and the natural environment

Education: methods for evaluating learning potential, progress, and achievement

Psychology: methods for detecting causal mechanisms based on experimental data gathered from complex biological systems

These various perspectives do have a common link: all require a core of mathematical techniques which efficiently model distributions of numbers. Consequently, professional statisticians tend to have an educational background in mathematics. Yet mathematics alone is certainly not a sufficient background to provide expert advice in all the possible areas of application of statistical techniques. This is why it is so important that basic statistical concepts be learned by students of a wide variety of disciplines.

For the purposes of this text, the word **statistics,** when used as a collective noun describing a field of study, may be defined as: the discipline concerned with the collection, description, and interpretation of data which exhibits unexplained variation. The precise meaning of the phrase "unexplained variation" will become clear in Chapter 7.

Although it is true that statistical methods are used in almost every discipline, it must be said that only a small proportion of people in each discipline actually use statistical methods frequently. But this is changing. The proliferation of computers for the summary and analysis of coded information (numerical or otherwise) is markedly increasing the opportunity for the use and abuse of statistics. Statistical know-how is increasingly valued in this "age of uncertainty."[10]

1.4 ANSWERS TO EXERCISES

Section 1.1

1. (a) The determination of sample size is discussed in Chapter 8.
   (b) Chapter 4, and possibly Chapter 5 as well. Once the logical status of a data set has been established or ascertained, summary of the data by descriptive methods is the next step.
   (c) Chapter 9.
   (d) Chapter 8 discusses the precision of estimates.
   (e) Chapter 7 discusses options for methods of sampling.
   (f) An experiment would have to be devised, requiring the techniques of Chapter 3 for the design of this experiment. Chapter 2 is also relevant.
   (g) One would require some techniques for describing the revenues before and after the initiation of the secondhand bookstore. The time-series methods of Chapter 6 would be helpful here.

2. (a) The person who buys one ticket is more rational (unless the person's aim is to buy publicity or to contribute to government revenues). With a single ticket, one purchases a chance at a large prize, and this seems more rational than guaranteeing a large loss.

   (b) They get more pleasure out of owning a ticket that has a chance of returning a large prize than in having $1 in their pockets. The fact that the chance is very slim does not affect this pleasure. That winning is a possibility is more important than how probable it is. Another aspect is the social pleasure people get in discussing with friends their plans for using the winnings. (Historically, lotteries have always had their detractors and supporters. The issues are psychological, economic, political, and religious, and only rarely statistical. It is not clear whether greater statistical sophistication of the populace, which seems inevitable, will alter this ongoing controversy.)

   (c) About 350 km (or about 220 miles) high.

3. The selection of students for the subgroups must be done in a way that makes the groups similar; otherwise, the differences in outcome might be attributable to initial group differences rather than to the different teaching strategies.

   If some students slip into the tutorial subgroup rather than the lecture subgroup to which they were assigned, or vice versa, any real difference in the effect of the two teaching methods will be masked: that is, the measured differences in the outcome of the two original subgroups will appear smaller than if crossovers were prevented.

Section 1.2

1. Fat people are continually trying to lose weight. Yogurt has a reputation of being low in calories but still nutritious. The condition of fatness apparently causes the eating of yogurt rather than the other way around. Of course, it is possible that the tendency to become fat and to desire yogurt is due to a third factor, such as diet in childhood.

   With regard to the Framingham study, it is unlikely that the onset of heart disease causes people to desire high-cholesterol diets. (However, this is a possible explanation for the correlation observed.) Perhaps a predisposition to heart disease and a predisposition to a desire for high-cholesterol foods have a genetic basis or some other unknown common cause.

   Thus the yogurt example and the Framingham example are both correlational studies (i.e., studies where there is an interest in whether or not two factors occur together in people), giving, by themselves, inconclusive evidence about causal paths.

PROBLEMS

1. There are many different electronic games in the entertainment arcades throughout North America. If you were interested in documenting the relative popularity of these games, based on actual usage, how would you do it? To simplify this question, suppose that you have decided to do a "drop-in" survey on 10 of the arcades in your city, that you do the observations yourself at selected times on successive Saturdays, and that for each arcade you note the games that are in use and those that are not. There are still several questions to resolve before the study can be completed:

(a) How will the 10 arcades be selected?
(b) In what order will they be visited?
(c) How will the data be recorded?

(*Note:* These are difficult questions, but your attempt to answer them at this stage will prepare you for the material of subsequent chapters.)

2. (Continuation of Problem 1.) Refer to the setting of Problem 1.
   (a) How will the data from different arcades be aggregated?
   (b) What summary numbers would you use to describe the results of the study?
   (c) What feature of the data will indicate whether the ranking of the games in terms of popularity would be reproduced if the study were repeated (on a different group of arcades)?

3. Many people monitor their weight with daily measurements. These measurements usually vary a pound or so from day to day, and there may even be some difference in two measurements taken a few seconds apart. Discuss the various sources of variation in these measurements. What practical problem does this variation present to a person who is attempting to lose weight?

4. Your college library will have a list of periodicals. Select five periodicals in a way that will make these five "representative" of the entire list, to the extent that this is possible.
   (a) Describe the rationale of your selection procedure.
   (b) Scan each article in the most recent issue of each periodical selected. Categorize each article under the following headings:

   - The conclusions of the article depend critically on data reported in the article.
   - The conclusions of the article are partially related to data reported in the article.
   - No data are reported in the article, or the data are irrelevant to the main thesis of the article.

   (c) What conclusions can you make about the potential for the use of statistical methods in the analysis of studies such as those that appear in periodicals? Elaborate.

5. An important part of the style of medical practice, at least in the developed countries, is the confidential and personal nature of the doctor–patient relationship. The entry of the state into health insurance schemes has coincided with the development of computerized information systems, and the information files of health insurance agencies contain some details of every chargeable activity of many doctors. For example, every diagnosis of pneumonia would be recorded, every appendectomy, every case of consultation for depression.

    What use might an insurance company (or government agency) make of these data in aggregate? That is, other than for checking on particular claims for payment made by a doctor, what might an insurance company do with such data to control its expenditures for insurance claims? Discuss the implications of this for the future privacy of the doctor–patient relationship.

6. The Sure Thing lottery sells 1,000,000 tickets for $1 each. One of these tickets will gain its owner a $100,000 prize and nine other tickets will gain their owners $50,000 each. Another lottery, called the Charity Special, also sells 1,000,000 tickets for $1 each, but 500 tickets provide prizes of $1000 each. From which lottery would you be more inclined to buy a ticket? Justify your choice. (For the purpose of this question, concentrate on the comparison of the two lotteries, not on whether you would or would not buy a ticket from such

lotteries. Note also that there is not necessarily a "right" answer here, but your justification should have some rational basis, consistent with your choice.)

7. In the survey of higher-education intentions described in Section 1.1.1, the questionnaire is filled out during class time and is administered by the particular teacher who has the class at the time chosen for the survey. All members of selected classes are asked to complete the questionnaire. What precautions should be taken to ensure that the responses given by the students are as truthful and accurate as possible?

Study Design I: Logical Keys to Statistical Thinking

It ain't so much the things we don't know that get us into trouble. It's the things we know that ain't so.

Artemus Ward (1834–1867)

People in general have no notion of the sort and amount of evidence often needed to prove the simplest matter of fact.

Peter Mere Latham (1789–1875)

The theme in this chapter is that logic plays an important role in making sense out of data. We also point out that the logical skills needed for statistical thinking are also very useful in everyday living. If you were never to do a statistical calculation in your life, the logical keys discussed here would still be useful to you. This chapter is a necessary prerequisite for the rest of the book. Chapter 3, which is also concerned with study design, is optional in that it is likely to be most important for those actually involved in planning experiments. Such experiments are routinely done in scientific work but only occasionally in the social science context, since social "experiments" are rarely true experiments.

There are just a few big ideas in this chapter:

1. Surprises are often informative.
2. It is possible to learn about a big thing from appropriately selected little pieces of it. But the selection has to be done properly, and this usually requires careful planning.
3. It is easier to assess something when it can be compared to something known.
4. For valid comparisons, it is important to know what is the same and what is different among the things compared.
5. Association does not imply causation.
6. Sources of variation in measurements can sometimes be identified separately.

By the time you finish reading the chapter, you should make sure that you have absorbed these points and understand fully their meaning in a statistical context. These points appear again in a slightly different form in the chapter summary (Section 2.9).

Key words are sample, population, random digits, systematic sampling, simple random sample, sampling with replacement, sampling without replacement, stratified sample, cluster sample, comparison, experiment, survey, confounding factor, Simpson's paradox, survey, nonresponse bias.

2.1 SEARCH AND RESEARCH IN EVERYDAY LIVING

One way that we cope with the many stimuli we encounter daily is to ignore most of them. Try to recall a single face in the crowd on a bus, or the location of your car in the parking lot after a long day. The things we focus on are the anomalies, the bits and pieces that are unusual or incongruous. We may notice a family with 10 children on the bus, or that we had to park in the parking spot farthest away from our destination. The method of "management by exception" appears to handle our everyday inputs of information unless we consciously dictate otherwise.

A closely related system of information management applies to our more deliberate observations. Suppose that we are studying socialization of preschool children in a day-care center. After observing these children for an hour or so, we realize that almost every new activity has been initiated by just two of the 20 children in the group. If we are surprised by this, we not only take note of it but may also wish to speculate why this is so, whether it is normal in this age group, whether these two children would be initiators on other days, and so on. In other words our "research" into this situation is guided by what seems unusual to us. The fact that none of the children are eager to begin their afternoon nap arouses less interest; we expect this.

Unfortunately, many of the "anomalies" that we observe do not lead to new theories or helpful explanations: they are merely accidents of nature, a confluence of several uncommon events. Although passive observation can lead to a multitude of intriguing hypotheses, it rarely establishes anything unequivocally. For example, suppose that we were to observe another group of children in a different day-care center. We may be surprised by the many incidents of fighting among the children. We may wonder whether the difference between the centers is related to the physical setting, the staff, the children, or the time of day. Or perhaps it is a combination of a great many differences each of which would have a negligible effect by itself. We clearly need to observe many more centers, or to devise a more carefully controlled comparison, or both. Although passive observation is often the starting point for many more formal investigations, the achievement of unequivocal results from a study requires very careful assessment of the design of the study, *before* the observation process begins.

The point of these examples is that although the basic modes of discovery are the same in both everyday living and formal investigations, the latter require more rigorous logical designs since the results are expected to be reproducible. Casual observation tends to be very productive of theories, probably more so than formal studies; however, most of these theories will have no enduring basis in fact. As T. H. Huxley has expressed it: "the great tragedy of science: the slaying of a beautiful hypothesis by an ugly fact."[1]

In the remainder of this chapter we explore the structure of formal research studies.

2.2 LEARNING THE LOGIC OF NUMERICAL ASSESSMENT

It is a popular misconception that the assessment of numbers is an arithmetic task. It might similarly be claimed that the writing of a play is a typing exercise. Numerical statements can be just as subtle and complex as verbal statements, but the precision of numerical statements leads us to expect an equivalent precision in the verbal text surrounding them. When our government announces that the unemployment rate has dropped 10 percent over the last 12 months and now stands at 8.0 percent, what are we to think?

- Was the former unemployment rate 18, 8.8, or 8.9 percent?
- Is the unemployment rate seasonally adjusted, or is this fact not relevant to interpretation of this figure?
- Is this reduction large enough to lead one to believe that an important reduction of unemployment has been achieved, or is 12-monthly variation (increase or decrease) usually of this magnitude?

The point is that the interpretation of numerical statements is not really common sense no matter how "common" the subject matter. What is needed by the discriminating consumer of numerical information is training in a certain kind of logic.

If we accept that these questions are relevant to the assessment of the government declaration, we may ask what kind of training would suggest these questions. In the first place, the assessor has to have seen many examples of apparently straightforward statements that are ambiguous on closer examination. He would have to have experienced the confusion between percentage decrease and percentage points decrease. He would have to have seen examples of a set of things getting larger but being, at the same time, a smaller proportion of some larger set (to understand the possible anomalies of "seasonally adjusted" data). He would have to understand that when two "shaky" numbers are compared, the interpretation can depend on just how shaky the numbers are compared to the size of the observed difference. Finally, he would have to have had enough experience in solving statistical problems to recognize that the unemployment statement had all these elements to be questioned. The conjuring up of the principles that apply to this particular instance is a process that is difficult to learn other than by working through statistical problems.

The logic of numerical assessment includes several abstractions. Some of the more important abstract concepts are:

1. Generalization from a part to a whole
2. Measurement qualities of coding systems
3. Independence and dependence
4. Correlation and causation
5. Separability of multiple causes
6. Accuracy and precision
7. Randomness

Such abstractions appear and reappear throughout the description of statistical methods. It is these logical principles that enable one to apply statistical methods intelligently to novel situations. The best way to learn the principles is to study the methods in artificially simple situations. This may explain why textbook examples in statistics seem unrealistic, even though the real-world applications of these methods are widely used; realistic applications of statistics are hardly ever simple. It also explains why undertaking the process of solving statistical problems is at least as important to statistical training as learning the answers to the problems.

2.3 A FIRST LOOK AT SAMPLING STRATEGIES

In common usage, the word *sample* means a part of something that may be inspected as evidence of the quality of the whole. For example, one does not usually feel compelled to eat a whole barrel of apples in order to assess their quality; a small sampling would usually be deemed adequate. In fact, if you believe that the whole barrel is of uniform grade but you cannot tell what that grade is just by looking, a taste of a single apple may be sufficient to reveal the quality of the entire barrel. On the other hand, if uniformity of the grade cannot be assumed, a larger sample would be required, but even then most of us would expect that an excellent assessment of the quality of the whole barrel of apples could be made on the basis of a small portion of the barrel. But what if all the rotten apples are at the bottom of the barrel? We must pay attention to the method of sampling when we discuss the logic of generalization from the part to the whole.

The technique of sampling is widely used:

- Politicians spend both public and party funds to determine the public's attitude to proposed legislation or foreign affairs initiatives. This is done by posing questions to a relatively small number of people.
- Manufacturers often test small portions of large batches of their products to ensure that they are meeting quality or reliability targets.

- Marketing firms may obtain the response of a small group of people to a proposed new product before large-scale distribution is undertaken.
- A pharmaceutical company must demonstrate the effectiveness of a new drug before making public claims that will help to sell the drug. This testing is usually done with a small number of volunteer subjects.
- Stock market analysts will sometimes monitor the trading activity in a small, manageable number of stocks to infer timely information about price trends in a large sector of the market.

When is it possible to learn something about a whole, which we will call a **population,** from a part, which we will call a **sample**? To say that the sample has to be representative of the population does not really help much. Suppose, for example, that we wish to estimate the average height of trees in a particular forest, and that we want to do this based on the observed heights of a sample. A *representative* sample would be one for which the average height were the same as the average height of the forest, but there is no method of selecting such a sample that avoids observing the whole population first. Since our object is to see what we can learn from only a part of the population, we may have to settle for an approximate solution. Perhaps we can devise a method of sample selection that, while involving some risk of resulting in misleading estimates, will have the virtue of usually resulting in estimates of the population average which are quite close to the true values. Of course, it would be nice to have a method that always produced representative samples, but such a method does not exist. It is possible, however, to specify a method of sampling that will not only tend to produce a representative sample, but will provide information about how representative the sample is likely to be.

2.3.1 Simple Random Sampling

To illustrate the mechanics of sampling and the implications for the validity of the information derived from a sample, we introduce an example typical of surveys undertaken by newspapers and magazines. Suppose that you are interested in the salaries of chief executive officers (CEOs) of North American companies with 100 or more employees. You happen to have a complete list of all 10,000 companies and their CEOs. You decide that a personal letter followed by a telephone call would enable you to obtain the information on salary income for these persons. Also, you decide that you have the time and resources to contact only 100 of the CEOs. How would you pick the sample of 100 names?

One method would be to number the listed names from 1 to 10,000 and pick the names numbered 1, 101, 201, 301, ... This method of sampling is called **systematic sampling** and is widely used because it is so simple to do. But you should worry a little about the order in which the 10,000 names are listed.

For example, if the list is ordered according to the number of employees in the company, the CEO of the company with the largest number of employees is guaranteed to be one of the 100 selected. Do you want this atypical CEO in your representative sample? Or, if the order in the list is the membership number of the CEO in a national association of CEOs, it is just possible that the prestige membership numbers, 100, 200, . . . , 10,000, are awarded to special members. Your systematic sample would miss all of them. These contingencies seem unlikely, but there are many situations where systematic sampling produces misleading samples. You should be on the lookout for this source of sampling anomalies.

The quality of the systematic sample will be affected by any factor affecting the order of the list. Let us assume that the ordering criterion is unknown and that we seek some sampling method that would produce "good" samples for any ordering. One method, although a bit laborious, would be to write the 10,000 names on cards, shuffle the cards thoroughly, and select 100 cards from the shuffled pile. This is the "picking numbers out of a hat" strategy. Such a sample would have the merit of being "unbiased"; that is, having no tendency to favor any particular group of names from the 10,000 in the selection of the 100. Would you expect such a sample to be representative? The answer to this question depends on what is meant by "representative." We will examine this question in more detail in the exercises. Meanwhile, let us assume that such a sample of the 10,000 is required and consider more convenient ways to select this kind of sample.

There is an easier, essentially equivalent way of accomplishing the foregoing sampling method—by making use of *random numbers*. There are tables of random digits, generated by computer, which have the property that each of the digits 0, 1, 2, . . . , 9 occurs with the same sequences and frequencies as would be obtained from repeated drawings from a shuffled deck of 10 cards labeled 0, 1, 2, . . . , 9; it is assumed in this description that each time a card is drawn it is replaced in the deck and the deck reshuffled before the next draw. This procedure is known as **random sampling with replacement.** We will select numbers consisting of four digits from the table of random digits. Each block of four digits could be the identification number for one of the CEOs in the list, so a sequence of 400 random digits would enable us to pick 100 CEOs.

There is one little hitch in this procedure that we will return to several times. If the same four-digit number occurs more than once while we are selecting our sample, the sample of size 100 will in fact contain fewer than 100 different names. What we really wanted the random numbers to do for us was to indicate how to **sample without replacement:** once an item is selected, further items are selected from the remaining population. This kind of sampling can easily be obtained from the same sequence of numbers by skipping over repeated numbers. See Figure 2.1 for a detailed explanation of this process.

We have not stated explicitly the properties of the computer-generated sequence of random digits; it will suffice for our purposes to claim that the sequence, used as directed, will in fact simulate the shuffling/selecting process. When the sampling is done without replacement, we refer to the process as **simple random sampling** (SRS). Had we imposed some additional constraint on the sampling, such as that we wanted to have an equal representation of eastern and western CEOs, the random sampling would no longer been "simple." Had we used some method other than

Table of random digits

```
4 1 7 4 9    9 3 2 8 3    3 0 0 0 7    3 8 2 5 1
0 3 4 8 5    6 9 3 7 2    3 2 1 1 1    7 8 4 2 8
9 5 4 2 7    0 5 4 7 7    0 3 8 6 1    4 3 7 9 1
```

Method 1: We wish to select five numbers, without replacement, from a population of 15. Because "15" has two digits, rewrite the random digits in pairs:

```
41  74  99  32  83  30  00  73  82  51  03  48  56  93  72  32  11
                                        03                      11  ← Select
17  84  28  95  42  70  54  77  03  86  14  37  91
                                        14  ← Select
```

Select items 03, 11, 14, . . . because these are all in the range from 01 to 15. We would need more random digits to get the full sample of five. Note that 00 is outside the range 01 to 15, and that since we wish to sample without replacement, the second 03 is ignored.

Method 2: As in method 1, we write the digits in pairs. But in order to use more of the numbers, we interpret numbers greater than 15 as being equivalent to a number in the range 1 to 15 by subtracting off multiples of 15. If the tabulated number is a multiple of 15, it is interpreted as selecting the fifteenth item. The first number 41 becomes 41 − 30 = 11. 74 becomes 74 − 60 = 14. The list becomes

```
41  74  99  32  83  30   . . .
11  14      02  08  15   ← Select
```

and the indicated sample of five consists of the items numbered 11, 14, 02, 08, and 15.
 Note that 99 was ignored because the full range 01 to 15 was not possible with two digit numbers beginning at 91. We should therefore ignore all numbers in the range 91 to 99 in addition to 00, to ensure that each number in the list 01 to 15 has an equal chance of being selected.

If sampling with replacement is desired, the same method is used except that repeated selections are allowed (i.e., the second 03 in method 1 is allowed, so the sample is 03, 11, 03, 14, . . .)

Figure 2.1 Use of random numbers to perform random sampling with and without replacement. In the example illustrated, a random sample of five items is selected from a population of 15 items.

random digits or a physical process equivalent to shuffling, the sampling process would not be called "random."

Those unfamiliar with formal sampling methods such as simple random sampling may be baffled by the introduction of artificial curiosities such as random numbers and sampling with replacement. The concept of examining a part to learn something about a whole seems simple enough, but what is gained by these further abstractions? The answer is that to assess the degree of representativeness of a sample of the population of interest, we need some mathematical results, and for these mathematical results to apply, we must be very exact in our specification of the method of sampling. We must try to say exactly what is meant by "sample 100 names 'at random' from the list of 10,000 names." A vague requirement of lack of bias is not enough; for example, it does not distinguish between sampling with and without replacement, yet this distinction will affect any reasonable measure of "representativeness" of a sample. A random sample selected without replacement will tend to be more informative of a population than a random sample with replacement of the same size. This will be explained in more detail in Chapter 7. The main point to remember here is that the phrase "sample at random" is ambiguous, and the different ways of sampling can make a difference to the usefulness of our sample.

2.3.2 More Complex Sampling Strategies: Clusters and Strata

A subject that is too large to cover in this text is sampling design: the strategies available to make a sample more representative than samples provided by the simple random sampling method. We will give only a qualitative idea of the basic issues by discussing examples where two alternative sampling methods, stratified sampling and cluster sampling, would be appropriate.

Stratified sampling and cluster sampling are alternatives to simple random sampling—they can include the use of random sampling, but for each technique the population is first divided into subpopulations. A *subsample* is selected from each of several subpopulations of the population, and then the subsamples are analyzed together (but in a different way than if the aggregated data were from a simple random sampling procedure) to describe the whole population. In **stratified sampling** the whole population is comprised of the subpopulations, but only a small sample from each subpopulation is observed. In **cluster sampling,** the whole population is comprised of the clusters, but only a few clusters are selected for the sample; all the individuals in a selected cluster are observed.

We begin with an example in which stratified sampling would be preferred to simple random sampling. If one of the political issues prior to an election were the access to abortions, it is likely that marital status would have some influence on the response. If so, we should consider the adverse effect on the quality of our survey of the accidental inclusion in our sample of too many or too few unmarried respondents. Most people are willing to reveal their marital status to a survey worker, so it would be feasible to include in the sample predetermined proportions of subjects of each marital status. Since these proportions are probably known for the population as a whole, we could thus guarantee that our sample was representative at least with respect to marital status. But note that this balance would be achievable only if we knew the population proportions beforehand.

In this technique, stratified sampling, the various categories of marital status are called *strata*. Stratified sampling is used to improve the representativeness of samples; the method is possible when the characteristics that are to be represented fairly in the sample occur with known frequency in the population sampled.

The use of stratified sampling would not be helpful unless the different strata really do provide different responses. Put another way, each stratum should have responses that are more similar within the stratum than the responses among different strata. In this case we say that each stratum is more *homogeneous* than the whole population. When this is true, it will be possible to get good information about the whole population from only a few representatives of each stratum. This is why a stratified sample can be more informative than a simple random sample of the same size.

The other sampling strategy to be discussed is cluster sampling. The difference between stratified sampling and cluster sampling is presented schematically in Figure 2.2. Cluster sampling is also described below, using a polling example.

Stratified Sampling   (A specified number of randomly selected individuals from each stratum)

Strata:	A1	A2	A3	A4	B1	B2	B3	C1	C2	C3	C4	C5	D1	D2	D3
Sample:		A2	A3		B1		B3		C2		C4		D1	D2	

Cluster Sampling   (All individuals from a specified number of randomly selected clusters)

Clusters:	A1	A2	A3	A4	B1	B2	B3	C1	C2	C3	C4	C5	D1	D2	D3
Sample:					B1	B2	B3	C1	C2	C3	C4	C5			

Figure 2.2 Stratified sampling versus cluster sampling. In stratified sampling, a few representatives of each stratum are selected into the sample. In cluster sampling, all representatives of a few clusters are selected into the sample. In both, the selection is random, and usually without replacement.

If we were to conduct a political opinion survey by visiting all households on a few city blocks, we would be doing a cluster sample. The households of a city block form a "cluster." An interviewer can easily visit every house on one block, and this will be much less time consuming than visiting an equal number of houses at randomly selected addresses.

Cluster sampling has an obvious advantage over simple random sampling: it provides a sample of a certain size at lower cost than does simple random sampling of households (or simple random sampling from voters lists). On the other hand, it would not be very helpful if people living on the same city block tended to have the same political opinions: for then we are getting the same opinions over and over again from one block, whereas we are interested, usually, in describing opinion from a larger constituency. But if each city block has a variety of opinion comparable to the target population, cluster sampling can be very effective.

The "fairness", or unbiasedness, of cluster sampling can be guaranteed by selecting the *clusters* by simple random sampling. That is, a random sample of city blocks is selected from the target constituency of all the city blocks, and then these selected city blocks are surveyed completely.

There are many variations on these ideas which are used for actual surveys. The main points to remember are:

1. All these methods introduce randomness at some point to ensure objectivity and to improve the representativeness of samples with respect to factors unknown to be related to the survey responses.

2. Information about the population that is known before the sample is collected, and is related to the survey responses, can be used to improve the representativeness of the sample by structuring the sample accordingly.

In this book we usually assume that ordinary random sampling, with or without replacement, is used. Much more detail of the consequences of random sampling is presented in Chapters 7 to 9. Some further discussion of cluster sampling and stratified sampling appears in Chapter 7.

Anyone undertaking a real survey should consult with those experienced in survey design. The description above is only a hint of the sophistication required to design and interpret, properly, a survey of a human population, or of any population.

EXERCISES

1. Use the table of random digits (Appendix A3) to select at random, without replacement, three letters from the alphabet. Using the letters in the order you generate them, is this an English word? Guess the proportion of such samples that would result in an English word. (In this exercise your guess should be based on your intuition and knowledge of word structure. It is not intended here that you repeat the process to determine the correct proportion.)
2. A woman discovers a case of unlabeled wine in a dark corner of her vast home cellar. She cannot remember whether she set aside the wine because it was so poor and she hoped the aging would make it drinkable, or because it was so promising and she wanted it to mature to its full potential. How many bottles should she sample to assess the quality of the whole case?
3. A merchant wishes to sell you a large box of firecrackers. He allows you to ignite three of them to prove their quality. (See Figure 2.3.)
   (a) Should you select the three by sampling with or without replacement?
   (b) If the "testing" of the firecrackers were merely a nondestructive inspection, would you select the sample with replacement?
4. Figure 2.4 displays a population of 49 squares. Select a simple random sample of 10 squares from the population of 49 squares. Is your sample representative of the population? Explain in what ways it is or is not representative of the population.
5. An auditor wishes to check the 1000 invoices of Veritas Publishing Company for the current fiscal year. The company publishes a monthly magazine. The auditor proposes to proceed chronologically through the file of invoices, picking out every eightieth invoice for closer scrutiny. Would you have any compelling reason to recommend that the sample of invoices be obtained in some other way? (Assume that the sample size is adequate.)
6. In sampling the voting intentions of residents of a city in an upcoming municipal election, which of the following subpopulations would be appropriate as a stratum in stratified sampling, and which would be better considered as clusters in a cluster sample: city blocks, the elderly population, apartment dwellers, households, males?
7. A population consisting of the numbers 1, 2, ... , 9 may be divided into subpopulations

Figure 2.3 The procedure says that we should test this one again.

in various ways. Two possibilities are indicated below as subpopulations A and subpopulations B. A sample of size 3 is to be selected to estimate the population average. [We know that the population average is $5.0 = (1 + 2 + \cdots + 9)/9$, but let us see what happens if we try to determine this using the sample alone.] Which definition of subpopulations (i.e., choose A or B) would be most appropriate for estimation of the population mean using

(a) Cluster sampling?
(b) Stratified sampling?
Explain your reasoning.

$$\text{Subpopulations } A: (1, 2, 3), (4, 5, 6), (7, 8, 9)$$

$$\text{Subpopulations } B: (2, 4, 9), (1, 6, 8), (3, 5, 7)$$

[*Hint:* Try methods (a) and (b) with the subpopulations defined as in *A* or *B* and see which gives the best (i.e., least variable) estimate of the population average.]

8. Select a simple random sample of size 10 from the numbers (0, 1, 2, 3, . . . , 9999). Is the average of the sample of size 10 within 1000 of the true population average? Repeat the whole sampling procedure five times to enable you to see if the sample average will have this property most of the time.

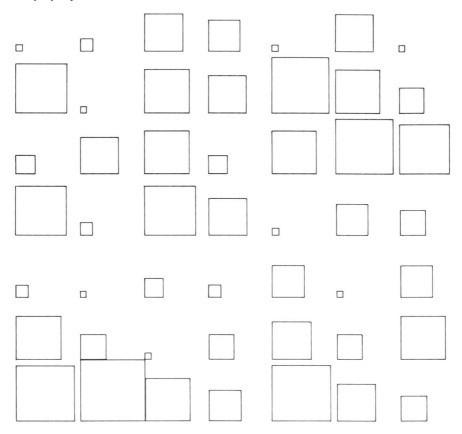

Figure 2.4 Population of 49 squares. See Exercise 4.

2.4 INFERENCES FROM COMPARISONS

In Section 2.1 we claimed that a common stimulus to new knowledge is the observation of something unexpected. Such observations often arise as the result of a **comparison.** If while a tourist to San Francisco you notice that it is cool downtown but hot in Palo Alto 30 miles away, you are made aware of the important influence of the ocean on local temperatures. You may have had no particular prior knowledge of what temperatures to expect in either location, but the comparison was revealing nevertheless.

The medical profession has tried to establish absolute norms for some of the measures they use frequently in assessing patients, such as with blood pressure assessments. For example, "normal values" for diastolic blood pressure are often said to be in the range 70 to 90 mm. The purpose of defining normal values is to make a single reading interpretable. But physicians often find a change in a patient's blood pressure from one visit to the next to be more revealing than the absolute level itself. Again, what is being used is a comparison in a situation for which norms are not specific enough to be relied upon.

If you were asked to assess a particular method of studying for tests, such as cramming the night before, an examination of the scores on the test of a group of students who used this method would not tell you very much. At the very least, you would want to compare these scores with the test scores of students who had used other study methods. To see why, suppose that the cramming group averaged 75 percent— would you be impressed by the results from this study method? What if the others averaged 100 percent?

Comparison can make a humdrum observation interesting; and it can also make a fascinating observation humdrum. We need some good habits to protect ourselves from these reversals.

When a claim is being made which is based on numerical evidence, it is a good idea to ask the following questions:

1. Do the quoted numbers refer to measurements that have well-established norms (i.e., standard values) with which to compare the measurements?
2. Are there two or more groups which are being compared, so that at least the relative sizes of the measurements have some meaning?
3. If there are two or more groups being compared, in order to support a claim, are there factors that differentiate the groups other than the factor mentioned in the claim?

The third question involves an issue that deserves a section of its own: the difference between experiments and surveys. As will be explained in Section 2.5, experiments have a structure that simplifies the interpretation of comparisons; unfortunately, this structure is not often attainable in practice.

EXERCISES

1. At the local high school, the five basketball players over 6 feet 6 inches tall averaged 20 or more points per game. If you were the coach, would you take this as an indication that you should recruit more players of this height next year?

2. Refer to Exercise 1. You are given the additional information that the other players (6 feet 6 inches or less) averaged 5 points per game. Does this change your answer?
3. Refer to Exercises 1 and 2. What alternative explanation is possible that would explain the better performance of the five tall players, other than the apparent advantage of height itself for basketball?

2.5 EXPERIMENTS AND SURVEYS

In assessing study design, an important distinction is whether the study is an "experiment" or a "survey." Before describing this distinction, let us point out the importance of study design even for the most pragmatic critics. The cover story in *Time* magazine (March 26, 1984) was entitled "Hold the Eggs and Butter" and subtitled "Cholesterol is proved deadly, and our diet may never be the same." The article reported several details of a $150 million study to determine whether reducing cholesterol in the bloodstream can reduce the incidence of fatal heart attacks. A drug was used by some of the subjects, on a long-term basis, to reduce cholesterol levels in their bloodstream, and sure enough it turned out that these subjects had a lower fatal heart attack rate than that of the subjects who did not get the drug. The article continues:

> There is no longer any doubt that lives can be saved by lowering cholesterol levels in the blood, but can this be achieved just by improving diet? If so, would healthier eating habits benefit all Americans? According to Columbia University cardiologist Robert Levy, who directed the study, the answer is yes on both counts. Says Levy: "If we can get everyone to lower his cholesterol 10% to 15% by cutting down on fat and cholesterol in the diet, heart attack deaths in this country will decrease by 20% to 30%."

The title of the article certainly seems to suggest that diet is the key, and several statements in the article, attributed to researchers, support this view. However, the article continues:

> Other doctors are not so sure, and urge a stricter interpretation of the study. Says Dr. Edward Ahrens, a veteran cholesterol researcher at Rockefeller University: "Since this was basically a drug study, we can conclude nothing about diet; such extrapolation is unwarranted, unscientific and wishful thinking."

This sort of equivocal result about the importance of cholesterol in diet has been appearing in the medical literature for many years. The same could be said of studies on the effects of smoking and heart disease, salt and hypertension, vitamin E and heart disease, exercise and longevity, and so on. What makes these issues so slippery? There is one important practical difficulty with all these studies: an "experiment" is needed to really decide the issue, but for practical and ethical reasons, only a "survey" is possible. Let us clarify the meaning of these terms in order to understand the requirements of definitive research studies. As the issue is complicated we will discuss some artificial studies first.

Studies involving comparisons are often of the following type: one group of items (or subjects) has factor A, whereas in the other group factor A is absent. The investigators are interested in the effect of factor A on some outcome variable *Result*. In other words, the investigators would like to determine the effect of A on *Result*. Note the implied interest in a causal mechanism in such a study.

For example, Professors Able and Willing want to determine which teaching method is better: using prepared overhead transparencies or creating the transparencies during the lecture. They split a large statistics class into two sections. Both sections are taught by alternating the same two professors. In section I, Able teaches lectures 1, 3, 5, . . . and Willing teaches lectures 2, 4, 6, ... In section II, Able teaches lectures 2, 4, 6, . . . while Willing teaches lectures 1, 3, 5, ... However, both instructors, when teaching section I, use prepared transparencies on their overhead projector, while in section II both instructors make up transparencies as they teach the class. The question is: Which teaching method is more effective? Effectiveness is to be measured by the students' average score on the final exam.

In terms of the general scheme we described for such studies, factor A is the use of prepared transparencies, and *Result* is the average score on the final exam. An important feature of such studies is that the only difference between the groups is the strategy being evaluated. Other differences should be eliminated if possible.

Why do you suppose that the two professors decided to both teach both sections? Surely, an attempt was being made to eliminate possible differences in teacher effectiveness so that any outcome differences could be ascribed solely to the difference in teaching method. Note that the time of day of the lectures, and the location and type of lecture hall, should be made as alike as possible for this same reason.

There is still one very large hole in the design of this study. The results would be completely worthless unless we could somehow ensure that the two groups of students are of about the same caliber when they begin the course. Ideally, we should ensure that both groups of students experience similar life-styles during the course, since for example if one class had a high proportion of students who were part-time, perhaps working full-time except for this course, it might be expected that their performance would differ from the others and confuse the interpretation of the results. One could think of many similar pitfalls. Is there a way around this difficulty? If not, all comparative studies would be on rather shaky ground.

The best way to deal with this problem is to resort to random selection methods. If there is a list of 300 students that is to be divided into two sections, select a simple random sample of 150 from the 300 to make up one group; the remainder would be the other group. One way to do this would be to number the 300 names and draw 150 numbered tickets from a mixed batch of 300 tickets, identifying group I. Another way that would accomplish almost the same thing would be to flip a coin 300 times, or use a computer to simulate this laborious sequence of flips; then a head would assign the next person on the list to group I, a tail to group II. (Note that there is a slight difference between assigning subjects at random to treatments and assigning treatments at random to subjects. Suppose that there are two treatments and 10 subjects. The assignment of subjects to treatments could be done by numbering the subjects 0, 1, 2, . . . , 9 and then allowing a sequence of random digits to assign five of the subjects to a particular treatment. This would guarantee five subjects in each treatment group. On the other hand, the assignment of treatments to subjects could be decided by the flip of a fair coin, heads indicating one treatment and tails indicating the other. Of course, in this case the resulting assignment would not necessarily produce equal numbers of subjects in the two treatment groups. For our purposes in this text, it is not necessary to focus on this distinction. Either method would lead to a valid experimental procedure, since any systematic differences between the subjects of the two groups could not be attributed to the experimenter or to prior differences in the subjects.)

The result of any of these random sampling methods is to let a chance mechanism balance the groups. Actually, there is no guarantee of balance with respect to caliber of student, or full-time/part-time split, or other characteristics, but there is a *tendency* toward balance *for each characteristic,* whether foreseen or not. This is very useful for simplifying the interpretation of comparisons between groups, for we can usually validly assume that the groups themselves are "comparable" except with respect to the factor under study. Furthermore, this random method of allocation has the very useful feature that the potential for imbalance can be assessed quantitatively, using probability methods. This point will be elaborated in Chapters 7 and 8.

Once we have similar subjects in the two groups, at least in the sense just described, and the treatment of the two groups is as similar as possible except for the factor of interest, any difference in the outcome measure can reasonably be ascribed to the factor. Except for one thing. Even if the teaching methods are identical in effectiveness, we do not expect the average scores on the final exams to be identical for the two groups. The groups are nearly balanced, but not exactly, by randomization. So we will have to know how much imbalance in the two groups can be attributed to the sampling process and how much indicates an effect of the factor. Clarification of this issue will be postponed until Chapter 9.

What we have described in the example is an **experiment.** The essential ingredient in an experiment is that the investigator decides, using an objective procedure such as the flip of a fair coin, which subject is to be assigned to which treatment, and then undertakes this assignment. Had the study omitted this step, that is, if students registered in group I or group II as they wished (even if we cannot see how this would bias the study), the study would be called a **survey** rather than an experiment.

Surveys are sometimes called *observational studies* to describe the passive role of the experimenter in forming the comparison groups. Surveys never establish a cause–effect relationship unequivocally because of the possibility that the comparison groups differed with respect to factors other than the one of interest. In this situation, a difference between two groups in the response variable cannot be attributed solely to the factor of interest.

Comparative studies examine the differences between two or more groups of subjects or items. In experiments, the groups are designed to be similar except for the application of a known treatment that is under study. In surveys, the groups are formed according to the presence or absence of a certain trait. The explanation of observed differences between the groups in a survey is more complex than in experiments. Group differences in surveys may include traits not under study, and possibly not even recognized, whereas group differences in experiments can only be explained as being due to the treatment of the groups.

The unwanted group differences between the comparison groups that occur in surveys are called **confounding factors.** If students in group I of a course have higher past grades on average than those of group II, "past grades" will be a confounding factor in comparison of the teaching effectiveness of the instructors of the two groups. We are "confounded" in our comparison because the groups are not similar in all relevant aspects. Confounding is a very serious problem in the analysis of survey data; sometimes its effects can be very misleading. This important and perplexing issue will be discussed in detail using a new example.

The Student Services office at a college has a counselor who offers advice to students wishing to improve their study skills. The counselor interviews students in order to suggest the best course of action. Two common suggestions are that students register in a noncredit short course, either in reading skills or in problem-solving skills. Let us call the groups of students sent to one or other of these courses the Read group and the Prob group. The counselor follows up on these students and notes whether or not they have passed all their exams (Pass group and Fail group). The data for an academic year are summarized in Table 2.1. Thus 405 of 625 (or about 65 percent) of the students taking the reading skills course passed the examinations, but only 220 of 625 (or about 35 percent) of the students taking the problem-solving course passed the examinations. So it looks as if the Read course is the better one since a higher proportion of students in this group subsequently pass than is true of the Prob course. But we must be cautious in jumping to this conclusion. What if the high-GPA students tended to be sent to the reading course and the low-GPA students tended to be sent to the Prob course? Let us call the group with lower GPA Low, and the group with the higher GPA High. Table 2.2 displays the data broken down by the GPA factor. Note that Tables 2.1 and 2.2 are consistent with each other. (For example, using the lower right-hand cells of each part of Table 2.2, 200 + 20 = 220, which is the lower right-hand cell of Table 2.1.) Yet it seems clear from each of these tables that the problem-solving course is more effective. For the High group, 400 of 600 or

TABLE 2.1 NUMBERS OF STUDENTS WITH EACH OUTCOME (FAIL OR PASS) AMONG THOSE RECOMMENDED TO TAKE A REMEDIAL COURSE (READ OR PROB)

		Short course recommended	
		Read	Prob
Examination outcome	Fail	220	405
	Pass	405	220

TABLE 2.2 EXAMINATION OUTCOME RESULTS FOR STUDENTS PARTICIPATING IN ONE OF TWO REMEDIAL COURSES[a]

High GPA group only

		Short course recommended	
		Read	Prob
Examination outcome	Fail	200	5
	Pass	400	20

Low GPA group only

		Short course recommended	
		Read	Prob
Examination outcome	Fail	20	400
	Pass	5	200

[a] The data are presented separately for two groups of students who came to the counseling service. The two groups were formed from the 1250 participants according to their prior grade-point average; the two groups are called High and Low. This table reveals that the students taking the Prob course are more successful no matter what GPA group they are in. This is in contradiction to the apparent indication of Table 2.1.

67 percent of the Read group passed all their exams, compared with 20 of 25 or 80 percent of the Prob group. Similarly, in the Low group 5 of 25 or 20 percent of the Read group passed, but 200 of 600 or 33 percent of the Prob group passed.

At first this seems baffling. How can the Prob course look better in each group of students (in High and in Low) but look worse in the combined group? You should assure yourself that this has in fact been demonstrated. An explanation is needed.

It is clear that the Read course students were better students since 600 out of 625 were in the High group, compared with only 25 out of 625 of the Prob group. So the comparison suggested by Table 2.1 really is comparing not only the Read and Prob courses but also the high- and low-GPA students. The large number of high-GPA students in the Read course made the course look good. (Check that the pass rates for the High group really are greater than those for the Low group.)

The true relative effect of the courses is revealed only when the comparison used students with similar GPAs. (Of course, our conclusion could still be erroneous for the same reason that we were misled the first time. There may still be important features differentiating these smaller comparison groups which have not been taken into account.)

A graphical explanation of the paradox is shown in Figure 2.5. The paradox arose because similar groups were not assigned to the two "treatments," Prob and

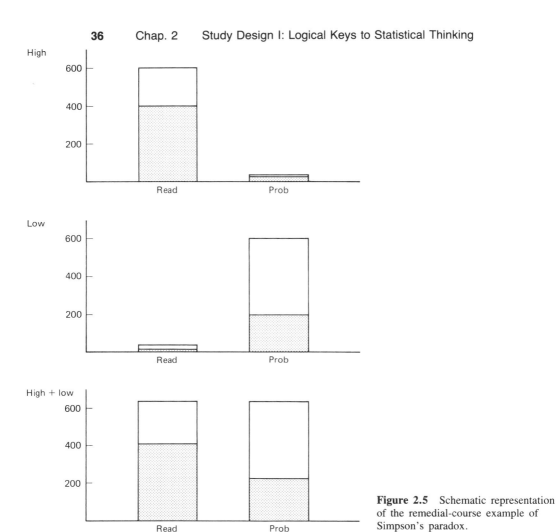

Figure 2.5 Schematic representation of the remedial-course example of Simpson's paradox.

Read. The effect of the course was "confounded" with the quality of the student (measured by Low versus High GPAs). The paradox that arises in this sort of situation is known as **Simpson's paradox.** If you think about Simpson's paradox carefully, and understand how it comes about, it will cease to be a paradox to you. A fading paradox signals a new insight.

The interpretation of survey data must deal explicitly with the effect of known confounding factors. Surveys, however, can still be misleading because of confounding factors that are either unknown or not measured. Because it is often not feasible to design a study as an experiment, we have to be on the lookout to avoid being misled by a "Simpson's paradox" in interpreting the results of a survey.

To keep the effects of causal factors, confounding factors, and common causes clearly distinguished, you may find the schematic diagram in Figure 2.6 helpful.

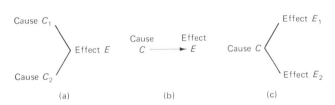

Figure 2.6 Schematic diagram showing three common causal paths. In (a), if only C_1 and E are measured, the importance of C_1 will be wrongly assessed. C_2 will be a confounding factor. In (b) is shown an uncomplicated situation that is usually the initial explanation of a relationship between C and E. Note that the statistical design may not distinguish this from the reverse path, with E causing C. In (c), it may be erroneously concluded that E_1 causes E_2, since the existence of a common cause will result in a relationship between E_1 and E_2.

Now let us return to the real-world issue of dietary cholesterol and its effect on the incidence of fatal heart disease. To study this proposed cause–effect relationship, we have to do an experiment. Several hundred subjects, preferably representative of some identifiable population (such as U.S. Caucasian males aged 20 to 30 years) would have to volunteer to participate in a long-term study, say of 30 years' duration. The investigator would assign dietary guidelines to the volunteers according to a random assignment mechanism, some guidelines having more cholesterol than others. The subjects would have to comply with their assigned guidelines for the 30 years. At the end we could then compare the incidence of heart attacks in relation to the diet assigned. Is it any wonder that the issue has not been resolved? It should be clear that even the best studies will only be able to produce suggestive evidence, and that confusion about this issue is likely to persist for many years, no matter how much money is spent on its resolution.

EXERCISES

1. Why are studies on the effects of smoking in human beings almost always surveys rather than experiments?
2. Suppose that you have 300 marbles, of which 100 are large and 200 are small. For each of the 300 marbles, you flip a coin to assign it to a group I (heads) or group II (tails). Would the two groups of marbles have a similar mix of large and small marbles? What relevance does this question have to the design of the teaching experiment described in this section?
3. Five greyhounds and five whippets are entered in the same dog race to decide which is the faster breed over a 250-m course. Is this a survey or an experiment, or could it be either one depending on how it were organized? Explain.

4. Construct a new example of Simpson's paradox starting with the following table.

		Short course recommended	
		Read	Prob
Examination outcome	Fail	100	200
	Pass	200	100

2.6 SPECIAL PROBLEMS OF STUDIES WITH HUMAN SUBJECTS

Sampling human populations is not quite like sampling apples from a barrel or mice from an experimental stock. Apples and mice are more manageable. Some of the major problems with human subjects are listed below.

1. Many people object to being categorized and listed and consider this an invasion of their privacy. When lists are made of certain categories of people, such as all subscribers to a certain trade magazine, these lists have commercial value. However, in exploratory studies, population lists are often not available at any cost, so it becomes very difficult to describe the target population explicitly. In such cases it is usual to define the target population implicitly by describing the method of selection of the sample. This is not very satisfactory from a logical point of view, but it is sometimes necessary.
2. Even when a complete list of the target population is available, prospective subjects can choose not to participate (see Figure 2.7). Worse still they can agree to participate and once included in the sample, change their minds. If the subjects who change their minds are unlike those who do not, the remaining subjects may present a biased response relative to the intended target population.
3. Some human participants may be too enthusiastic and unwittingly bias the results of the study in this way. Or they may interact in complex ways with the investigator so that the investigator's bias affects the subject's response.
4. Another feature of human populations is their heterogeneity: small samples can be quite different from the population they are supposed to represent.
5. Human populations are particularly subject to change with time. They are similar to other animals in their diurnal and seasonal rhythms, but they have the additional influences of political and intellectual forces. An example of a politically induced time influence on most human activities is the weekly cycle; an example of a more complex influence would be time trends in the sympathy toward social welfare induced by economic cycles. Thus when a human population is studied, if it cannot be observed at a particular instant, these time influences can complicate the interpretation of the aggregated results.

Sec. 2.6　Special Problems of Studies with Human Subjects　39

Figure 2.7 The understandable reluctance of human subjects to participate in certain experiments.

Typical studies with human subjects do not eliminate these problems entirely, but there are several techniques that minimize their influence on the results of the study.

We now consider these various problems in connection with a survey of opinion about a proposed property tax increase in a particular municipality. Let us suppose that we have been retained as consultants by the incumbent political party in preparation for an election campaign. The party would like to know what new programs would be popular enough to offset the disfavor incurred by increased taxes. Whose opinion should be sought? If we find that we cannot define the target population exactly, this is a good indication that we have not clearly defined the study objectives. If we say that our target population is the collection of people who will vote on election day, we have a crisp but useless definition, since it is this definition that must help us to select the sample *before* the election. A compromise might be to use voters lists, which are usually posted well in advance of an election. Of course, these lists will contain many who will not vote, and who are thus, in a sense, of no importance to the political candidates. This sort of compromise is inevitable in social studies, and must be accounted for, informally at least, in the interpretation of results.

When subjects have some control over their degree of involvement in a study, we have to consider what effect this is likely to have on the results. This effect is not entirely amenable to analytical methods; some of it will have to be subjective. For example, in the political opinion survey, if prospective subjects sense that the poll is being conducted for the use of the incumbent party, and if they do not support that party, they are unlikely to participate. So we must judge how many prospective subjects are likely to react in this way. Certainly, this special subgroup of the population, if they were to respond, might be expected to have response tendencies different from the rest of the population, so the potential for bias is present. This bias is called **nonresponse bias.** A possible technique for assessing the seriousness of this bias is to select out the responses of those who agree to participate but claim not to support the incumbent party. This group may be somewhat akin to the nonrespondents in their political affiliation, and yet we would have their responses to the survey.

Another phenomenon exhibited by human subjects is the placebo response: many investigators find that subjects will tell the investigator what they think the investigator would be pleased to hear. ("Placebo" in Latin means "I will please." Originally, the word was used in medical contexts for a medication that would please a patient. The most common modern usage of the word *placebo* is to refer to a pill that has no biochemical effect and is used in medical experiments for comparison with a pill supposed to have some biochemical effect. This is discussed further in Chapter 3.) The investigator can be misled in this way. The objectivity of the investigator, or at least of the agent who is doing the data collection, is necessary. If you have ever been approached to answer such a survey, it is likely that you have been assured that the agency is objective and nonpartisan so that your response will not be biased. This assurance may be truthful even though the survey is commissioned by one party. Clearly, a polling organization must be very concerned about its reputation for objectivity if it wishes to obtain responses that are free of the placebo bias.

One final word about studies with human subjects. The example used in the discussion above was a survey rather than an experiment. There are only a few contexts where human experiments are feasible. The practical difficulty in experimenting with human beings, other than the obvious ones of contacting willing subjects and subjecting them to treatments that could be harmful, is that human subjects will not inconvenience themselves very much in the interest of research. Of course, many proposed experiments have moral difficulties as well; it can be argued that even when subjects are willing to participate in an experiment, the investigator is not absolved from the ethical issues in the manipulation of other human beings. But to address the simpler practical issues, we will consider in more detail an example of a proposed medical experiment.

What would be necessary for an experiment to determine whether or not drinking large amounts of coffee, over a period of five years, causes bladder cancer? Volunteers would have to drink the prescribed dosage of coffee for the five-year period whether or not they felt like it. Another group of volunteers would have to not drink more than a certain amount for the same length of time. Since this is a proper experiment, the assignment to the high-dose group or to the low-dose group is made on a random allocation basis, with no allowance for the coffee preference of the subjects. This study would clearly fail because of noncompliance of the subjects to the treatments allocated. Many issues of importance could be resolved but for this unfortunate pattern of behavior of human beings. Experiments with human subjects are very difficult to perform so that they are both ethical and informative. This is why most studies of human populations are of the observational type.

EXERCISES

1. List very briefly the special problems involved in studies with human subjects.
2. Outline the methods available to assess the seriousness of nonresponse to a survey.
3. A television ad claims: "More dentists recommend Black Magic toothpaste than any other brand." You have never heard of this particular brand and wonder whether dentists are deliberately keeping this marvelous product to themselves. Are there any ways that you can think of that the manufacturer of the toothpaste could be making a technically truthful statement even if the toothpaste does not have widespread acceptance among dentists?
4. To begin a study on hyperactivity in preschool children, it is proposed that a random sample of hyperactive children in your school district be surveyed. What would be the major practical difficulty that you would have in executing this survey?

2.7 SOURCES OF VARIATION IN MEASUREMENTS

By *variation in measurements* we mean the tendency of measurements to result in different values. The variation may be due to differences in the items measured, but even if the same item is measured repeatedly, there will be some variation in the measurements. This variation could be due to the inconsistency of the measurer, to instrument imprecision, or to variation over time in the aspect measured. In this section we study the relative importance of these sources of error and how they affect study design.

Let us begin by supposing that you have just taken a test designed to measure your IQ and the result is reported to you as 125. IQs are usually between 70 and 130, so 125 is quite a high score. Does this score mean that your IQ is exactly 125, between 123 and 127, or between 110 and 140?

How much stability is there in this number? You would probably not get exactly the same score if you were to be tested again. Such a retest would involve a different set of questions, at a different time, and the test might well be administered under slightly different conditions. Your ability to perform on IQ tests may actually change over time. There are several reasons why this score of 125 is considered only a rough indication of your ability to answer the types of questions posed in such tests. (We will avoid the interesting question of what "intelligence" really means and whether the IQ test really measures it: this is something for the psychologists to clarify.) What can we do to make this IQ measure less "rough"? In other words, how can we organize the measurement process so that the potential variation in the score is reduced? Variation reduction is one of the common aims of study design techniques.

To reduce variation due to the particular selection of questions on a test, we could increase the number of questions selected at random from the question bank; in this way the test itself can be made to have the same difficulty each time it is offered, even though it is not feasible to allow much repetition of questions. To reduce variation due to the conditions of writing, we could attempt to conform to set standards (time of day, level of background noise, weather conditions). To reduce variation due to changes in the ability of the subject (i.e., the writer), we could specify certain criteria which must be met by the subject, such as: has been able to read for six years, is of chronological age 12, is female, and so on.

However, it is possible to go too far in our elimination of variation. Some variation may be the object of study and should not be reduced or eliminated. For example, does IQ at age 12 depend on the number of years a child has been reading? This question would be impossible to answer if only children who had been reading six years had been tested. This example points out one of the great dilemmas of study design: How can we eliminate the variation we do not want to study without eliminating the variation that we *do* want to study? This is why it is crucial to examine the possible sources of variation in measurement when considering the design of a study.

Let us next consider a simpler example. A manufacturing firm wishes to check the quality of their product, computer disk drives, before shipping them to a customer. Let us suppose that their test of quality is simply to count the number of errors made by the drive in reading a large test file. The number of errors made will depend in part on the quality of the product and in part on the circumstances of the test. If the humidity is higher than usual, this might increase the error rates even though the product, when received by the customer, would be unaffected. When we are trying to decide whether to ship the product, or to return it for further calibration and checking, we should be aware of the size of the measurement error (in this example, due to ambient conditions of the test). If the measurement error were known to be large, and if the error rate during the test were high, we may have to repeat the testing procedure to decide whether the product conforms to our quality standards. If the measurement error were small, we could make the decision right away. The definition of "large" or "small" here would be made in terms of the relative size of the variation in actual product quality and the variation due to measurement error.

The important thing to remember is that variation in measurements does not necessarily mimic actual variation in the items measured; the latter is often much less. This is obviously true when the same item is repeatedly measured. The effect is less obvious, but still present, when each measurement is on a different physical item. For example, consider the following problem of clinical chemistry. The measurement of the concentration of cholesterol in blood is often said to be an unreliable measurement. The measurement of cholesterol in the same person on successive days can be quite different. Measurements on each of several persons can be quite different even if all are in "normal" health. Such variation reduces the value of the measurement to the clinician. To improve this situation, one would have to know whether the variation is in the measuring process, among individuals, or changing over time in individuals, and also to know the relative importance of each source of variation. We would not naively assume that the variation exists in the persons themselves.

For applications of statistical methods, the consideration of sources of error is usually very important. (See Figure 2.8.) We deal explicitly with modeling measurement error in Chapter 7.

EXERCISES

1. "Normal ranges" for biochemical tests of human serum are often determined by examining the values obtained for a group of people who are in apparently good health. These ranges are used in the assessment of new patients and in the monitoring of patients undergoing some treatment. One such measure is serum glucose. Would you expect a healthy person whose serum glucose was measured 10 times over one year to have measurements as variable as those of a group of 10 apparently healthy people who were measured once? Explain.

2. Grocery stores frequently sell their meat in packages that have been weighed and this weight is automatically converted into a package price depending on the unit price (price per pound or per kilogram). The butcher scoops a quantity of ground beef that he judges to be approximately 1 kg and the weigh scale produces a label with the measured weight and the associated price printed on it. Let us suppose that the scale reports the weight to the nearest 0.02 kg (i.e., 0.96, 0.98, 1.00, . . .). Is there any way you could choose a package of ground beef that would tend to save you a few cents per package? Explain.

3. At a carnival, an expert gypsy will, for 25 cents, guess your weight and pay you 10 cents for each pound she is in error. (If she errs by three or more pounds, you will have won at least 5 cents.) The gypsy keeps a list of her guesses and a list of the actual weights as measured by an accurate scale. Which list would you expect would have the greater spread of values?

The Unexpected Tiger

Princess: You're the king, father. May I marry Michael?
King: My dear, you may if Mike kills the tiger behind one of these five doors. Mike must open the doors in order, starting at 1. He won't know what room the tiger's in until he opens the right door. It will be an *unexpected* tiger.

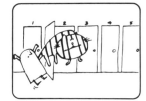

Having proved there was no tiger, Mike boldly started to open the doors. To his surprise, the tiger leaped from room 2. It was completely unexpected. The king had kept his word. So far logicians have been unable to agree on what is wrong with Mike's reasoning.

When Mike saw the doors he said to himself:
Mike: If I open four empty rooms I'll *know* the tiger's in room 5. But the king said I wouldn't know in advance. So the tiger *can't* be in room 5.

Mike: Five is out, so the tiger must be in one of the other four rooms. What happens after I open three empty rooms? The tiger will have to be in room 4. But then it won't be unexpected. So 4 is out too.

By the same reasoning, Mike proved the tiger couldn't be in room 3, or 2, or 1. Mike was overjoyed.
Mike: There's no tiger behind *any* door. If there were, it wouldn't be unexpected, as the king promised. And the king *always* keeps his word.

Figure 2.8 An example of the failure of common sense.[3]

2.8 ANSWERS TO EXERCISES

Section 2.3

1. Your answer should consist of three different letters of the alphabet. Your guess at the proportion can be checked by doing Problem 3 or 14. (You may be surprised at the small percentage of three-letter combinations that are real English words.) The proportion will be different for each person, since the comprehension of a "word" will depend on the vocabulary size of each person.

2. One bottle should be sufficient. See the discussion in the first paragraph of Section 2.3.
3. (a) Sampling should be without replacement. Sampling with replacement would include the possibility of testing a firecracker twice, which surely would be senseless.
   (b) No. There is little to be gained by inspecting the same firecracker twice, and this is a real possibility if sampling is done with replacement. It would be better to ensure that three different firecrackers are inspected. Of course, if the box of firecrackers is large enough, the difference between the two methods would be negligible in practice.
4. The sampling procedure is fair in the sense that each square has an equal chance of being selected into the sample. But whether it is representative in any sense depends to an extent on luck. (One possible meaning of the word "representative" would require that the sample be the same *size* as the population. The sample is certainly not representative in this sense. We will use more permissive meanings of "representative." The fact that the term has more than one meaning is one of the lessons in this exercise.) But the proportions of squares of each size in the sample might, one would hope, be something like the population itself. In this case this is a vain hope, because the sample size is quite small. It is probable that some square sizes will be missed entirely. About all that we could realistically hope for in this small sample is that the average square size (side length or area) might be well approximated by the sample, and the sample would therefore be representative in this sense.
5. Since 1000/80 is approximately 12, there is a good chance that the monthly nature of the business will make these selected invoices unrepresentative of the whole 1000. To avoid this difficulty the auditor should choose his invoices according to some random sampling scheme.
6. Clusters are supposed to be similar in subject mix to the populations so that a few of them will represent the whole population well. Strata are chosen to be quite different from each other, and hence different from the total population of which they are a part. The idea with strata is that one tries to get a few subjects from each of every strata, with the strata being as homogeneous as possible so that they are well described by the few subjects chosen. Thus city blocks and households would be reasonable clusters, while the elderly population, apartment dwellers, and males would be possible strata.
7. (a) *B*
   (b) *A*
   Any subpopulation in *B* gives the same average as the entire population, 5.0. So the subpopulations in *B* would make good clusters. On the other hand, any one cluster in *A* could give estimates with a large error. In population *A*, all the subpopulations need to be represented in the sample in order for the sample averages to be similar to each other (and to 5.0). These subpopulations of *A* make good strata. You can easily see that stratified samples from *B* could produce quite widely varying sample averages.
8. Your sample average of the 10 random numbers (each random number in the range 0 to 9999) should be within 1000 of 4999.5 (the actual population average) more than 70 percent of the time. The theory underlying this is covered in Chapter 7.

Section 2.4

1. No. It is not certain that the short players averaged fewer points than the tall players, based on the information given.
2. This does make a comparison possible, and does suggest that height is an important factor

in scoring performance. However, the evidence is not conclusive: there may be characteristics of tall basketball players, other than height, that are the key to good performance. For example, if tall athletes tend to be less well coordinated than short athletes, they may be induced at an early age to play more basketball than other sports, and as a result have more basketball experience than other sports, and as a result have more basketball experience at a later date than short players of the same age. (This is not a serious proposal, only a logical possibility.) Another consideration is that a basketball team may be more successful with a mixture of short and tall players.

The new information changes the answer from no to maybe.

3. See the example in the preceding answer.

Section 2.5

1. To perform an experiment on people which would detect the effects of smoking, one would have to assign, at random, a regimen of smoking or nonsmoking to subjects who were willing to participate in the experiment and capable of complying with the regimen assigned. Considering that the effects of smoking may take many years to become noticeable (as with cancer or high blood pressure), it would be almost impossible to enlist subjects in a true experiment of this issue.

2. It is likely that the proportion of small marbles in each group would be close to 100/300. Yes, they would be similar except in rare instances. The allocation of large and small marbles is a model for any characteristic of the students in the teaching experiment that is possessed by 100 students but not by the other 200 students. The actual split into 100 and 200 is unimportant in this model. Thus any characteristic possessed by some students and not by others is likely to be sorted into the two comparison groups in approximately equal proportions. So the model is relevant to the example. (*Note:* To appreciate the ingenuity of random allocation, try to think of another method of allocation that tends to balance the two comparison groups with respect to *all* characteristics, including those not yet thought of.) So the marble experiment is logically equivalent to the allocation of students to two groups, with the identical goal of balancing the groups with respect to any particular characteristic.

3. It is not an experiment, since the breed of dog cannot be assigned to experimental animals by the experimenter. The animals available for the study are greyhounds and whippets, and this cannot be changed at the outcome of the flip of a coin. The practical problem here is that we do not know whether the dogs available for the study are representative specimens of their breed. If the five whippets were a random sample of all whippets, and similarly for the greyhounds, some firm decision about the breed speed could be made.

4. The original table given is the following:

		Short course recommended	
		Read	Prob
Examination outcome	Fail	100	200
	Pass	200	100

Two tables that add to the one given are

		Short course recommended	
		Read	Prob
Examination outcome	Fail	5	195
	Pass	0	85

		Short course recommended	
		Read	Prob
Examination outcome	Fail	95	5
	Pass	200	15

Using these, the explanation is the same as in the example described in Section 2.5.

Section 2.6

1. - Population lists not available
    - Sample dropouts
    - Placebo effect
    - Heterogeneity
    - Variation over time
2. - Subjective evaluation of the likely reason for dropping out and its relationship to the information sought in the study
    - Comparison of information available about nonrespondents (or noncompleters) with the same information about respondents (such as sex determined from name, and address)
    - Study of a select group of respondents thought to be similar to the nonrespondents, and adjustment to the respondent data to reflect the entire population sampled

   The manufacturer could have asked the dentists: "Here is a list of ingredients in our Black Magic brand toothpaste. Would you recommend this brand?" All that is needed is one positive response for the television claim to be "truthful."
4. It would be difficult to obtain a list of the population of interest, preschool children that were hyperactive, and in this case the selection of a random sample would be impossible.

Section 2.7

1. The 10 people will have serum glucose values that vary more than the 10 readings from one person, because there is an extra and independent source of variation, the variation among people.
2. Choose the lightest package, as indicated by the label. It will tend to include a negative rounding error, since the weight has turned out to be the smallest.
3. The list of guesses has an extra source of variation: the error of the gypsy relative to the accurate weight. This list will probably have more variability than the list of true weights. (The exceptional instance would be the one where the heavy people are underestimated and

the light people overestimated. In this case, if the tendency is strong enough, the list of guesses would have the lesser variation. These phenomena are discussed in Section 9.5.1.)

2.9 SUMMARY AND GLOSSARY

In this chapter we have touched on many complex but fundamental issues in the area of study design. Even if you were never to design a data-based study, it would still be advisable to know the basics of study design simply so that you can evaluate this aspect of studies in which the conclusions are important to you. What, then, are the most important points to be retained from this chapter?

1. Unexpected observations may ultimately lead to useful discoveries, but usually a formally designed study is required to decide whether the cause of the unexpected observation is as suspected.
2. The best way to learn how to assess critically reports of studies is to practice the process, either under supervision or with some other form of corrective feedback.
3. When a study focuses on sample information, it is crucial to examine the method of sampling before generalizing to the population. Random sampling of some form has more predictable properties than other forms of sampling.
4. The query "Compared to what?" is often a useful one in the assessment of numerical information. A useful follow-up question is "What other factors might influence the comparison?"
5. The establishment of a causal mechanism in the presence of unexplained variation is almost impossible unless the investigator can control the assignment of the suspected causal agent to the study subjects.
6. Group comparisons from surveys can give misleading results. The phenomenon called Simpson's paradox occurs when in addition to the factor of interest, confounding factors contribute to the observed group differences. When these confounding factors are unknown or not measured, misleading conclusions can result.
7. Studies involving human subjects tend to be equivocal concerning the establishment of causal mechanisms since human subjects are rarely willing to be manipulated in the way the investigator decrees. Even the simple process of description of human populations can meet with resistance, a fact that complicates the design of surveys with this objective.
8. The presence of unexplained variation complicates the process of inference from data, and investigators try to reduce this as much as possible by appropriate design strategies; but in doing so the investigator must be careful not to reduce the variation that is the object of the study.

In the next chapter we deal with study design strategies in more detail.

Glossary

A **population** is an aggregate of items, or of numbers that represent measurements on the items. Usually, a population is the object of interest of a study, and usually the information about the population must be gleaned from a sample of the population. (2.3)

A **sample** is a subset of a population. (2.3)

Random digits are digits (0, 1, 2, . . . , 9) produced in such a way that the patterns of digits that are sequentially produced are like those that would be produced by successive draws (with replacement) of tickets from a hat with thorough mixing between draws. (2.3)

A **random sample** is a sample selected from a population in a way that uses some chance mechanism to determine the items selected. Usually, the chance mechanism is simulated (or mimicked) by following a sequence of random numbers. (2.3)

A **simple random sample of size n** is a sample of n items selected without replacement from a population in such a way that each possible distinct sample of size n has the same chance of being selected. Using random numbers to identify the items selected from the numbered population list is the usual way to do this. For a simple random sample, note that it is impossible to select a single population item more than once, since sampling is without replacement. (2.3)

A **systematic sample** is a sample selected from a population by arranging the population in some sort of order, and then selecting every kth item from this ordered arrangement. The starting position is often selected at random. (2.3)

Stratified sampling is sampling from a population by selecting a simple random sample from each of several subpopulations, called **strata**, and combining the information from all the stratum samples. The stratified sample will include representatives from every stratum. (2.3)

Cluster sampling is sampling from a population by first dividing the population into subpopulations, called clusters, and then forming the cluster sample from a simple random sample of these clusters. Typically, the cluster sample includes all representatives from a few clusters. (2.3)

An **experiment** is a comparative study in which the comparison groups are formed from a single pool of study subjects by the investigator. This assignment to comparison groups by the investigator is usually done according to some random mechanism or table of random digits. (2.5)

A **survey** is a comparative study that is not an experiment. Usually, a property of the subjects themselves determines which comparison group the subject joins. (2.5)

Simpson's paradox: When proportions are compared between two groups, the group with the larger proportion in the raw data can become the group with the smaller proportion in adjusted data if the adjustment is made for the effect of a hidden confounding factor. (2.5)

Nonresponse bias: When a survey of a population is performed by attempting to access a sample from the population, and when the persons sampled do not all provide the information requested, the summary of the respondents may be systematically different from a summary of the entire sample, and hence give a biased indication of the population. (2.6)

PROBLEMS AND PROJECTS

1. In the CEO salary survey described in Section 2.3, does a simple random sampling method guarantee that a representative sample will be obtained? Explain why or why not.
2. In the CEO income survey described in Section 2.3, you wish to estimate the average income of the 10,000 CEOs based on the 100 reported incomes.
   (a) How would you do this?
   (b) Is there any information in the 100 reported incomes that would give you a hint about the accuracy of your estimate? Explain. (*Hint:* What would you conclude if all 100 incomes in the sample were identical?)
3. Repeat Exercise 1 of Section 2.3 until you think you have a rough estimate of the proportion of "words" that make an English word. Choose your answer from among the following:

   $$0.1\%, 1\%, 5\%, 10\%, 20\%, 30\%$$

   (This problem is continued as a computer project in Problem 14.)
4. Simpson's paradox occurs in the data set described below. Exhibit the paradox clearly and give a verbal explanation of it in the context of this example. (*Hint:* Examine the data in the two tables below and compare them with a table for the two plots combined.)

   Description of data set: A farmer is comparing the quality of two kinds of tulip bulbs that he is planning to produce commercially. He has both kinds of bulbs planted in two garden plots. The bulbs are classified as type I or type II and are evaluated as being of High quality or of Low quality. The number of bulbs in each category is shown below.

   Plot A

Quality	Type I	Type II
High	55	5
Low	70	10

   Plot B

Quality	Type I	Type II
High	13	75
Low	7	60

5. Describe how you would conduct a survey of students in a statistics class to determine their impressions of the competence of the instructor. Your description should include consid-

eration of the proportion of the class responding to the survey, the format of the student responses (checklist, prose, verbal, etc.), the method of data summary, the applicability of the results for comparison of instructors in different classes, and the confidentiality of the students' responses.

6. The university administration is seeking the opinion of the student body concerning the relative desirability of more on-campus residences versus greatly improved transportation facilities to the surrounding community. The university has 10,000 students. Suggest methods for obtaining this information at modest cost. Explain the rationale for your choice.

7. A survey of college freshmen is based on a simple random sample of 1000 students out of 5000 in the freshman category. In a personal interview with the surveyor, the key question asked of the students sampled is: What was your GPA (grade-point average) in the most recent year that you were in high school? The average GPA for the 1000 responses was 2.5. The next year a similar survey is done using mailed questionnaires. There were 500 completed questionnaires returned and the answers to the key question averaged 3.0. Should we conclude that college freshmen this year have higher GPAs from high school, on average, or is there a better explanation? Discuss.

8. Over a period of two years, a doctor treats 200 cases of a certain disease with one of two possible treatments. One treatment is the standard treatment (S) and the other is an experimental (E) treatment. Both treatments are used throughout the two-year period. The patients are able to be classified as "cured" (C) or "not cured" (N) within a week of the administration of the treatment. The doctor's data are summarized below.

		Treatment	
		S	E
Result	N	50	90
	C	50	10

After analyzing in detail the data from the two-year study (the contingency table above and related variables such as severity of the illness, age of the patient, etc.), the doctor decided to use the experimental treatment all the time, since she believed that this had a higher cure rate. What rational explanation can you suggest for the doctor's belief? (*Hint*: Your answer should be based on a general concept covered in this chapter. You may wish to expand the example given to illustrate your answer.)

9. What is the principle involved in Exercise 4 of Section 2.6? What does this principle imply for the selection question in Problem 1(a) of Chapter 1?

10. The student society at a college is conducting a study of student attitudes toward the co-operative education option at the college. The most active co-op programs at the college are in the science and business departments, with only small or nonexistent programs in the humanities and fine arts. The purpose of the survey is to decide whether or not the student society will become active advocates of the expansion of the program. One important question in the survey is: Do you agree that the student society should encourage the administration to shift additional financial resources to the co-op program? A simple random sample of 100 students is selected from the registration lists, and the following data

are noted concerning the sample itself compared to the student body as a whole, based on information from the registrar's files. (The figures in parentheses should be interpreted as, for example, 213 students of the 1253 are in co-op.)

	College	(Co-op)	Sample	(Co-op)
Science (including Computing Science)	1253	(213)	30	(4)
Business	984	(345)	29	(6)
Humanities and Fine Arts	2145	(17)	37	(1)
Other	57	(0)	4	(0)

This information is available to the student society organizers before the survey is conducted, since the data are available from the registrar.

(a) What do these presurvey data tell you about the representativeness of your sample?

(b) How would you use these data to improve the accuracy of the information provided by the survey when it is undertaken? (Suppose that your objective is to estimate, from the sample of 100 students, the proportion of all 4439 students that answer "yes" to the question about increased support of co-op education.)

(c) How would you treat the information from the 11 co-op students in the sample—separately or included with the rest? (State what considerations would help you to decide.)

(d) If you were in charge of this survey, how would you design the sample selection? Justify your choice.

11. Taxis in a large city are issued special license plates with numbers 00001, 00002, 00003, . . . up to however many numbers have been issued so far. These numbers are issued to the taxi owners in the order in which they are requested, so that, for example, license number 00343 is the 343rd license issued. As a visitor to the city you notice three taxis numbered 00287, 01012, and 01311. You would like to estimate the number of taxi licenses that have been issued. You guess that there are about 2000, but you realize that any number 1311 or greater is a possibility. But what is an intelligent estimate?

Since you have time to kill and a microcomputer under your arm, you decide to investigate this problem using the Monte Carlo technique. Begin with a tentative assumption that the number of taxis is 2000, say, and use a table of random numbers to see what kinds of samples of size 3 you would expect to get if the assumption is correct. The process of going from the sample (such as the given one of 00287, 01012, 01311) to the population size (2000?) could then be studied.

To get a feeling for the Monte Carlo method, try generating samples of size 3 using a random number table. If you try 10 samples of size 3, for a total taxi population of 2000, and do all this again for a total population of 3000, you should get some idea of how to proceed from the sample to the population size. Try out your method on the three-number sample given. (This problem is continued as a computer project in Problem 15.)

12. With reference to Figure 2.4, the population of 49 squares, consider the following sampling procedure. Throw a 12-in. ruler onto the figure; select the squares intersected by the ruler edge. (Decide before the throw which edge you will use.)

(a) Is this selection procedure equivalent to a simple random sample? Explain why or why not.

(b) Does the selection procedure give each square an equal chance of being selected?

(c) Perform the experiment and comment on the sample obtained.

13. At Aeio University, Statistics 202 is a second course for which Statistics 101 or Statistics 102 are adequate prerequisites. Students cannot take both 101 and 102 for credit. It is proposed to examine the records of all people at Aeio University who have taken Statistics 202 this year and to compare the 202 grades obtained by those who have taken 101 with the 202 grades of those who have taken 102. The aim is to see which course provides better preparation for Statistics 202. Comment on the logical problems inherent in the proposed study design.

Computer Projects

14. (Continuation of Problem 3.) See Exercise 1 of Section 2.3. Using random numbers to print out "words" of three letters, generate three lists of 50 words each. Inspect each list to detect the number of English words included. Note the similarity, or lack of it, in the three proportions. If each list produced about the same proportion, you would probably believe that you knew the proportion you would get from a large sample. Does a 50-word list yield a useful estimate of the desired proportion? (For example, would it allow you to pick reliably the best answer from among those given in Problem 3?)

15. (Continuation of Problem 11.) Enlarge the investigation suggested in Problem 11 by studying the relationships among:
    - The population size
    - The average license number in a sample
    - The maximum license number in a sample

    for samples of size 3 and for populations of size 1000, 1500, 2000, 2500, and 3000.

Study Design II: Principles of Experimentation

Garbage in, garbage out.

Anonymous (*Often with reference to computer-based data processing*)

Observation is a passive science, experimentation is an active science.

Claude Bernard (*1813–1878*)

In Chapter 2 we saw that the unambiguous detection of causal paths required an experimental design. The crucial step was the use of randomization in the formation of comparison groups. When an experiment is being planned, it is often desirable to assess several experimental strategies simultaneously; this leads to more complex design issues. We touch on three principles of design of experiments in this chapter:

1. The use of averages to avoid the confusion produced by uncontrolled variation
2. The use of blocking to eliminate irrelevant variation from comparisons
3. The use of factorial designs to assess several proposed causal factors at once

As a preliminary to these topics we introduce some terminology and the setting for a simple experiment.

Key words are experimental units, factor, level, independent variable, dependent variable, treatment, replication, randomization, blocking, crossover design, factorial design, and interaction.

3.1 THE TERMINOLOGY OF EXPERIMENTATION

Let us first describe a very simple experiment. Its objective is to determine if the consumption of sugar prior to a 1-mile run improves the performance on the run. We say that the amount of sugar given to a runner is the value of the *independent variable*

Sec. 3.1 The Terminology of Experimentation   55

(since the value is set by the investigator prior to the experiment itself), and the time taken to run 1 mile is the value of the *dependent variable* (since this value may depend on the amount of sugar consumed).

We need a group of volunteers who are willing to accept whatever regimen the investigator chooses for them. These volunteers form the *experimental units* (or experimental subjects as they are usually called when the units are people).

The causal *factor* under study is the consumption of sugar, and let us say that 4 oz, the amount in about two chocolate bars, is the amount with which we wish to experiment. Some subjects will receive the 4 oz of sugar, and some will receive none. We say that the 4 oz of sugar is one *treatment* and the 0 oz of sugar is the other *treatment*. Thus we have two treatment groups.

Ideally, we would give the "no sugar" treatment group something that looked and tasted like sugar so that there would be no possibility of a psychological factor contributing to a difference in performance of the two groups. A way to do this might be to present the potion as a clear syrup and use saccharine or a similar sugar substitute to sweeten the sugarless potion. When the two treatments are indistinguishable to the subjects, it is customary to call the no-action treatment (the saccharine treatment in our example) a *placebo* treatment. As noted earlier, the idea behind the use of a placebo is that a treatment that the subject thinks is helpful will be pleasing to the subject even if it has no effect at all. It is obviously quite important that the placebo does not have any chemical influence on the performance. That is, we would hope that the only effect of the placebo treatment is the psychological one; otherwise, it will be difficult to interpret the outcome of the experiment.

The 1-mile course will be the same for all subjects. We may wish to start all the subjects at one time to ensure equality of conditions for the experiment. The measure of performance will be the time taken to complete the course.

Now with the experiment as described, we should be able to detect the effect of the sugar, if any, on performance. Suppose that the data are as follows:

Group	Runner	Time (sec)
Sugar	1	320
No sugar	2	285

Before drawing any conclusions, we would want to know whether the two runners would have had more equal times if they had received the same treatment; in other words, are the runners of comparable ability? The investigator may say that he used an unbiased method to allocate the "sugar" treatment to runner 1 and the "no-sugar" treatment to runner 2. But this really does not clarify the situation at all. If the runners are different in ability, there is no way to tell whether the difference is due to a difference in ability or due to the treatment. If we observed a similar time difference for five such two-runner experiments, in which each treatment allocation was done with the flip of a coin, this would be much more revealing. Repetition of a part of an experiment is called *replication*. Replication helps us to sort out which differences are due to chance or uncontrolled variations, and which are due to the deliberate differences applied by the experimenter.

In discussing experimental design, it is common to use the term *randomization*. This term has a specific meaning in this context: it defines a way that treatments are allocated to experimental units—when this is done in such a way that all possible allocations are equally likely, the treatments are said to be randomized to experimental units. There is a hidden constraint on this process; usually the number of experimental units receiving each treatment is specified in advance so that the "possible assignments" are then understood to mean the assignments that result in the desired number receiving each treatment.

Note that this process is quite different in principle from the selection of a random sample—random allocation is a different process from random selection. Yet the mechanics of random allocation are the same as the mechanics of random selection. The assignment of "sugar" or "no sugar" to the 10 runners, in such a way that the two treatment groups were each of size 5, was done by selecting a random sample of 5 from the 10 and using this split of the 10 runners for the assignment of treatments.

We will explore the implications of replication and randomization further in Section 3.2. Let us first summarize the terminology we have introduced so far.

Experimental units are the objects on which the experiment is performed. (When the experimental units are people, it is customary to call them experimental subjects.)

A **factor** is a deliberately imposed variation in the conditions of the experiment; it is usually suspected to have a causal influence on the outcome of the experiment.

A **level** is a specific value of the factor.

An **independent variable** is a variable whose values indicate the levels of the factor to be used in the experiment. Experimental units that are allocated the various values of the independent variable form the comparison groups.

A **dependent variable** is a variable whose values measure the outcome of the experiment; group differences in the dependent variable measure the effect of the various levels of the independent variable.

A **treatment** is the application of a particular level of the factor to an experimental unit. The experimental units receiving a particular treatment compose a treatment group.

Replication is repetition of an experiment under exactly the same conditions except with different experimental units.

Randomization is the assignment of two or more treatments to experimental units in such a way that specified treatment group sizes are achieved, and such that each possible resulting allocation of treatments has an equal chance of occurring.

EXERCISES

1. Use the terminology of Section 3.1 to describe an experiment for determining which of two brands of suntan lotion produces the greater user satisfaction.
2. Describe the design of an experiment to determine the durability of two brands of outdoor paint. Use the terminology introduced in Section 3.1.
3. Two different month-long aerobics training programs are to be compared. The outcome measure is: increase in lung expiration volume over the duration of the program. Describe the design of the experiment.

3.2 USE OF AVERAGING TO REDUCE CONFUSION

In this section we discuss the necessity for relying on averages when we are comparing the effects of treatments in an experimental framework.

If we use only two runners for our experiment, as initially proposed in Section 3.1, we will have difficulty in separating the differences in runner ability from the effect of the sugar treatment. Even if the two runners usually clock similar times for this race (before the experimental difference is imposed), their performance in the experimental run will probably show some difference. It will not be clear whether the difference is large enough to interpret it as a consequence of the treatment, or whether it is simply a result of imperfectly controlled conditions of the race.

In summary, there are two problems with the two-runner experiment:

1. The random allocation of runners to treatments is "fair" in a sense but will not eliminate the unequal abilities of runners in the comparison groups when each group consists of a single runner.
2. The conditions of the race cannot be perfectly controlled—even if runners of equal ability run the same race, they will traverse slightly different paths, and there will be cycles of performance that are unlikely to be in phase for the two runners. We must have some way of anticipating the size of this effect, so we will know when the treatment effect is responsible for observed differences.

Happily, there is one solution to these two problems: increase the number of experimental units—runners in this example. Suppose that 10 runners are available for the experiment. A method of random allocation of the 10 runners to the two treatments is needed: one method to ensure five runners in each group is to select a simple random sample from the population of 10 runners—methods for this were discussed in Section 2.3. The five chosen could receive the "sugar" treatment (S), while the other five could receive the "no-sugar" treatment (NS). This method effectively "allocates treatments at random to the experimental units." Hypothetical data are displayed in Table 3.1.

To get an intuitive feeling for the information in this table about the effect of the

TABLE 3.1 HYPOTHETICAL DATA FOR AN EXPERIMENT ON 10 RUNNERS TO ASSESS THE EFFECT OF SUGAR ON RUNNING TIME (*S*-SUGAR, *NS*-NO SUGAR)

Treatment	Runner	Time (sec)
S	1	320
S	2	280
S	3	250
S	4	315
S	5	290
NS	6	275
NS	7	310
NS	8	330
NS	9	295
NS	10	320

treatment, look at the graphical display of Figure 3.1. What should be apparent from this picture is that:

1. There is considerable overlap in the times of the two groups.
2. There is some evidence that the sugar group tends to have lower times.
3. The evidence for conclusion 2 is too weak to be definitive—we would need an even larger number of subjects to decide the issue.

Figure 3.1 Graphical representation of experimental outcomes which displays a mild suggestion of a treatment effect.

Now the issue of unequal ability of runners is becoming less important. Each runner in the study has the same chance of ending in the *NS* group as in the *S* group. With five in each group we feel there is a chance that the running ability should be distributed fairly equally between the treatment groups. Of course, there is still a chance that the five best runners end up in one treatment group, but this chance becomes remote with the larger group size. Certainly, with treatment groups of size 5 we have less chance of being subject to bad luck in the allocation of subjects to treatment groups than we would with treatment groups of size 1.

The influence of uncontrolled variation is suggested by the spread of running times in each treatment group. In fact, unequal ability can be considered as just one more uncontrolled factor. We can see from Figure 3.1 that there is considerable variation due to uncontrolled factors. We hope that the differences in times between any two runners in different treatment groups is large compared to the difference among the individuals in each treatment group; otherwise, we would not be sure that the treatment difference would be reproducible—a difference that would be observed consistently among runners such as these under similar experimental conditions.

Why do five measurements per group give a better estimate of a treatment effect than one measurement per group? The reason is that averages of many measurements are often more consistent indicators of group differences than are single measurements. To see this, consider the treatment group averages: 291 seconds for the *S* group and 306 for the *NS* group. The average measures the "center" of the spread of running times, and we are better able to guess the difference between the two treatment averages for a very large number of subjects, based on the 10 in our experiment, than if we just use one representative from each treatment. This intuitive idea will be clarified in Chapter 7.

The main point for our present purposes is that replication of experimental designs produces more reliable estimates of the treatment effects than does a single measurement at each level in a design.

Qualitatively, the 10-runner design that has been used to study the effect of sugar on runners' performance has two advantages over the 2-runner design: the treatment differences can logically be attributed to the effect of the treatment, and the uncontrolled factors that still introduce variation to our measurements can be controlled by replication and averaging.

EXERCISES

1. In the experiment evaluating a sugar dose on runners' performances, how important is it that the runners participating in the experiment have equal running ability? Does inequality in ability of the experimental subjects present an insuperable problem for the experimenter?
2. Every runner experiences variation in performance from day to day. What strategy can be used to overcome the inconsistency that such variation can produce, in the comparison of performances of runners with the two treatments?

3.3 BLOCKING, FOR MORE PRECISE COMPARISONS

In the experiment described in Section 3.2, the problem of forming two groups of runners of comparable ability was encountered. The comparison of the groups after the treatments were allocated was always complicated by the fact that the groups may have been dissimilar before the treatments were allocated. One way to avoid this complication is to try the different treatments, sugar (*S*) and no sugar (*NS*) on the same runners. On day 1, runner 1 might use no sugar, and on day 2, this same runner could use sugar. The difference in the runner's times would indicate the effect of sugar. This could be done for each runner.

There are some obvious problems with this design: the weather on day 1 may not be as beneficial as that on day 2. There may be a carryover effect from one day to the next; a longer "washout" period might solve this, but then the runners might be in better shape on the later day. There might be a psychological effect of always having one treatment on the first day: for example, if the sugar treatment does tend to produce faster times and is always used on the first day, the runner may be encouraged to run with even more intensity on the second day, thereby obscuring the real difference. Such considerations suggest a new design, shown in Table 3.2.

TABLE 3.2 DESIGN THAT CAN COMPARE *S* AND *NS* TREATMENTS FAIRLY, EVEN IF THE RUNNERS ARE OF UNEQUAL ABILITY BEFORE THE ASSIGNMENT OF TREATMENTS

Runner	Treatment	
	Day 1	Day 2
1	S	NS
2	S	NS
3	S	NS
4	S	NS
5	S	NS
6	NS	S
7	NS	S
8	NS	S
9	NS	S
10	NS	S

If the two days' conditions are different enough to affect the outcome, or if the simple ordering of the days makes a difference, it seems reasonable that it should affect the two groups in the same way. (When this assumption is violated, we say that the day number and the treatment "interact"; this is discussed in Section 3.4.) If we assume that the "day effect" is to change the race time by a constant number of seconds, no matter which treatment is used, this constant amount will contribute equally to the average times for the two treatments, averaging over all 10 runners. This will allow us to estimate the effect of the sugar treatment in a way that is not affected at all by a day effect. We can similarly eliminate the possible influence of the runners being more fatigued on the second day, since we have equal frequencies for the two orderings of treatments.

The design just described is called a *crossover design*. Note that the treatments are being compared for each experimental unit, so the differences among experimental units do not complicate this comparison.

The crossover design is a particular design that uses the principle of *blocking*. The runners are "blocks" and the treatments are compared "within blocks." Blocking is commonly used when several treatments are being compared. The block is chosen such that treatment differences within blocks can be ascribed to the treatments themselves since the experimental units within a block are so similar. Suppose, for example, that we had a third treatment called "water" (*W*). (We might be interested to see if the obvious absence of sugar from the treatment has any effect on the runners, especially if they believe that a dose of sugar before the race is a good thing. If the sugar treatment is offered as a drink, with the sugar dissolved in water, the no-sugar treatment could be offered as an apparently identical drink, but using an artificial sweetener.) A design that might be used to compare the three treatments is shown in Table 3.3. Note that we need six runners (or blocks) to achieve all possible orderings of the three treatments. In the earlier design we needed only two runners for this; the reason we used 10 before was because we wanted to repeat the two combinations so as to protect ourselves from being misled by variation due to unknown or uncontrollable causes. For the same reason, we may wish to increase the number of runners in this three-treatment design. However, we should increase it in multiples of 6, to maintain equal representation with respect to the ordering of treatments. So 6, 12, 18, . . . runners could be used with this design.

TABLE 3.3 EXPERIMENTAL DESIGN TO COMPARE THREE TREATMENTS THAT WILL NOT BE UNFAIRLY AFFECTED BY DAY EFFECTS: COMPLETE BLOCK DESIGN

Runner	Treatment		
	Day 1	Day 2	Day 3
1	NS	S	W
2	NS	W	S
3	S	NS	W
4	S	W	NS
5	W	NS	S
6	W	S	NS

When each block compares the same number of treatments, and each treatment is applied to the same number of experimental units, the design is said to be *balanced*. When the number of treatments per block is equal to the number of treatments that are compared in the experiment, the design is said to be *complete*. So the six-subject design above is called a *balanced complete block design*. An example of a balanced *incomplete* block design is shown in Table 3.4. The advantage of this design over the complete block is that it uses half the number of subjects. In some situations, where there are a large number of treatments and only a few experimental units, incomplete block designs may be more feasible than complete block designs. Note that even in the incomplete block design, we still have a way of eliminating the day effect as long

TABLE 3.4 INCOMPLETE BLOCK DESIGN

Runner	Day 1	Day 2
1	NS	S
2	S	W
3	W	NS

as it is the same for each treatment. That is, the averages for each treatment may be compared without worrying about the influence of the day effect.

Note the status of the day variable in the designs we have described so far. The aim of the analysis has been to ignore the day effect. In this situation the day is called a "nuisance" variable. However, if we were interested in its effect, we might try to estimate this from our balanced designs. For example, in the incomplete block design above, the difference between the day 2 time of runner 1 and the day 1 time of runner 2 would estimate the day effect, but it would also be affected by any differences in the abilities of runners 1 and 2. Now, if we have enough such estimates based on a large number of runners, and if the runners have been allocated at random to the various orderings of the treatments, the influence of the variation in runners' abilities should average to a negligible amount. But we have lost the advantage of blocking for this estimate, since our day comparisons are not made "within" runners.

Let us review the definitions introduced in this section.

> **Blocking** is the experimental strategy of using a set of similar experimental units for more than one treatment, thus enabling a comparison of treatments that is not affected by differences among experimental units. The set of similar experimental units is called a block, and two or more treatments are compared in each block. The blocks may be physically divided into experimental units, as would be done in agricultural experiments: the experimental units are the subplots, and the blocks are the plots. Or they may be divided in time, as in the crossover design: the block consists of the identical experimental unit at different times.
>
> **Crossover design** is a blocking strategy in which the various treatments to be compared are applied to the same experimental unit at different times. Usually, the order of the assignment of treatments is arranged so that any time-dependent effect can be averaged to a negligible amount by the analysis.

EXERCISES

1. For the complete block design displayed in Table 3.2, the experiment on two treatments, explain why the day effect will not influence the relative sizes of the treatment averages.
2. (a) What is the minimum number of experimental units needed for a balanced complete block experiment that compares four treatments, A, B, C, and D?
   (b) What is the minimum number of experimental units in a balanced incomplete block design that compares these four treatments in pairs?
3. Consider the following two designs:

    Design I: Two runners are tested using two different treatments (S and NS) on two days, one getting S first and then NS, the other using the reverse order of treatments.
    Design II: Four runners are tested once, two with treatment S and two with treatment NS.

Assume that treatments are allocated at random to the runners, using the structure described. Which design, I or II, would best assess the relative effect of the two treatments? Explain.

4. A square field is to be used for an experiment to compare the yield and disease resistance of three varieties of wheat. Because the field varies considerably over its extent with respect to its moisture-holding ability and nutrient concentrations, the field is divided into three rectangular plots for the experiment. Each rectangular plot contains, in three square sub-plots, each of the three varieties of wheat.

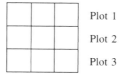

(a) What procedure would you recommend for the placement of wheat varieties in the subplots, assuming that each variety will occupy one subplot?
(b) What is the advantage of this multiplot design over the use of three large rectangular plots that use the whole field, with one variety in each large rectangular plot?
(c) Identify the experimental units, the treatments, and the blocks in the experiment originally described.

3.4 FACTORIAL DESIGNS

Let us return to the experiment assessing the effect of a dose of sugar on the time taken to run a mile. Suppose that we are also interested in the influence of the time of day. We have a theory that morning (*AM*) runs should be slower than evening (*PM*) runs. Of course, we could do a second experiment of the *AM–PM* effect. We would need two subjects for a balanced complete block design, so four subjects would be the minimum number for the two experiments. Note that the effect of each factor would have to be estimated from just two measurements. However, it would be better to have each factor's effect estimated from all four runners, allowing some of the uncontrolled variation to average out close to zero. The factorial design can achieve this.

A **factorial design** is one in which each experimental unit receives one level of each factor under study. A factorial design for the two factors "sugar" and "time of day" would be set up as indicated in Table 3.5. A more usual way to report these data

TABLE 3.5 FACTORIAL DESIGN

Runner	Sugar factor	Time-of-day factor	Observed time (sec)
1	S	AM	260
2	NS	AM	310
3	S	PM	295
4	NS	PM	325

are shown in Table 3.6. Each level of one factor appears with each level of the other factor. The effects of both factors can be estimated: the sugar factor would be estimated by comparing runners 1 and 3 with runners 2 and 4. The effect of the time-of-day factor would be estimated by comparing runners 1 and 2 with runners 3 and 4.

TABLE 3.6 DATA FROM A FACTORIAL DESIGN SHOWING THE CAPABILITY OF THIS DESIGN TO REVEAL THE EFFECTS OF EACH FACTOR SEPARATELY[a]

		Sugar	
		S	NS
Time of Day	AM	260	310
	PM	295	325

[a] There appears to be both a time-of-day effect and a sugar effect. (When the data set has more than one measurement in each cell, averages may be used.)

More exactly, the sugar treatment appears to reduce the running time by 50 sec in the morning and by 30 sec in the afternoon. Of course, we suspect that some of these differences are due to the differences in ability of our runners, but we still estimate that a reduction of about 40 sec (a compromise between 30 and 50 sec) is attributable to the sugar treatment, at least as a tentative conclusion.

Note the assumption that the reduction is the same for both morning and evening. If this is not true, we would not be entitled to average the two observed reductions. The reason we would want to use the average is that it is a more reliable estimate of the actual effect than is either of the differences alone. (This claim will be verified in Chapter 7.) The assumption we have made is described as the assumption of the absence of an **interaction** between the factors. When two factors interact in an experiment, the effect of one factor on the dependent variable will differ depending on the particular level of the other factor being used. For example, if there were a large interaction between the sugar factor and the time-of-day factor in their effect on running time, the use of the average of 40 sec as an estimate of the sugar effect, regardless of the time of day, would be nonsensical. It would be necessary to report separately the sugar effect for each level of time of day. One final advantage of factorial experiments is important to experimenters. When interaction is absent, a fact that can be inferred from the data, any single-factor effects which are present must have been demonstrated under more than one set of conditions. An experimenter will be more confident that the treatment effect is a real one if he or she observes it under more than one set of conditions. A factorial experiment has this feature.

Factorial designs may be used with a large number of factors and with a large number of levels for each factor. It is an efficient way of assessing several factors at once, provided that the interactions between the factors are not too complicated. It allows the experimenter to observe the effects of treatments under a variety of conditions.

EXERCISES

1. Three blends of gasoline are to be compared using two models of car, to see which combinations produce the best mileage. The following data on mileage (miles per gallon) are collected for the six possible experiments.

		Model	
	1	20.5	21.0
Gasoline blend 2		19.5	20.0
	3	19.5	19.0

   Assuming that these results are reproducible without error, describe the outcome of the experiment. Use the terminology of this chapter where appropriate.

2. A runner wishes to examine a possible relationship between the duration of his warm-up, the duration of his previous night's sleep, and the clocked time of his usual morning run. He wishes to do a preliminary study, with himself as the only subject.
   (a) Suggest a factorial design for an experiment to assess the effect of these factors on running time. (*Hint:* What are the experimental units for the experiment? How will you randomize treatments to experimental units?)
   (b) What is the most serious problem with the experiment (i.e., the problem that could prevent any useful information from being obtained)?
   (c) If the experiment were to be repeated several times, what aspect of the experiment would be improved? What important aspect is not improved?

3. A woman wishes to settle an argument with her neighbor about the most beneficial method of brewing tea. She feels that the teapot should be heated first with boiling hot water, whereas her neighbor feels that this is unnecessary unless a larger-than-usual quantity of milk is to be added to the tea. Design an experiment in which an impartial panel of four will decide the issue. Be sure to mention safeguards that would prevent the "impartial" panel from inadvertently adding their biases.

3.5 ANSWERS TO EXERCISES

Section 3.1

1. Volunteers at a public beach would be asked if they would be willing to participate in a comparative test of two suntan lotions. The two brands would be randomized to the volunteer experimental subjects. These subjects would be given two packages, marked A

and B but not otherwise identified as to brand. The packages would be handed out to volunteers on their way to the beach. The names and phone numbers of the volunteers could be obtained for a follow-up call to see which brand they liked the best.

The factor being examined is the brand of suntan lotion, the "values" of the independent variable are A and B, the dependent variable is the preferred brand as obtained from the follow-up phone call, the treatments are the two brands, and each volunteer is a replicate of the experiment. Labeling of the two brands as A and B could be done at random, according to the flip of a coin, but the code for which volunteer was given which label would have to be recorded when the lotions were handed out. In other words, the labels would not have the same meaning for all the volunteers, but a volunteer's preference could be identified properly by the investigator.

It can be argued that the effect of the packaging on user satisfaction has wrongly been omitted from this experiment. However, information on brand preference can be obtained more simply by asking sunbathers directly, if the actual use is not to be assessed. In this case the experiment would be partly testing the effectiveness of advertising. The purpose of the study would have to be spelled out more carefully to decide which type of study is most appropriate. In any case, if an experiment is to be done, as suggested by the exercise, presumably it should be done as described above. Of course, there are many variations that might reasonably be suggested.

2. The experimental units would be a number of houses in a community typical of the area in which the paints were to be marketed. The owners of the houses would have to agree to accept either brand of paint and to choose from among a small list of colors. The allocation of brand of paint to a house would have to be done in a way that ensured that among the houses with a certain color, there was a fairly even representation of both brands. This could be done by allocating the brands to pairs of houses requiring the same color, using the toss of a fair coin to decide the allocation. The comparison of brands could then be made without confusing brand quality with color choice.

The brands are the treatments, and the dependent variable for this experiment could be an index from 1 to 10 which describes the condition of the painted surfaces after three years. The values 1, 2, ..., 10 could each have a description attached to it to explain the assignment of the index values.

The houses should all be painted during the same season.

3. Experimental subjects would be people willing to attend an aerobics class and to attend the one chosen by the experimenter. Experimental subjects would be randomized to the two programs. The dependent variable is the change in lung expiration volume over the one-month program. The treatments being compared are the two programs.

Section 3.2

1. The inequality in the ability of the runners is not important as long as the investigator can obtain a large number of runners who are willing to participate in the experiment. If both treatment groups are large, the randomization of treatments to runners should balance the groups with respect to ability, even though each group could contain runners of widely varying ability. If only a small number of runners is available, wide variability in these runners' abilities would present a problem for the interpretation of the results. One way to deal with this problem is to attempt to form pairs of runners of like ability and randomize the treatments within these pairs. (Thus the pairs would be the blocks; see Section 3.3.)

2. Replication of the experiment, either with different runners or with the same runners on different days. In the latter case the randomization should be repeated on each day. (Or use a crossover design; see Section 3.3.)

Section 3.3

1. The average time for the *S* treatment includes the same mix of day 1 and day 2 times as does the average time for the *NS* treatment. Whatever the day effect is, it will be present to the same degree in both averages. The treatment effect is assessed by looking at the difference in the averages, and the day effect will subtract out.
2. (a) 24:

ABCD	BCDA	CDAB	DABC
ABDC	BCAD	CDBA	DACB
ACBD	BDCA	CADB	DBAC
ACDB	BDAC	CABD	DBCA
ADBC	BACD	CBDA	DCAB
ADCB	BADC	CBAD	DCBA

   (b) 4:

   AB
   CD
   BC
   DA

3. Design I would be best, since the relative effects of the two treatments will not be affected by variation in runners' abilities. Note that both designs result in comparison of treatment averages based on two measurements for each treatment.
4.

A	B	C	Plot 1
B	C	A	Plot 2
C	A	B	Plot 3

   (a) The advantage of this design is that each variety has an equal amount of "northness" and "eastness"; if there are trends in soil conditions across the field, this design has a chance of equalizing the treatments with respect to these trends.
   (b) With only one plot per treatment, the differences between treatments could not be separated from the differences due to fertility across the field.
   (c) Experimental units are the subplots. Treatments are the varieties of wheat, and the blocks are the plots.

Section 3.4

1. Blend 1 is best for model 1 and blends 2 and 3 are equivalent. For model 2 the ordering of the blends (best first) is 1, 2, 3. There is an interaction between blend and model since the effect of blend is different for each model.

2. (a) The experimental units are the one person (yourself) at different points of time. Let us consider a categorization of the sleep and warm-up variables so that these can be considered as "factors" with "levels" (this step is just to comply with the requirement that the design be a factorial one).

    Sleep: 8 hr (call it Long) or 6 hr (Short)
    Warm-up: 5 min (Short) or 10 min (Long)

Treatment	Sleep	Warm-up
1	Short	Short
2	Short	Long
3	Long	Short
4	Long	Long

    Using treatment combination 1 on day 1, and so on, would be one possible design except that the assignment of treatment combinations to days should be determined by a random process. One could select four digits in the range 1 to 4 from a table of random digits (without replacement), and the order in which the four digits occurred would determine the design exactly. For example, if the order generated by the table of random digits were 2, 4, 3, 1, the design would be

    Short Long on day 1
    Long Long on day 2
    Long Short on day 3
    Short Short on day 4

    Probably this procedure should be repeated a few times; thus the experiment might last 16 or 20 days.

    (b) A big problem is that the runner knows which treatment has been applied: his prior expectations about the treatment may affect his performance. This could possibly lead to misleading results.

    (c) The transient effects of the particular conditions of a particular day could be "averaged out" of the treatment comparisons. But the problem mentioned in part (b) would not be avoided.

3. A factorial experiment could be arranged with one factor being Teapot Temperature—Hot or Cold, and the other Milk—Lots or Little. The judges should be prevented from seeing which cups have the Hot Teapot Temperature, and ideally they should not know which cups have a lot of milk. A blindfold might achieve the latter if the amounts of milk were not extremely different. (Otherwise, taste would tell.) The subjects should receive their cups in either a random order (if there are a large number of subjects), or in an order determined by one of the block designs suggested in Section 3.3.

3.6 SUMMARY AND GLOSSARY

1. In an experiment, the investigator will assemble a set of experimental units and assign the treatments to these experimental units at random. The purpose of such an experiment is to compare the effects of the treatments on some outcome. The treatments consist of specified levels of the factors of interest. The factors are sometimes referred to as independent variables and the outcome as the dependent variable. This terminology and structure is common to all experiments.

2. Replication of experimental designs allows the reduction of the effect of extraneous factors on the comparison of treatments. It does this by enabling the replacement of single measurements by averages whose values are determined mostly by the actual effects of treatments and only slightly by extraneous factors. Randomization achieves a similar purpose: it reduces the effect of variation in experimental units on the treatment comparisons by distributing this variation more or less equally to the comparison groups. The averaging process helps to reduce the effect of extraneous factors in comparing the effects of two or more treatments.

3. Blocking reduces the variation induced by randomization of subjects to treatments. Blocks are formed in such a way that experimental units in blocks are more similar to each other than is the case among the entire set of experimental units. When blocking is possible, the treatments should be compared within blocks, since differences in outcome within blocks are likely to be ascribable to the treatment differences alone.

4. A factorial design allows the assessment of the effects on an outcome measurement of several possible causal factors with a minimum of experimental units. The effect of each factor is tested under a variety of conditions described by various levels of the other factors.

Glossary

Experimental units are the objects on which the experiment is performed. (When the experimental units are people, it is customary to call them experimental subjects.) (3.1)

A **factor** is a characteristic whose effects are to be explored by the experiment. (3.1)

A **level** is a specific value of the factor. (3.1)

An **independent variable** is a variable whose values indicate the levels of the factor to be used in the experiment. Experimental units, which are allocated the various values of the independent variable, form the comparison groups. (3.1)

A **dependent variable** is a variable whose values measure the outcome of the experiment. Group differences in this variable measure the effect of the various levels of the independent variable. (3.1)

A **treatment** is the application of a particular level of the factor to an experimental unit. The experimental units receiving a particular treatment form a treatment group. (3.1)

Replication is the repetition of a part of an experiment under exactly the same conditions, except with different experimental units. (3.1)

Randomization is the assignment of two or more treatments to experimental units in such a way that specified treatment group sizes are achieved, and such that each possible resulting allocation has an equal chance of occurring. (3.1)

Blocking is the experimental strategy of using a set of similar experimental units for more than one treatment, thus enabling a comparison of treatments that is not affected by differences among experimental units. The set of similar experimental units is then called a block, and two or more treatments are compared in each block. The blocks may be physically divided into experimental units, as would be done in agricultural experiments: the experimental units are the subplots, and the blocks are the plots. Or they may be divided in time, as in the crossover design: the block consists of the identical experimental unit at different times. (3.3)

A **crossover design** is a blocking strategy in which the various treatments to be compared are applied to the same experimental unit at different times. Usually, the order of the assignment of treatments is arranged so that any time-dependent effect can be averaged to a negligible amount by the analysis. (3.3)

A **factorial design** is one in which each experimental unit receives one level of each factor under study. (3.4)

Interaction describes a relationship between two or more factors in their effect on a dependent variable. When the effect of one factor depends on the level of the other factor, there is said to be an interaction between the factors in their effect on the dependent variable. (3.4)

PROBLEMS AND PROJECTS

1. Describe an experimental design that would be appropriate for the taste testing of three blends of coffee, using, at most, 10 subjects.
2. A rectangular field is known to contain increasing moisture from its east side to its west side, and an increasing proportion of sand to clay from its north side to its south side. Design an experiment that will allow a simple comparison of the average yields of the three varieties such that the comparison is not affected by the variation in soil conditions across the field. You should assume that the effects of moisture and soil consistency are to increase yield regularly as one crosses the field from east to west or from north to south.
3. Discuss the role of replication in a situation such as that described in Exercise 1 of Section 3.4, if the car models available for the experiment were of different ages (i.e., if several years of model 1 were available and several years of model 2 were available).
4. There is good evidence that physical stress, even stroking the leaves of a house plant for a

minute each day, inhibits plant growth. Some claim (without good evidence) that speaking kindly to house plants encourages growth.
   (a) Discuss the design of an experiment to assess the effects of physical contact, talking to plants, or both, on growth. You must carefully describe the treatments and other aspects of the protocol.
   (b) Carry out your experiment and write a report describing the results.
5. A common parlor game is the following. A holds a dollar bill by one edge just above B's hand. B holds her hand so that there is room for the dollar bill to slip between her thumb and index finger, unless she is able to close the thumb and finger together before the bill slips to the floor. A releases the bill without warning. After a number of trials, B's reflex time is measured by the height at which A must drop the dollar bill so that B has time to catch it.
   (a) Design an experiment to determine the effects of coffee consumption on reflex time, where reflex time is measured as suggested by the parlor game.
   (b) Perform the experiment and report your results under the following headings:
   Objective
   Design and methods
   Results
   Conclusions
6. "Blocks," "clusters," and "strata" are all names given to subpopulations. But each term refers to a subpopulation defined for a different purpose.
   (a) Is it likely that the subpopulations called blocks, clusters, or strata would actually be the same subpopulations? Explain why or why not.
   (b) Is a properly defined block more like a properly defined cluster or more like a properly defined stratum?
   (c) In sampling surveys, the collection of all clusters (for a given specification of the clusters) comprises the target population. The same is true of strata. In experimental designs, does the collection of all blocks comprise a population of interest? How important is it that the blocks be representative of the target population to which one wishes to generalize the conclusions about the treatment comparisons? Discuss briefly.
7. An experiment is planned to decide which of two trivia games has the more difficult questions. (These games all have a similar format, where players are asked questions and they advance in the game if they answer the question correctly. When a large number of people are participating, it is usual to form teams, all members of each team collaborating in arriving at their official answer to any question.) Twenty people from the neighborhood agree to participate in the experiment. It is planned to split the group into four teams of five players each. This is done by selecting simple random samples (without replacement of course) of size 5 until the 20 players are all assigned. Game A is played by teams 1 and 2 and game B is played by teams 3 and 4. The two teams that are assigned to a game alternate in trying to answer successive questions from that game. The proportion of correct answers obtained by each team is as follows:

Game A:	Team 1	40%
	Team 2	95%
Game B:	Team 3	90%
	Team 4	80%

Interpret these results. Comment on the adequacy of the design for the purpose stated. Be sure to mention both the advantages and disadvantages of this design over other designs that are possible.

8. Suppose that of the 20 players in Problem 7, 10 can answer 75 percent of the questions in either game and the other 10 can answer 25 percent of the questions in either game. Call the 75 percent players X (for expert) and the 25 percent players Z (for sleepers). How often will the allocation scheme described in Problem 7 result in one or more teams consisting entirely of Z players?

    To answer this, use a table of random digits, ignoring all digits other than 1, 2, 3, 4. The sequence of these four digits will tell you how to assign the 20 players to teams. (The first digit will assign player 1, the second digit will assign player 2, and so on.) Once five players have been assigned to a team, ignore all future digits relating to this team. Eventually, all teams will be selected. (That is, when 15 of the 20 have been assigned, the fourth team will be the five people who are left.)

    Repeat this process 10 times to estimate the desired proportion of allocations. (This problem is continued as a computer project in Problem 9.)

Computer Project

9. (Continuation of Problem 8.) Write a program to accomplish the procedure described in Problem 8. Use it to estimate the proportion from 100 repetitions (or more, if this is feasible) of the assignment of the 20 players.

Descriptive Methods for One Variable

> *There are three kinds of lies: lies, damned lies, and statistics.*
>
> *Benjamin Disraeli (1804–1881)*
>
> *He uses statistics as a drunken man uses lampposts—for support rather than for illumination.*
>
> *Andrew Lang (1844–1912)*

This chapter and the next two introduce certain aspects of descriptive statistics. In this chapter we introduce descriptive strategies for summarizing several measurements of a single variable: for example, the starting salaries of all business school graduates from a particular university. In Chapter 5 we introduce methods for describing two variables: for example, the relationship of starting salaries of new business school graduates to their undergraduate grade-point averages. In Chapter 6 we introduce methods for describing time-series data: for example, the average starting salary for business school graduates for each of the last 20 years.

Chapters 4, 5, and 6 all concern descriptive statistics and are conceptually simpler than the rest of the text. Chapters 7 to 9, and to some extent Chapters 1 and 2, involve techniques for relating a sample to a population, whereas in descriptive statistics the aim is simply to summarize the sample itself. As we shall see, the descriptive techniques introduced in this chapter become a starting point for many of the inferential methods of the later chapters. The descriptive methods are not only useful by themselves but form a basis for further study of statistics.

In this chapter we begin with some ideas about measurement in Section 4.2, and then introduce the central idea of the chapter, the frequency distribution, in Section 4.3. Sections 4.4 and 4.5 present ways to summarize frequency distributions.

Key words are measurement, instrument, qualitative, quantitative, bias, precision, accuracy, validity, frequency distribution, histogram, mode, skewness, average, median, range, standard deviation, mean absolute deviation, invariance, standard units, normal distribution, percentile, and interquartile range.

4.1 INTRODUCTION

We will be introducing a variety of techniques for summarizing data. A noteworthy characteristic of statistical summaries is that they often reveal more information than do the raw data they summarize. (Of course, the information in the summary is present in the raw data, but it is usually not apparent until the data are summarized.) For example, a page of numbers representing the number of students in each class at a college will not, without further statistical summary, reveal facts such as "no classes with fewer than 5 students" or "class sizes tend to be either in the range 20 to 40 or in the range 100 to 200 students, with relatively few in the range 40 to 100." It is simple facts such as these that the techniques of descriptive statistics are designed to reveal. The widespread use of these techniques may be explained by their success in particular application areas, rather than by a fascination with the techniques themselves. It may be helpful for self-motivation to ask yourself repeatedly, as new techniques are being described, what applications you foresee in an area of interest to you.

4.2 MEASUREMENT

4.2.1 Measurements and Instruments

Measurement may be thought of as the process of assigning a number or numerical code to something. If we wish to ascertain the width of the page in a book, we use a ruler in the usual way and read off the width. The result of our measurement, 15 cm, is assigned to the book by the measurement process. The reason for describing measurement in this way is that we need a word that includes every conceivable way in which we might produce data to be subjected to statistical analysis.

For example, if we have the responses of 1000 delegates at a political leadership convention, in which each delegate's leadership preference is recorded, we have 1000 measurements. The measurement process consists of the delegates entering a voting booth, marking an X on their ballots, and the vote being recorded as a numerical code by a statistical clerk. Let's say that a delegate is "measured" to be "3." The measurement 3 need not have the usual numerical significance; in this example, the 3 is just a label, providing a convenient way to computerize the tallying of the ballots. But we will still refer to the process of determining this code as a measurement. Many questionnaires request nonnumerical responses, such as "agree," "disagree," or "neutral," which are then coded for computerized analysis as "1," "2," or "3."

An **instrument** is the device used to assist in the production of a number from the thing measured. A ruler is a simple instrument; so is the combination of ballot and statistical clerk in the voting example. It is common to refer to survey questionnaires as instruments.

A good starting point for the statistical analysis of a data set is to determine the instruments and the measurement procedures that are used. In this section we suggest some important characteristics to look for; the options for statistical description and analysis depend on these characteristics.

4.2.2 Measurements of Quantity and Quality

If you were to describe the houses in your neighborhood, you would probably use a convenient set of categories which capture the major distinguishing features, such as "Tudor," "colonial," "cedar and glass contemporary," "ranch," and so on. The assignment of such a category to a house is a measurement of the house; it is a purely qualitative measure, since there is no natural notion of quantity in these categories. Of course, if you had personal preferences that allowed you to rank these categories, you might not consider this measurement as purely qualitative; but even then there would be no natural way to quantify the description. On the other hand, if you were interested in the sizes of the houses, your measurement might well be quantitative, such as assessed market value or total square feet of living space. The main point here is that some measures are quantitative and some are not.

Suppose that the data shown in Table 4.1 are collected for a neighborhood. A rough summary might be "50 percent colonial, 50 percent miscellaneous styles" and "average size about 2100 ft^2." Qualitative data, such as architectural style, are usually summarized using percentages for each category. Quantitative data, such as living space, are usually summarized using averages. Do you think that such statements as "the average architectural style is colonial" or "12.5 percent of the houses contained 2100 ft^2 of living space" are appropriate summary statements?

TABLE 4.1 HOUSE DATA, QUALITATIVE AND QUANTITATIVE

House	Architectural style	Living space (ft^2)
1	Tudor	2000
2	Colonial	2150
3	Colonial	2800
4	Cedar and glass	1850
5	Colonial	2100
6	Ranch	1600
7	Tudor	2200
8	Colonial	1900

In summary, **qualitative measurements** are those that result in categorical information (without any connotation of size), whereas **quantitative measurements** are those that result in numbers whose size indicates the aspect measured.

We will see that not only the description of data, but also their statistical analysis, depends on the type of data (i.e., quantitative or qualitative) that are recorded.

[A purely quantitative scale is sometimes called a *cardinal scale*. If it also has a natural zero (such as age or the number of tickets sold, it is a *ratio scale;* otherwise (date of the month, Fahrenheit scale of temperature), it is called an *interval scale*. A scale that describes an ordering of things (rank in class, list of video sets in order of preference) is quantitative in a weaker sense and is called an *ordinal scale*. A purely qualitative scale (brand of cereal, type of music, breed of dog) is called *nominal* or *categorical*. It is important to be aware of these distinctions, but for a first course the qualitative/quantitative distinction is the most important.]

4.2.3 Precision, Bias, and Accuracy

It is sometimes a little disconcerting when a repeated measurement fails to give the same value as the original measurement, especially when the thing being measured is thought to have a fixed size. Consider, for example, the measurement of distance from your garage at home to the parking lot spot where you usually park during the day. Suppose you consider that the distance you are trying to measure is the minimum distance you could drive, legally (see Figure 4.1), and arrive at your destination. Each day you drive this same route. On Monday, you measured the distance on your odometer to be 12.50 mi. On Tuesday, it turned out to be 12.35 mi. You are clearly experiencing some measurement error and you may have no single explanation for it. Many factors that were not under control during your drive may have combined to accumulate to the errors contained in the 12.50 and 12.35 readings. But you do know that there is a number which is the true minimum legal distance that it is possible to trace out with your car. Thus your measurement is comprised of:

1. The true minimum distance
2. The difference between the true minimum and the average value you get on your odometer over a large number of trips
3. The difference between your odometer average over a large number of trips and the actual reading on a particular day

That is,

> today's measurement = true minimum distance
> + (average measurement − true minimum distance)
> + (today's measurement − average measurement)

This equation is obviously true, but it helps to have names for these separate components of a measurement. A shorter version of this equation is

> measurement = true value + bias + unexplained variation

The term **unexplained variation** has several synonyms in statistics: imprecision, random error, chance error, chance variation, and random variation. Most often it is simply called *error*. The problem with this term is that bias is also error of a kind, but it is very different from unexplained variation. Unexplained variation will eventually average to zero—this is part of its definition.

Figure 4.1 Occasionally one has to forego the smooth comfort of the freeway in the interest of a minimum path.

The term **bias** is sometimes called *systematic error*. The word "systematic" means "according to a system" or "regular," but its use in the phrase "systematic error" is more specific. A bias is a constant deviation that appears in every measurement; that is, every measurement contains the same systematic error. In using "bias" as a replacement for the component (average measurement − true minimum distance) in the expository example, the average we are describing here is the long-run average and is not itself subject to unexplained variation. It is the amount by which our measuring method is wrong, not counting unexplained variation, that is called bias.

In the example concerning driving distances, since we were trying to measure the minimum driving distance possible, the bias is likely to be positive (i.e., the measurements are slightly inflated, on average, relative to the true value). Or the odometer itself may be missing a tooth in its contact cog and thus sytematically underestimate the distance. These effects are bias. In general, bias can be positive, zero, or negative, but for a given measurement context, it is constant. On the other hand, unexplained variation will sometimes be positive and sometimes negative even within the same measurement context. Unexplained variation will average out to zero if we repeat the measurement the same way a large number of times. But notice that no matter how many measurements we take, the bias portion will still influence our average measurement. In fact, since the bias is in every measurement, the average measurement will contain just as much bias as will each individual measurement.

When we measure something more than once, it is usually because we want to use all the measurements to estimate the true value of the quantity measured, and we expect that this combined estimate will be better than any single measurement. The usual way of combining several measurements of the same thing is to average them. What do we gain by this? Using the equation above shows that the average measurement will be very close to the true value *plus* the bias, because the unexplained variation averages to approximately zero. Thus a large number of measurements reduces the influence of unexplained variation in measurements; however, it does not help to eliminate bias.

To summarize, measurements tend to give "wrong" values for two reasons: bias and unexplained variation. Bias is the amount the measurement tends to be wrong, on average. **Imprecision** is a word that describes the amount of unexplained variation. We define imprecision as the amount of variation in a measurement from the average measurement. When *both* bias and imprecision are small, the measurement is said to be **accurate.** A measurement for which the bias is zero is called **unbiased**. Figure 4.2 presents a visual description of the difference between bias and imprecision.

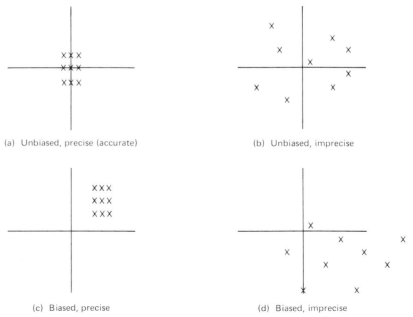

Figure 4.2 Unbiasedness, precision, and accuracy.[1]

In addition to being unbiased and precise, it is desirable that measurements be valid. The **validity** of a measurement is the degree to which it measures the quality or quantity it is supposed to measure. If a caliper is used to measure skinfold thickness, the assumption is that this will measure the percentage of a person's weight that is fat. The validity of the skinfold measure as a measure of percentage of body fat would have to be established using other more accurate techniques for measuring body fat.

If an exam is to measure problem-solving skills, to be a "valid" exam it should not contain questions whose answers rely only on rote learning.

The validity of economic measures is often tricky. It would be convenient to measure unemployment level by the number of people who claim unemployment insurance benefits, but the validity questions we would have to ask are: Are there unemployed people who are not collecting the unemployment insurance benefits? Are there people claiming benefits who are not really unemployed? (The economist's definition of the unemployment level is the number of persons actively seeking work but unable to find "suitable" work. To determine this level, many countries conduct a monthly labor force survey. However, the definition of unemployment is tricky to apply in practice.)

Validity is important in the design of measures for statistical analysis, but there are no universal rules for how to do this. The construction of valid measures requires familiarity with the subject or aspect measured: in other words, factual knowledge. Valid measures of economic activity require the expertise of an economist. Measures of pollutants require expertise in chemistry or biochemistry. Unfortunately, it is not within the scope of this book to guide the reader in the construction of valid measures, but we hope to deal effectively with bias and imprecision.

EXERCISES

1. (a) Which of the following house measurements are purely quantitative and which are purely qualitative?

    Appraised market value
    Number of levels
    Exterior facing material
    Rank based on size of lot
    Rank based on personal preference
    Age of house
    Architectural style
    Number of bedrooms
    Landscaping (poor, middling, good)
    Distance to shopping center
    Type of roofing material

(b) For the measurements that do not seem to fit these two categories, suggest a way to summarize the data indicated.

(c) For each of the measurements listed, suggest the type of instrument necessary to obtain the data values.

2. The word "representative" has several possible meanings in statistics. We can use the phrase "unbiased measurements," "precise measurements," and "accurate measurements" to clarify our meaning. Replace the word "representative" with the appropriate phrase in the following sentences:

(a) Large samples usually yield averages that are more representative of population averages than those of small samples.

(b) Large random samples yield averages that are representative of the population average.

(c) Small random samples will yield averages that are representative of the population averages if these samples are selected according to the simple random sampling method.

3. A farmer wishes to assess the fertility of his soil so that he can order the necessary amounts of fertilizer for next season's crop. His fields are now lying fallow and only weeds and volunteer grains are growing in them. He has two ways of measuring fertility. One is simply to observe visually the health of the weeds and volunteer grains that are growing in the fallow fields. The other is to send a soil sample to a laboratory to determine the nitrogen concentration in the soil. Which do you think is the more valid measure?

4.3 FREQUENCY DISTRIBUTIONS AND HISTOGRAMS

W. Edwards Deming is an international expert in the area of quality control. He has long advocated the creative use of descriptive statistical methods in industry. It is said that his influence on Japanese quality control methods has been an important factor in the amazing success of Japanese industry. We will describe two examples used by Deming,[2] which happen to illustrate the material of this section: frequency distributions and histograms. Although the context of the examples may seem a bit mundane, it should be remembered that the people who make high-quality hardware also make a lot of money. Perhaps this fact will help you maintain your motivation through this section.

The first example is based on a data set consisting of measurements on 500 steel rods. The most demanding requirement of these rods was that their diameter be very close to 1.000 cm, since these were to be part of a bearing assembly. Actually, the value 1.000 cm was to be the lower specification limit, which was the most important one, since a rod with too large a diameter could be machined down to 1.000 cm but one that was too small had to be scrapped. A first step in the analysis of these data was to construct Table 4.2. Table 4.2 displays the recorded diameters of each of the 500 rods; the only information that is left out of this table is the order in which the measurements were recorded. We should keep in mind that this information has been omitted and proceed with the summary table as given; it is certainly much easier to comprehend than a list of 500 diameters.

TABLE 4.2 FREQUENCY DISTRIBUTION FOR THE ROD-DIAMETER DATA[2]

Rod diameter (to nearest 0.001 cm)	Number of rods
0.996 or less	0
0.997	10
0.998	30
0.999	0
1.000	80
1.001	60
1.002	100
1.003	90
1.004	60
1.005	40
1.006	20
1.007	10
1.008 or more	0

The table itself is a summary of the data, and it has a name: **frequency distribution.** The frequency distribution of a list of numbers consists of the list of the possible different values that the original list contains and the frequency with which each of these possible values occurs. It reveals how the frequencies, which are the number of repetitions of a given measurement, are distributed over the observed values of the measurement. The tabulation of a frequency distribution is one of the most basic strategies of descriptive statistics. The concept of a frequency distribution pervades all statistical theory and applications. Let us see what it tells us in this instance.

In our example, the great majority (86 percent) of the diameters are in the range 1.000 to 1.005. Some are as small as 0.997 and some are as large as 1.007. There is another feature of this distribution that is most easily seen from a graphical version of the frequency distribution, which is called a **histogram.** See Figure 4.3. A histogram

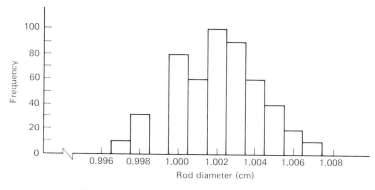

Figure 4.3 Histogram of rod-diameter data.[2]

is a graphical representation of a frequency distribution in which a rectangle represents the possible values and their frequencies: the position of the rectangle indicates the possible values being described, and the size of the rectangle represents the frequency of the values being described. (Sometimes words are a lot more cumbersome than pictures.)

One cannot help noticing the gap between 0.998 and 1.000 in Figure 4.3. In this gap there are zero rod diameters, whereas judging from the adjacent frequencies, one might expect to find about 40. Note that the frequency for diameter 1.000 cm is a bit higher than one might expect from the rest of the distribution. The histogram seems to suggest that inspectors measuring the rods decided that 0.999 cm was close enough to 1.000, the minimum acceptable diameter, and so recorded these as 1.000 so that these borderline cases would not be scrapped.

The practice of accepting marginally inferior rods was unfortunate in this case because the person responsible for maintaining the calibration of the machine was using "percent rejected" as his guideline. Because the percent rejected was only about 8 percent, which was close to the lowest rate that had been attained in previous months, there was no thought of recalibration of the machine. When the 0.999 diameters were recorded faithfully, the percent rejected increased to almost 15 percent, indicating that recalibration was necessary.

The example above illustrates the power of the histogram as a descriptive technique. However, there is one feature of the data that is revealed better by the table than by the graph; it is that the frequencies used in Deming's example (i.e., 0, 10, 30, 0, 80, . . .) are probably not the actual frequencies recorded at the plant. It is unlikely, to say the least, that all the frequencies were multiples of 10. This is of no importance for our purposes except that it illustrates again the effectiveness of these summary displays for revealing anomalies in the data. As we have seen, tracking down the reasons for anomalies can be productive work.

The examination of a histogram for the detection of anomolies is most useful when the examiner is familiar with the context of the data. Special circumstances in the generation of the data will always be relevant to the interpretation of the histogram. However, there are some things that are generally useful to watch for:

1. *More than one mode:* A **mode** of a histogram is an interval over which the height of the histogram is greater than for values just outside the interval. It refers to the "humps" of the histogram. A histogram that has just one hump is called **unimodal.** The presence of more than one mode usually represents a group that can usefully be identified and separated for further description. For example, the distribution of kernel size for samples from a wheat field may indicate the presence of more than one species (or contamination from a wild variety). (A term that is often taught but seldom used is *the* mode. It is the most frequent value occurring in a data set.)

2. *Skewness:* **Skewness** is the property of a histogram to have one "tail" much longer than the other—a histogram is said to be skewed right when the long tail is to the right. A short tail may reveal some selection bias—kernels of wheat that

are too small to handle may have been left out by the sampling procedure. A long tail may indicate "contamination" by a few sampled items that really do not belong in some sense—for example, a few large kernels may have been selected from a part of the wheat field near an outhouse.

3. *Gaps in the histogram:* These are intervals of the histogram in which there are very few observations, even though the nearby frequencies are not so small. (The Deming example illustrated this.)
4. *Abrupt tails:* One or both tails of the histogram become zero very quickly, as if "cut off." This may be a sign that some extreme values have been eliminated from the data set just because they are extreme. A conscientious data analyst will track down the reasons for omissions of data values.

The histogram is widely used because it is a simple as well as effective technique. However, there can be complications. Some of these are listed below, first in summary form, then in more detail.

1. Histograms must be based on grouped data when the frequencies of individual values are too small.
2. Histograms used in comparisons should be based on relative frequencies.
3. Grouped data histograms based on unequal grouping intervals should use the area of the rectangles to convey the frequency of data in these intervals.
4. When the vertical and horizontal axes do not intersect at the origin, axes "breaks" should be used to highlight this fact.
5. Open-ended intervals cannot be properly graphed with a histogram and require special representation.

These details will be needed when you are constructing your own histograms. Moreover, an awareness of them is also necessary for a proper interpretation of histograms prepared by others.

1. Grouped data histograms. Suppose that a data set which consists of 100 numbers has 50 different values. Simply recording the frequency of each possible value would not give a smooth histogram that could be remembered or described any easier than the data itself. In this situation, the data are grouped into intervals; the values are simply noted as being in a certain interval or not. Thus frequencies can be calculated for each interval, and if the intervals are large enough, the frequencies will be substantial. The 100 data values with 50 different values might boil down to 10 intervals of values. The choice of the widths of the intervals for the purpose of accumulating frequencies depends on the application at hand: one must balance the desire for adequate detail in the features of the histogram with the desire for a simple curve that is easily remembered and described. (See Exercise 6 at the end of this section.)

84 Chap. 4 Descriptive Methods for One Variable

(a)

(b)

Living histograms, showing that bimodality can be an indicator of the existence of distinct subpopulations. (a) Heights of female students; (b) heights of students. (Courtesy of B. L. Joiner, "Living Histograms," *International Statistical Review*, Vol. 43, No. 3, 1975, pp 339–40.)

If in the quality control example given above, the data had been recorded to the nearest 0.0001 cm instead of the nearest 0.001 cm, the histogram would have been too onerous to draw and too complicated to assess visually or to describe. In this case the grouping procedure would produce a few intervals, and we would record the frequencies for each interval of values. In this example an easy way to accomplish this would be simply to round the data to the nearest 0.001 cm, and the result would be to produce a histogram identical to Figure 4.3.

2. Relative frequency histogram. The vertical axis of the histogram shown in Figure 4.3 is labeled "frequency." In some cases, *relative frequency,* which is the frequency expressed as a percentage of the total frequency, is preferred. In the rod example, if the relative frequencies were used, the labels on the vertical scale would be 4, 8, 12, 16, and 20 percent instead of 20, 40, 60, 80, and 100 (since 20/500 is 0.04 or 4 percent, and so on). The advantage of this variant of the histogram is that it can be used for comparison; if in another month there were 339 measurements instead of 500, we could make the graphs for two months comparable by using relative frequencies.

Note that the relative frequency histogram looks exactly like the frequency histogram: without the change in the scale on the vertical axis, you would not notice any difference. The need for relative frequency histograms arises only when histograms are being compared.

3. Unequal intervals for frequency accumulation. It is sometimes necessary to draw a histogram based on a frequency distribution for grouped data (as suggested in point 1). This is straightforward when the grouping intervals are all of equal size. But suppose that in the data graphed above we did not know that there were 10 0.997s and 30 0.998s. Instead, we knew only that there were 40 values rounding to 0.997 or 0.998. How would we represent this in a histogram? If we were to draw the frequency 40 over 0.997 and 0.998, as in Figure 4.4a, we would be suggesting that *each* value occurred 40 times. The rectangle over 0.997 and 0.998 would be the same height as the one over 1.005. This is clearly a wrong depiction of the true frequency.

A rule that will keep us out of trouble in this situation is to construct the rectangles so that the *area* of the rectangle is proportional to the total frequency it represents. Thus the height of the rectangle over the double interval should correspond to a frequency of one-half the combined frequency of 40, that is, 20 (see Figure 4.4b); 20 is a reasonable number to represent the frequency of each of 0.997 and 0.998 if all you know is that 40 is the combined frequency.

Note that the frequency labeled on the vertical axis corresponds to the frequency for each 0.001 cm on the horizontal axis. We would have a different labeling of frequency if we chose to base our frequency calculations on an interval of 0.0001 cm instead of 0.001 cm. Then the scale labels would be 2, 4, 6, 8, and 10 instead of 20, 40, 60, 80, and 100.

The guiding principle is that the scale should enable the viewer to assess the

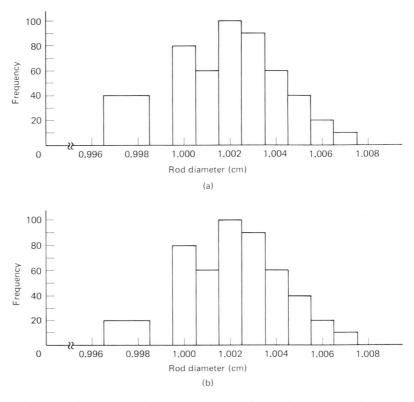

Figure 4.4 Histogram (a) is incorrect: the rectangle over the interval 0.997–0.998 should indicate that the values in this interval occur with the same frequency as 1.006, since this was the case in the original distribution (Figure 4.3). Histogram (b) achieves the proper visual effect. The area of the rectangle is proportional to the frequency for the interval on which the rectangle sits.

frequency (or relative frequency) of intervals of interest by counting up the units on the horizontal axis and mulplying this by the frequency (or relative frequency) indicated by reference to the scale of the vertical axis.

4. Axis break. Note the break in the horizontal axis near the vertical axis in Figure 4.3. This detail is simply to draw the viewers' attention to the fact that the vertical axis and the horizontal axis do not cross at the origin, which is what many people are used to. Without noticing this, a quick glance at the graph might suggest that diameters vary about 50 percent above and below their average value, reflecting poor quality control indeed. In fact, the variation is only about 1/2 of 1 percent. The graphs in Figure 4.5 indicate an example in which a similar oversight on the vertical axis can lead to an erroneous interpretation of a graph.

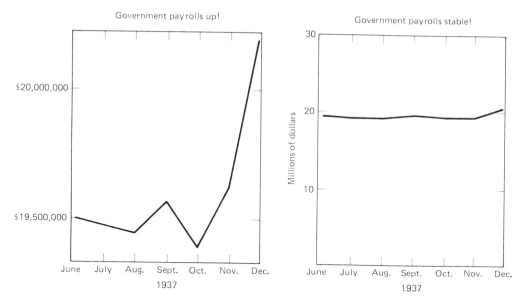

Figure 4.5 Graphsmanship.[3]

5. Open-ended intervals for grouped data. The original frequency tabulation had the category "0.996 or less." If this interval had not been empty, how would you draw its frequency on the histogram? For example, if there were three diameters in this interval, would one draw a rectangle of width 0.996 and height 3/500? This would clearly underrepresent the frequency of values in this interval. The answer here is that there is no good way to represent graphically these open-ended intervals when they have nonzero frequencies. If you must use a histogram to display a frequency distribution which has used an open-ended interval, some special notation or footnote should accompany the histogram so that it will not mislead its viewers.

This completes our discussion of the details of constructing histograms. The final point to be made in this section concerns the inappropriateness of histograms for summarizing ordered data.

In Deming's article which introduced the rod-diameter data, there is a second data set that illustrates another property of frequency distributions and histograms. These data are listed in order of the date of manufacture. Each value represents the elongation of a spring, in centimeters. The 50 numbers displayed below represent a sequence of test results, which we may think of as samples of size 1, from 50 successive batches of springs. The specification limits for this manufacturing process are 0.0004 to 0.0016 cm. For convenience, all the measurements have been multiplied by 10,000 in the sequence of measurements shown in Table 4.3.

A histogram would *not* be a good summary of these data. It would show that the

TABLE 4.3 SEQUENCE OF 50 SAMPLES OF MEASUREMENTS OF COIL-SPRING EXTENSIONS[2,a]

Week 1:	13, 13, 12, 12, 11, 12
Week 2:	11, 13, 11, 11, 12, 10
Week 3:	11, 10, 10, 9, 11, 11
Week 4:	10, 11, 11, 11, 10, 11
Week 5:	12, 10, 9, 10, 9, 10
Week 6:	10, 9, 9, 9, 9, 10
Week 7:	9, 8, 10, 9, 8, 10
Week 8:	8, 8, 7, 9, 8, 9
Week 9	8, 6

[a] The measurements recorded here are 10,000 times the original measurements, which were in centimeters.

measured elongations are within the specification limits, and therefore that no special maintenance of the machinery is necessary. Yet even a cursory scan of the data reveals that the diameters are decreasing over time and suggests that the lower specification limit will soon be crossed. Consequently, some defective material would be produced unless some corrective action were taken first.

The moral here is that histograms are useful when the order of the observations is not important. However, when measurements are used for monitoring, the order is usually an informative clue to the behavior of the system.

A more extreme example of the inadequacy of histograms to summarize ordered data may dramatize this point: A histogram of the frequency of each letter that appears in Hamlet's soliloquy "To be or not to be . . ." clearly fails to convey the essence of this famous set of "ordered data."

EXERCISES

1. Construct a frequency distribution and a histogram of the steel-rod-diameter data for the frequencies that result by grouping the frequencies for 0.997 to 0.998, 0.999 to 1.000, 1.001 to 1.002, and so on. Does the anomaly in the data described in the text still persist? What rule would you suggest for attaining the correct amount of grouping of data values in a histogram?
2. From the histogram shown in Figure 4.6, an executive wants to determine the proportion of people who have worked for the company for five years or more. Assume that the data on which the histogram is based have been rounded off to the nearest year.
   (a) Explain why it is not necessary to know the scaling on the vertical axis for the executive to accomplish her goal.
   (b) Guess the answer to the executive's query.
   (c) Check your guess by an exact method.

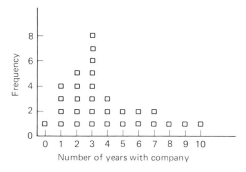

Figure 4.6 Distribution of employment duration, employees of Newwave Co., 1985.

3. A class of students attending a lecture in Statistics 1000 record their major fields as follows: psychology, economics, computing science, commerce, computing science, computing science, commerce, English, physics, commerce, biology, kinesiology, psychology, mathematics, chemistry, chemistry, computing science, computing science, commerce, biology, geography, chemistry, kinesiology, archeology, psychology, criminology, computing science, commerce, history, archeology, computing science, kinesiology, commerce, kinesiology, chemistry, commerce, kinesiology, and psychology.
   (a) Make a graphical summary of these data using a kind of histogram.
   (b) Do the gaps between the rectangles in this histogram and multiple modes have the same meaning as the ones in the histogram of the rod-diameter data? Explain.

4. The numerical grades from an examination in English Literature are to be assigned the letter grades A, B, or F.
   (a) Draw a rough histogram for the numerical grades and suggest a way to assign the grades, justifying your choice, if possible. The professor has begun to summarize the data by sorting it into 10s, 20s, 30s and so on, as follows:

   10s: 12, 19
   20s: 27, 24, 22, 25, 29
   30s: 32, 30, 33, 30
   40s: 48, 47, 49, 47, 45
   50s: 50, 57, 53, 55, 50, 53, 58, 56
   60s: 69, 60, 65, 62, 64, 64, 66
   70s: 74, 72, 75, 71
   80s: 84, 81, 86, 83, 81, 89, 81
   90s: 92, 90

5. A bank employee wishes to fill in an idle lunch hour by looking through rolls of pennies for rare-issue dates. The employee does not know which years are "rare" beforehand, but accumulates the following frequency distribution from nine rolls plus a few pennies from her pocket:

Year	Frequency	Year	Frequency	Year	Frequency
1983	47	1973	11	1963	3
1982	65	1972	10	1962	2
1981	103	1971	6	1961	6
1980	45	1970	10	1960	1
1979	30	1969	4	1959	1
1978	35	1968	1	1958	0
1977	15	1967	4	1957	0
1976	18	1966	3	1956	2
1975	12	1965	1	1955 or earlier	0
1974	17	1964	2		

What features of this distribution might be of interest to the novice coin collector?

6. (a) Construct a histogram of the life-span data (years)[4] displayed below. To do this, group life spans into five-year intervals before recording the frequencies. The first interval would include all life spans from zero years up to but not including five years, the second interval would include all life spans from five years up to but not including 10 years, and so on.

North and South American

Domestic cattle	30
Domestic dog	20
Domestic cat	28
Wild goat	18
Canadian beaver	20
Domestic guinea pig	8
Armadillo	10.3
Lemming	3
Horse	46
Canadian porcupine	7.5
Lynx	19
Golden hamster	4
Canadian meadow mouse	3
House mouse	3.5
North American mink	10
Little brown bat	24
Muskrat	6.2
Domestic sheep	20
Deer mouse	8
Harbor seal	34
Brown rat	3.3
Eastern American mole	1.2
Gray squirrel	15
Shrew	1.5

North and South American (cont.)

American badger	24
Human being (Caucasian)	75
White-tailed deer	17.5
Raccoon	14
Virginia opossum	8
Chinchilla	11
Cotton rat	3
Common marmoset	12
Great anteater	4

African and European

Chimpanzee	44
Camel	29
Elephant	70
European hedgehog	4.2
Giraffe	34
Monkey (*Speciosa*)	30
Kangaroo	19
Tiger	26
European wild bear	25
European red fox	14
Chaoma baboon	36
Gemsbok gazelle	18.5

(b) Comment on the effect of using one- or 10-year intervals in part (a) instead of five-year intervals.
(c) Construct a one-dimensional scatter diagram of the life-span data.
(d) Comment on the relative merits of the displays constructed in parts (a) and (c).

7. The body lengths of 1116 plaice (a kind of flatfish) are recorded in the following frequency table.[5]

Lower end of interval (cm)	Frequency
17.5	4
18.5	21
19.5	80
20.5	100
21.5	110
22.5	40
23.5	12
24.5	24
25.5	45
26.5	91
27.5	152
28.5	170
29.5	140
30.5	63
31.5	25
32.5	24
33.5	9
34.5	2

(a) Draw the histogram for these data using 1-cm intervals.
(b) Draw a new histogram using 4-cm intervals.
(c) Comment on possible interpretations of the histogram in part (a) and on the relatiave merit of the histograms in parts (a) and (b).

4.4 SUMMARY MEASURES: AVERAGE AND STANDARD DEVIATION

The main purpose of this section is to discuss the most commonly used summary measures that describe a frequency distribution. We shall be especially concerned with summaries of the center and spread of a frequency distribution.

A frequency distribution is a list of numbers and their associated multiplicities. It is a useful summary of data for which the order of observations in the list of data values is unimportant. We can extract or convey useful information about a list of numbers quite quickly by visually scanning its histogram. However, for some purposes, we will need numerical summaries of the important characteristics of a frequency distribution, and for this purpose we need some additional techniques. For the discussion of characteristics of frequency distributions, it will help to have a method slightly less formal than histograms to describe frequency distributions. Instead of the rod-diameter histogram displayed in Figure 4.3, we will instead use an approximation of this histogram as shown in Figure 4.7. Actually, if the histogram of Figure 4.3 were based on a very large number of measurements, where each measurement was recorded to many decimal places, it would look very much like the smooth histogram of Figure 4.7. We will want to become comfortable with comparisons of frequency distributions using these smoothed histograms, and that is why the relative frequency scale is used for the vertical scale. (With the absolute frequency scale, the larger set of numbers of two sets would tend to have higher frequencies for all values, making a visual comparison of the shapes of the two histograms awkward.)

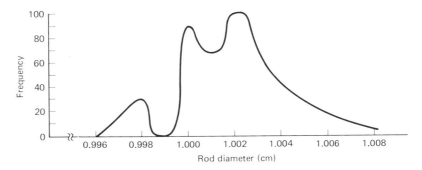

Figure 4.7 Smoothed histogram illustrating alternative graph of data originally graphed in Figure 4.3.

An interesting experiment undertaken by psychologist Robert Tryon[6] involved the selective breeding of two groups of rats, the selection criterion being the number of wrong turns (or errors) made by the rats while attempting to get through a maze to a supply of food. The result was a group of maze-bright rats, who tended to have low error scores, and another group of maze-dull rats having high error scores. The performance of the two strains of rats after several generations of selective breeding is depicted in the histogram in Figure 4.8. There is no question that the maze-dull rats "tend" to make more errors than the maze-bright rats. But note that it would be false to claim that a maze-dull rat will definitely commit more errors than a maze-bright rat. The two frequency distributions overlap. We can make statements about groups of rats that may not apply to any particular pair of rats. One way to make this idea precise is to refer to the differences in the average performance of the two groups. The **average** of a list of numbers is the sum of the numbers in the list divided by the number of numbers in the list. The "average" is one numerical characteristic of a frequency

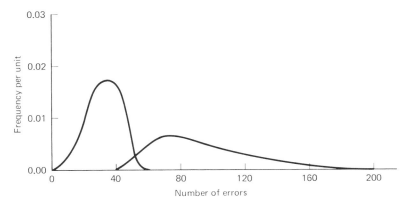

Figure 4.8 Histograms showing the frequency distributions of the number of errors made by two groups of rats: maze-dull rats and maze-bright rats. With respect to other measures of rat intelligence, the two groups were similar.[6]

distribution (or histogram) which summarizes the frequency distribution. The average is the most commonly used summary of the center of a distribution.

If you reexamine the histogram for the two groups of rats, you will notice another difference in the two distributions. The maze-dull rats have a more variable performance than the maze-bright rats. There is a difference in the spread of the two distributions.

The Tryon data reveal that the two strains of rats differ in both the center and spread of the distribution of wrong turns. Are these differences a property of the rats or a property of the groups of rats? To tell the difference, we would have to have individual rats repeat the experiment. Usually, we expect some variation for a single rat on successive tries (even ignoring a possible learning effect), and also some variation among the rat capabilities within one group, so the numbers describe a composite of the two influences.

The interesting part of the Tryon experiment from a scientific perspective is the comparison of the distributions of scores on other rat intelligence tests. The maze-bright rats were no better at these other tests than the maze-dull rats. Tryon felt that this was evidence against the existence of a general factor of intelligence, at least in rats. What do you think?

4.4.1 Measures of Center

We have seen that the concept of the center of a frequency distribution is a useful one for descriptive purposes. Just to focus temporarily on the calculation procedures for certain of these measures, let us consider a very simple data set based on the scores of only five maze-bright rats, which we will label B, and the scores of five maze-dull rats, labeled D.

94 Chap. 4 Descriptive Methods for One Variable

B: 31, 50, 25, 45, 29
D: 45, 170, 85, 110, 90

The average of a list of numbers is the sum of the numbers divided by the number of numbers in the list. Check that the averages for B and D are 36 and 100, respectively. The average is the most widely used measure of the center of a list of numbers. A word of warning: The word "average" is used by some people to include measures of center other than the one we have defined. These people would use the term "arithmetic mean" or simply "mean" for the calculated value we have just defined as "average." In this text we consider "average," "mean," and "arithmetic mean" to have identical meaning.

An alternative measure is the **median,** which is defined as the middle-ranked value in the list. If you order the five numbers in B (i.e., 25, 29, 31, 45, 50) the middle-ranked one is the one whose rank is 3: namely, the number 31. See Table 4.4. Similarly, the median of the five measurements in D is 90.

TABLE 4.4 CALCULATION OF THE MEDIAN AS THE MIDDLE-RANKED VALUE

Data value	Rank
25	1
29	2
31 ← the median	3
45	4
50	5

The median value is close to, but not the same as, the average value. To assess which measure, median or average, does a better job of describing the difference in the centers of the B data and the D data, examine the **one-dimensional scatter diagram** of Figure 4.9 (i.e., showing the scatter of values of a single variable; we will see similar two-variable plots in Chapter 4).

The choice between the average and the median is by no means clear cut in this instance: 100, the average, seems a bit high for D; 90, the median, seems more representative. The median, 31, seems a bit low for B; 36, the average, seems a better measure in this case. It would appear that sometimes the median is better and sometimes the average is better. The usual prescription is: The median is best for describing highly skewed (not symmetrical) distributions or when there are one or more outlying

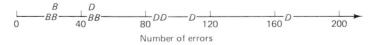

Number of errors

Figure 4.9 One-dimensional scatter diagram. The ten data values displayed are error totals for 10 rats that have run a maze. There are two groups of five rats; the maze-dull rats (D) and the maze-bright rats (B).

values whose validity is suspect. Otherwise, the average is best. (The reason averages are best for distributions that are close to being "normal" is a bit technical and will be omitted here.)

Of course, we have a real dilemma if we want to make a comparison between the two distributions in the example above because it would be desirable to use the same measure of center for both distributions for such comparison. There is no easy solution to this dilemma (see Exercise 1 at the end of this section).

One detail concerning the calculation of medians: When the number of measurements is even, there is no one middle-ranked value, because there is no middle rank. For the numbers 1, 3, 3, 4 we would like 3 to be the median, but which data value has rank 2.5 (the middle of the ranks)? When the size of the data set is an even number, the median is defined to be the average of the two middle-ranked values. The median of 1, 3, 3, 4 is $(3 + 3)/2$ or 3. The median of the numbers 25, 29, 30, 31, 45, 50 is 30.5.

The median and the average can be given a physical interpretation. First we have to give a physical model for a histogram. Imagine that a group of 60 sheep all weighing between 50 and 100 lb are to be weighed on a weigh scale. See Figure 4.10. The scale you must visualize for this example is the kind with a balance bar, and the sliding

Figure 4.10 Each contribution to a distribution counts.

marker moves from the far left side of the balance bar, at 50 lb, to the far right side, at 100 lb. Each sheep's weight is read at the marker when the balance bar is balanced. For each sheep, suppose that we place a small sugar cube at the final position of the marker. For two sheep of the same weight, we pile up two sugar cubes. When all 60 sheep have been weighed, we carefully spray a little water on the cubes to make them stick together. This gives us a 60-cube solid structure that looks like the picture in Figure 4.11. (We have avoided, for simplicity, the very real possibility that the distribution had gaps in it; our explanation of the average would have been more complicated because the structure would not be one solid piece.)

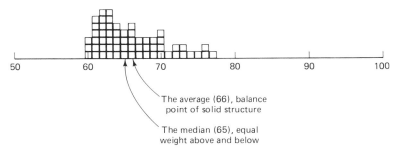

Figure 4.11 Sugar-cube histogram illustrating a constructive definition of average and median.

The average weight of these sheep is 66 lb; if we put a ruler under the structure at 66, it should balance. If anyone moves one of the cubes after the ruler is balanced, the structure would fall off one side of the ruler. Thus the average has a balance-point interpretation.

The median is 65: 28 cubes are less than 65, and 28 are greater than 65. Four have been centered at 65. If we cut these central four in half, we have the weight of 30 cubes on either side of the median. If a cube is moved to a new position but does not cross 65, the median does not change. As long as the weight below 65 equals the weight above 65, the median is unchanged.

Both the median and the average are balance points, but they balance different quantities. The difference is that the median balances the weight on either side without regard to the position of these weights, while the average balances the solid structure, the position of each cube being crucial. (Another way of saying this is that the moments of inertia on either side of the average are balanced. The average is the center of gravity. The median partitions the total "weight" into two equal parts.)

The median and the average measure the location of the center of a distribution. See Figure 4.12. Let us next consider measures of spread.

4.4.2 Measures of Spread

One obvious measure of spread is the **range,** the largest minus the smallest value. For group *B* (the maze-bright rats) it is 25 (50 − 25 = 25) and for *D* (the maze-dull rats) it is 125 (170 − 45 = 125). This measure of spread does seem to capture what we

Sec. 4.4   Summary Measures: Average and Standard Deviation

The Deceptive "Average"

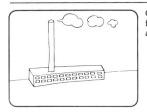

Gismo Products has a small factory where supergismos are manufactured.

Mr. Gismo: Here's what we pay out each week. I get $4800, my brother gets $2000, my six relatives each make $500, the five foremen each make $400, and the ten workers each get $200. That makes a weekly total of $13,800 for 23 people, right?

The management consists of Mr. Gismo, his brother, and six relatives. The work force consists of five foremen and ten workers. Business is good, and the factory needs a new worker.

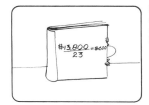

Sam: Okay, okay. You're right. The average is $600 a week. But you *still* misled me.

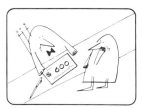

Mr. Gismo is interviewing Sam for the job.
Mr. Gismo: We pay very well here. The average salary is $600 a week. During your training period you'll get $150 a week, but that will soon increase.

Mr. Gismo: I disagree. You just didn't understand. I could have listed the salaries in order and told you that the middle salary is $400, but that isn't the average. It's the *median*.

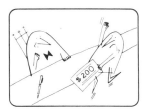

After working a few days, Sam asked to see the boss.
Sam: You misled me! I've checked with the other workers and not one is getting more than $200 a week. How can the average salary be $600 a week?

Sam: Where does the $200 a week come in?
Mr. Gismo: That's called the *mode*. It's the salary that *most* people are making.

Mr. Gismo: Now, Sam, don't get excited. The average salary is $600. I'll prove it to you.

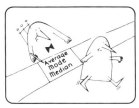

Mr. Gismo: The trouble with you, my boy, is that you don't know the difference between average, median, and mode.
Sam: Well, I know now. And . . . I quit!

Figure 4.12 The sensitivity of an average to an outlier.[7]

see in the histograms—that group D has a much wider spread of values than B. This measure has one serious defect, however: it is dependent on the number of values in our data set. If we had had only two values in group D and 100 in group B, it is likely that the range of the two values would be less than the range of the 100 values, which would convey an impression contrary to what the histogram clearly shows. Other measures of spread avoid this defect.

The most commonly used measure of spread is called the **standard deviation.** The definition and the calculation of the standard deviation is discussed below: its value for data sets B and D is 9.70 and 40.87, respectively. Are these numbers reasonable representations of the "spread" in the two data sets? To be able to answer this question, we have to be more specific about what is meant by "spread."

The standard deviation measures, in a special way, how far the numbers in a data set tend to be from the average of the data set. Let us return to data set B: 30, 50, 25, 45, 30. The average was 36, so the deviations from 36 for the five values are -6, 14, -11, 9, -6. Would the average of these deviations be a good measure of spread? No. The sum of these numbers is always zero, and so is the average. What about the average absolute value of the deviation, which is just the average value of the deviations once the minus signs are ignored? This average absolute deviation turns out to be 9.2 for these data. This average absolute deviation is certainly a reasonable measure, and simple to describe, but it turns out to be not as useful as the standard deviation for statistical work. For the standard deviation we get rid of the signs by squaring the deviations, and then we average these, and finally get back to the units we want by taking the square root. (The reason for this is that in descriptive statistics we are usually describing the distribution of an average, rather than of a population distribution per se. It can be shown mathematically that the variability of an average depends on the population standard deviation rather than the average absolute deviation of the population. This explanation should be more understandable after Chapter 7 has been completed.)

For the B data, the standard deviation is

$$\sqrt{\frac{(-6)^2 + (14)^2 + (-11)^2 + (9)^2 + (-6)^2}{5}} = 9.70$$

We will be using standard deviation frequently, so let's call it SD. The calculation procedure for SD may be described as follows. The standard deviation, SD, of a list of numbers is the square root of the average squared deviation of the list. "Deviation" in this definition means the difference between a number and the average of the list. Can you get SD = 40.87 from the D data set? Try it.

[This definition of SD seems preferable to the "$n - 1$" version popular in some texts (in which the sum of squared deviations is divided by $n - 1$ instead of n). The $n - 1$ definition has no rationale when the measure is to be applied to a population; in this case the definition given is the only reasonable one. In descriptive statistics, the data set is the population to be described, so the n definition is the appropriate one. When we get to inferential statistics in Chapter 7, we use the $n - 1$ definition when it is appropriate (i.e., in the estimation of variances). See the comments of Lindgren,[8] as well as Problem 28 of Chapter 7 concerning alternative estimates.]

As a descriptive measure of spread, the SD can be thought of as a "typical" deviation: that is, if one had to guess the distance of a randomly selected number from the average of the list, the SD value would be a reasonable guess. (More precisely, the SD is the root-mean-square deviation.) Data sets are often summarized as "average ± SD." The appropriateness of this procedure will be discussed further in Section 4.5.

The SD is quite sensitive to the extreme values; the addition or removal of a single value that is larger or smaller than the others can change the SD quite a bit. For example, the removal of 170 from the D data reduces the SD from 40.87 to 24.36. We must be aware of this when considering whether or not an apparently "wild" measurement is to be removed from a data set.

The *average absolute deviation,* or *MAD* as it is sometimes called (an acronym for "mean absolute deviation"), and the SD both measure the size of a typical deviation, but in two slightly different ways. If the measurement of spread is defined as the measurement of a typical deviation, the two measures would both be reasonable ones to use. However, there are two reasons for preferring the SD to other measures of spread. One is the mathematical convenience of the SD for describing the variability of averages (a point we will return to in Chapter 7), and the other is the practical advantage of the following widely known "empirical result" (which is only approximately true):

68 percent of the values in a list of numbers are within 1 SD of the average of the list.

95 percent of the values in a list of numbers are within 2 SDs of the average of the list.

The conditions under which this result is likely to work well are discussed in future sections. Basically, one requires that the distribution be "normal"—this property is described in Section 4.5.

The combination of the average and the standard deviation is often used to describe a whole list of numbers. There are many situations where this is an adequate substitute for a frequency distribution as a descriptive summary of the list. But note that this empirical rule does not always hold. For example, the list {1, 2} is entirely within 1 SD of its average, and only 33 percent of the list {1, 1, 2, 2, 3, 3} is within 1 SD of its average. The conditions under which it does hold will be outlined in Section 4.5.

We have discussed three measures of spread: the range, the SD, and the MAD. The range is the simplest conceptually but most awkward for comparative descriptions. The MAD is quite simple to describe and it can be used in comparative studies, but is less well known than the SD. The SD is more difficult to describe, to those unfamiliar with it, than the MAD, but is actually more widely known and can be used in comparative studies. The reason for the popularity of the SD over the MAD is based on its use in inferential statistics. For descriptive purposes the MAD seems more natural than the SD, but because it is unknown to many, its use poses a practical dilemma.

4.4.3 Some Invariance Properties of Descriptive Summaries

A very useful idea in many technical subjects is the idea of **invariance.** We will describe this idea by example before defining it. For an example of the use of this idea, we will join the Smith family on an outing to a "pick-your-own" berry farm. Each family member takes a different pot or pail to pick berries, but because each picker is charged for the berries by weight, the container is weighed empty, before the picking, and full, afterward. The weight of the berries picked is the difference between the "after" weight and the "before" weight. It is quite obvious that this difference does not depend on the weight of the pot. The difference is "invariant" with respect to pot weight.

The six members of the Smith family record the weight of berries they each have picked: 3.4, 3.7, 2.0, 4.6, 3.2, and 4.1 kg. These numbers average 3.5 kg and have an SD of 0.8 kg. But it is discovered that the scale used for the weighings was set so that even with no weight at all on the scale, the weight read 0.2 kg; therefore, all the weighings were 0.2 kg too large. How do we adjust for this in the calculation of average and SD? We could fix all six numbers and recompute. But surely we can predict the outcome of this exercise—the average of the fixed data should be $(3.5 - 0.2)$ or 3.3 kg, but what about SD? Will it be $0.8 - 0.2$ or just 0.8? If you recall how you calculated the SD, based on deviations from the average, you will realize that the SD stays the same, 0.8 kg. The SD is invariant to the addition of a constant to each data value.

When the Smith family arrives home, two of the pickers discover that they have left their pails of berries at the berry farm. So they want to recompute the average amount picked, based on the four remaining values. Now the two discarded weights are 2.0 kg and 4.6 kg—the two extreme values. Noticing that the four values would have the same median (3.3 kg) as the six values, they decide that the median value is good enough for recording the typical weight picked. The median is invariant to chopping off extremes as long as extremes are chopped off both ends of the distribution.

Even if one of the two abandoned pails is retrieved, the median would not change very much. For example, if the 2.0-kg pail is retrieved, the new median weight changes from 3.3 kg to 3.2 kg. The median is not invariant to lopsided chopping of values from the distribution, but it is almost invariant. We use **insensitive** to describe this "almost invariant" property. Contrast this with the sensitivity of the average and the standard deviation to the deletion of a single extreme value, 2.0 kg. The average increases from 3.5 kg to 3.8 kg and the SD drops to 0.5 kg from 0.8 kg.

The sensitivity of measures of center and spread to extreme values is an important practical issue. If an extreme value is farther from the rest of the data than is usual, it may indicate that a measurement "mistake" has been made. This could be an error in recording the measurement, or the inadvertent inclusion of a measurement belonging to a different group.

For example, if one measurement of the weight of berries picked were 35 kg, it would be senseless to summarize the data by an average and an SD. If it can be determined that this measurement is likely to be 3.5 kg rather than 35 kg, the change should be made. If no such explanation seems justifiable, the data should be summarized in two pieces: the average and SD of the data grouped near 3 to 4 kg, and the value at 35 reported separately. In either case, the value of 35 is called an outlying observation or **outlier.** An outlier is a data value that is so far from the other data values that it must be summarized as a special case in order to present a useful summary. The issue is not merely whether the wild observation is a mistake: even if one picker did pick that much, the combined measurements are not well summarized by the average and SD.

Averages and standard deviations are sensitive to outliers. The median is not. Since the most common summary method is to use the average and SD, we must always be on the alert for the distorting influence of outlying observations.

There is another kind of invariance idea that is important in statistics. A change in the units in which a measurement is recorded should not change the statistical summary in an important way. As an example of this new kind of invariance, we visit Japan. If a team of Sumo wrestlers has an average girth of 200 cm, they also have an average girth of 2 m. Changing the measurements from centimeters to meters (i.e., dividing by 100) had the effect of changing the average by this same factor. This is so obvious that it is possible to miss the point: changing the units before we calculate the average has the same effect as changing the units after we calculate the average. Our summary of the wrestler's girth is invariant in meaning, although not numerically. The average, the median, and the SD all have this kind of invariance because they all have the same units as the original measurements.

This second kind of invariance is fairly easily understood in terms of changes from centimeters to meters. The idea of invariance gets a bit more complicated in the description of "standard units"; standard units are covered in the next section.

So "invariance" is a slippery concept; it means different things in different contexts. The common aspect that justifies the use of the one term for two ideas is that in both cases changing one thing does not change some other thing, in some sense. The "other thing" is said to be invariant with respect to changes in the "first thing." This fuzzy summary indicates that the examples explain this concept best.

EXERCISES

1. In an elementary school, 40 percent of the students are girls and 60 percent are boys. The average age of the boys is 9.5 years and of the girls is 9.0 years. What is the average age of students in the school?

2. Noon temperatures in three locations in a desert are 110, 116, and 107 degrees on the Fahrenheit scale.
   (a) Compute the average and standard deviation of these temperatures.
   (b) Fahrenheit temperatures, F, are converted to Celsius temperatures, C, by the formula

   $$C = (F - 32) \times \frac{5}{9}$$

   Thus the three temperatures can be written as 43.3, 46.7, and 41.7, respectively, on the Celsius scale. Recalculate the average and standard deviation using the Celsius temperatures.
   (c) Which are more variable, Fahrenheit temperatures or Celsius temperatures? Or are they equally variable? (Variability is usually measured by SD.)
   (d) Let your SD answer to part (a) be called SDF and your SD answer to part (b) be SDC. Is it true that

   $$SDC = SDF \times \frac{5}{9} \ ?$$

   Suggest an alternative method to obtain your answer to part (b).

3. The following represent the annual incomes of the employees of a small company:

   $90,000
   $45,000
   $34,000
   $28,000
   $27,500
   $27,000
   $26,500
   $26,000
   $22,000
   $18,000

   (a) What percentage of the employees earn more than the average salary?
   (b) If you had to use a single number to describe a typical salary, would you prefer the average or the median for this purpose?
   (c) What percentage of incomes are within 1 standard deviation of the average? Recalculate this percentage after the $90,000 is eliminated from the list. Speculate on the property a data set must have in order that the empirical result (see Section 4.4.2) be a useful predictor.

4. Verify, using a numerical example, that the SD of two numbers is the range of the two numbers times 0.500.

5. Calculate the average and the median for each of the following data sets:

Sec. 4.5   Normal Approximation to Frequency Distributions   103

Data set 1:   1, 2, 2, 2, 3, 6, 7, 10

Data set 2:   1, 2, 3, 3, 4, 4, 4, 5

Data set 3:   1, 2, 3, 4, 4, 5, 6, 7

Which data set is skewed right, and which is skewed left? Suggest a way to tell whether the average or median is larger without having to do any calculations.

Data set 1 represents the number of responses to eight "for sale" advertisements placed in newspaper 1. Data set 2 gives the corresponding data when these same ads are simultaneously placed in newspaper 2. Which paper is best for the seller, based on these data? Discuss.

6. One hundred university professors are surveyed to determine the number of weeks of vacation taken over the previous 12 months. The survey reported the following data:

Number of weeks of vacation	Number of professors
0	4
1	8
2	21
3	45
4	12
5	4
6	0
7	1
8	5

(a) Calculate the average vacation, in weeks. Assume that "3 weeks" means exactly 3 weeks, and so on. (*Hint:* Four fives add up to $4 \times 5 = 20$.)
(b) Calculate the SD of the vacation duration. [*Hint:* Four quantities such as $(5 - 3)^2$ sum to $4 \times (5 - 3)^2$.]
(c) Does the empirical rule work well with these data?
(d) What is the median vacation time?

4.5 NORMAL APPROXIMATION TO FREQUENCY DISTRIBUTIONS

4.5.1 Introduction to the Normal Curve

We have discussed the use of the frequency distribution (or histogram) as a summary technique, and also the use of the average and standard deviation for this same purpose. Let us compare these two methods of summary.

With the availability of computers for calculations, both are easy to calculate. In terms of the space required for presentation, the average and SD method is a clear winner, but it conveys less information. For example, if we are told that Deming's rod diameters average 1.020 and have an SD of 0.0022, we have no hint of the important gap anomaly revealed by the frequency distribution. As a first step in the analysis of any data set, the investigator is wise to look at the frequency distribution before accepting the average and SD as an adequate summary. Thus the use of average and SD seems to be limited to situations in which the frequency distribution does not exhibit any interesting anomalies in its shape.

Figure 4.13a displays the shape of the histogram for many data sets. Notice that it has one hump and that it is symmetrical about its center. Figure 4.13b shows a two-hump histogram. Figure 4.13c shows a histogram that is not symmetrical; it is said to be "skewed to the right." We will focus in this section on a method for summarizing frequency distributions that look like Figure 4.13a. It turns out that it is these distributions that are well summarized by the average and the SD. From a descriptive point of view, these symmetrical distributions are unremarkable in shape, and we might expect that the average and SD are adequate summaries when this shape occurs.

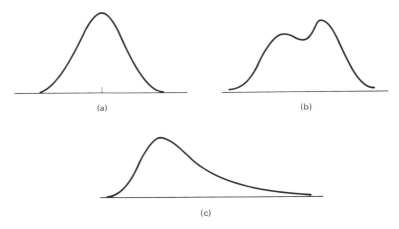

Figure 4.13 (a) Symmetrical one-hump histogram. (b) Two-hump histogram. (c) Asymmetrical one-hump histogram, skewed right.

A mathematical model for distributions of the shape shown in Figure 4.13a is based on the curve defined by the particular mathematical formula

$$h = \frac{1}{\sqrt{2\pi}} e^{-x^2/2}$$

The use of the mathematical results concerning the normal curve will require very little mathematics, and certainly no direct use of the formula shown above. The reason for quoting the formula is to emphasize that the normal curve really is a particular one, not actually defined by its properties of having one hump or being symmetrical: there are other curves with these properties. Methods to distinguish among these histograms that look normal are beyond the scope of this text. We will be content to treat all distributions that look like Figure 4.13a as if they were approximately normal.

This particular curve is called the **normal curve.** Histograms that look like

Figure 4.13a can often be shown to have a form approximated well by this formula, once the measurements are recorded in *standard units*. When this is the case we can apply to the observed histogram all the knowledge that mathematicians have generated about the normal curve.

In the application of the normal model to data sets, a key step is the expression of our measurements in standard units. Only one normal distribution is tabulated and that is the one with average 0.0 and SD 1.0. Yet real distributions have many averages and SDs, so to use the normal table to describe typical data sets we have to have a way of converting the data to units that have average 0.0 and SD 1.0.

Standard units for a measurement are the units that have the property that the average measurement in these units is zero and the standard deviation of the measurements in these units is 1. It will be shown that it is always possible to find units with this property. To understand the meaning of this definition, it is first necessary to see how we can make any measurement have any average and SD we wish, by appropriate choice of the units of measurement.

For example, if the weights of three people are 150, 170, and 130 lb, the average weight is 150 lb and the SD is 16.3 lb. If we convert these weights to kilograms (by dividing by 2.2), the weights are 68.2, 77.3, and 59.1 kg. In these new units the average is 68.2 kg and the SD is 7.43 kg. These are the same people with the same bodies but the average and SD have changed because the units of measurement have changed. Is there some unit of measurement that would make the average 0.0 and the SD equal to 1.0?

We want 150 lb to become 0 in these new units, and 16.3 lb has to become 1.0 in the new units. The simplest way to accomplish this, beginning with the weights in pounds, is to subtract the average weight 150 from each weight to give new weights: 0, 20, and −20. Divide these by the SD 16.3 to give 0.00, 1.23, and −1.23. Now these numbers have average 0.0 and an SD of 1.0. Yet they still measure weight, although in these new units the weight is measured relative to the average weight, and in SDs instead of in pounds. The numbers 0.0, 1.23, and −1.23 measure the weights of the three people in standard units.

Next we present a graphical explanation of standard units. Figure 4.14 illustrates two scales of measurement on which are recorded the three weights 130, 150, and 170. The visual representation of the three weights does not change at all; only the numbers that we use to describe the weights change when we express the weights in standard

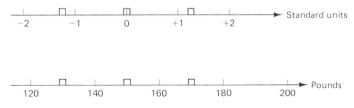

Figure 4.14 Scatter diagrams showing the visual equivalence for the distributions expressed in original or standard units.

units. This numerical conversion is what happens when we subtract the average and divide the result by the SD. The numerical advantage of this new scale is that the average is 0.0 and the SD is 1.0. If we use a normal approximation to the histogram (which in the case of the three numbers would be a poor approximation), the tabulated normal distribution, which also has average 0.0 and SD of 1.0, can be used to describe the observed data.

Perhaps the easiest way to describe standard units is to say that standard units are the units that describe how many SDs a measurement is from its average. If I weigh 1.55 in standard units, I weigh 1.55 SDs more than the average. Of course, the average and SD in this definition are understood to be calculated from some data set, and the standard units are therefore dependent on this reference set for their interpretation.

Sometimes an idea that seems simple when it is explained in one context does not seem so simple in another context. Let's work through another example. At a sailing school, 200 students take a test based on a course they have just completed. The instructor marks the exam and records the number of errors incurred by each of the 200 students. The average number of errors is 35, and the standard deviation is 7. If we are interested in one particular test result, that of a student who has made 50 errors, we might like to know what this is in standard units. This would allow us to get a rough idea of where this score of 50 ranks the student in this class. The score of 50 is $(50 - 35.0)/7.0$ standard deviations above the average for the group. This quotient is about 2.1 and we say that the measurement, 50 errors, is 2.1 standard units. So we see that standard units are achieved by first subtracting the average and then dividing by the SD. But do not get lost in the mechanics of the conversion: a measurement expressed in standard units is simply the distance from the average in SDs, with the sign indicating above (+) or below (−) the average.

It was claimed that the conversion to standard units is the key to the use of the

Sailing school practical exam. Applying theory takes practice. (Courtesy of Sea Wing Sailing School.)

normal model for frequency distributions. But how does this help us to evaluate the student who incurred 50 errors? It does not unless we are willing to accept that the histogram of the 200 scores is approximately normal. The decision to accept this is usually based on a visual inspection of the histogram. Assuming that we have accepted the shape as being close to that of the normal curve, an assessment of the relative size of a value 2.1 in standard units can be determined from a table of the normal distribution (see Table A4.1 in Appendix 4). From this table it can be found that a score of 2.1 exceeds 98.2 percent of the values. (The mechanics of this will be explained shortly.) Thus approximately 98 percent of the 200 students, or 196 of them, have fewer than 50 errors. We would expect to find in the list of 200 students' error scores about four that are greater than 50. (We ignore for now the complication that an appreciable percentage of the scores may equal 50, so that "greater than 50" is not the only alternative to "less than 50.")

Thus, starting from knowledge of the average and SD, and the assumption of normality, we can determine certain percentages that would otherwise require the frequency distribution, or histogram, itself. With the histogram at hand, we could easily read off the number of values greater than 50. But with the conversion to standard units and the normal curve table, we have done the same thing without the histogram. It was enough to have the average and SD.

When a data set has a histogram that is at least approximately normal, it does not make sense to describe it with a histogram. The average and SD, together with the comment that the distribution is approximately normal, are enough, since the normal table gives us the rest. Of course, it may be necessary to construct the histogram to discover that the distribution is normal, but one would not have to use this histogram to communicate the distribution to others (e.g., in a written report).

Normal distributions are conveniently summarized by the average and the standard deviation. Of course, not all distributions are normal; and not all significant numerical facts are distributions. Standard units may have no relevance with some numerical data. In the summer of 1983,[9] an Air Canada pilot apparently used the wrong units in assessing the fuel estimate given to him by a mechanic. The pilot was responsible for ensuring an adequate stock of fuel on board the Boeing 767 under his command. The flight from Ottawa to Edmonton required about 19,000 "units" of fuel. The confusion centered on whether 11,425 units of fuel on board were liters, pounds, or kilograms. The pilot thought that 11,425 ℓ (which is just over 20,000 lb) would be enough. The mechanic who measured the fuel on board thought the 20,000 figure calculated by the pilot was kilograms, and this seemed to him to be enough. Needless to say, the introduction of standard units to this situation would not have prevented the error. Happily, the pilot was able to make an emergency landing in Manitoba, but the airline was greatly embarrassed by the incident.

4.5.2 Table of Normal Percents and Percentiles

Let us next consider in a bit more detail the mechanics of using the the normal table. Table A4.1 relates the size of a measurement, which we denote z, to the percentage of values that are less than z, which we denote $A(z)$. It gives this relationship for a list of measurements that have a histogram that follows the normal curve exactly. z must be expressed in standard units to use this table.

The table may be used to find z that corresponds to $A(z)$, or to find $A(z)$ that corresponds to z. In the example of the sailing test scores, we mentioned that the $A(z)$ corresponding to 2.1 (a student's score of 50 errors expressed in standard units), was 98.2 percent. To determine what standard score, z, exceeds 90 percent of the standard scores, one looks for $A(z) = 90$ percent and reads off $z = 1.3$ standard units. Of course, for this $z = 1.3$ to be useful we must convert it back to the original units of the measurement, which was "number of errors." But 1.3 standard units is 1.3 standard deviations above the average, because measurements expressed in standard units have average 0 and SD 1. What measurement, in the original units, is 1.3 SDs above its average? Since 1 SD is 7.0, 1.3 SDs is 7.0 × 1.3 or 9.1. The average is 35 errors, so the score that is 1.3 SDs above its average is 35 + 9.1 or about 44 errors. Thus 44 is the error score that exceeds 90 percent of the error scores.

Now let us check that we know how to reverse this calculation: that is, if we were told the average and SD as before, but this time we were asked what percentage of scores are less than a score of 44 errors. The first step is to convert the question to an equivalent one in terms of standard units. What is the score of 44 in standard units? Well, how many SDs above the average is 44? The average is 35, so 44 is 9 errors above average, which is 9/7.0 SDs above average, that is, 1.3 SDs above average. Thus 44 errors becomes 1.3 standard units. Now we can look in Table A4.1 to find the percentage of values less than 1.3 standard units. For $z = 1.3$ we read off $A(z) = 90$ percent. The conclusion is that 90 percent of the error scores were less than 44. See Figure 4.15.

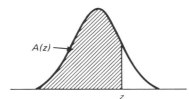

Figure 4.15 Percentages $A(z)$ correspond to the shaded area.

Getting used to these conversions with the normal curve takes a bit of practice. One may well think that the advantage of being able to summarize certain histograms with two numbers instead of a whole table, or graph, is not worth all the trouble and confusion in learning to use Table A4.1. However, this procedure is used throughout statistical work, not just for histogram description. Like learning to ride a bicycle, once you master the procedure, you will wonder why there was any difficulty at all.

There is more to learn about using Table A4.1. Figure 4.15 shows graphically how the table actually relates to the normal curve. It is the *area under the curve* that corresponds to percentages, just as with histograms as discussed in Section 4.3. The total area is defined to be 100 percent. Recall that the normal curve corresponds to a histogram for measurements with an average of zero, and that the normal curve is symmetrical about zero. What percentage of values should be smaller than zero when expressed in standard units? Since one-half of the area is to the left of zero, the answer

Sec. 4.5   Normal Approximation to Frequency Distributions   109

is clearly 50 percent. You can check this by seeing what value of $A(z)$ corresponds to $z = 0$ in Table A4.1.

We will also want to be able to deduce from Table A4.1 the areas such as those in Figure 4.16. These areas are related to tabulated areas, as indicated in Figure 4.17. For example, the area between standard unit values of -1.0 and $+2.0$ is $97.7 - 15.9$, that is, 81.8 percent. The area above the standard unit value 2.0 is $100 - 97.7$ or 2.3 percent. It is worthwhile to draw freehand histograms relating the area (percentage) you want to the area given by the table. Otherwise, it is easy to get confused.

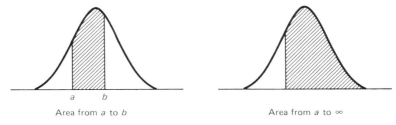

Figure 4.16 Typical area calculable from normal table.

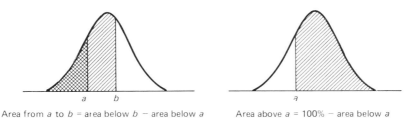

Figure 4.17 Calculation of areas under the normal curve from the tabulated areas of Appendix A4.1.

Here is a fresh example to see if you can put together all the machinery of using the table of the normal curve. One measure of "intelligence" is a written test for which the score is reported in such a way that the average score is 100 and the SD of all the scores for the population tested is 15. This intelligence quotient (IQ) has a histogram that follows the normal curve. (There should not be any mysticism read into this statement: the raw score on the intelligence tests may not follow a normal frequency curve. However, in the conversion of these scores to IQs, the distribution is forced to be normal. The IQ of a given raw score is the value whose normal-curve percentile exceeds the same percentage of normal scores that the raw score also exceeded. So the IQ has to be normally distributed itself, no matter what the distribution of the raw score.) Now suppose you find that your score on this test is 125. That would certainly be better than average, but how much better? One way to rephrase the question is: What percentage of IQ scores are greater than 125?

To find the answer to this question, the first step is to convert 125 to standard units. 125 is 25 above the mean, and this is 25/15 SDs above the mean. Thus 125 is 25/15 or 1.67 in standard units. From Table A4.1 we find that 1.67 corresponds to a percentage $A(1.67) = 95$ percent. This means that 95 percent of IQs are less than 125, and of course $100 - 95$ or 5 percent are greater than 125. This answers our question. You should try several of the exercises at the end of this section.

Suppose that we were to reverse the process in the last example. Then the question might be: What is the IQ value that just exceeds 95 percent of IQs in the population? The answer 125 has a special name: the 95th *percentile* of the distribution ("percentile" is defined below). Table A4.1 is essentially a table for converting percents into percentiles, and vice versa. The terminology is a bit confusing but it is widely used; the main point to keep straight is that a percentile is not a percent. The other little hitch in finding percentiles is that Table A4.1 does this only in standard units (i.e., it gives 1.67 as the 95th percentile). The 95th percentile for IQs was $100 + 1.67 \times 15$ or 125 (not 125 percent).

The next point to stress is that percentiles for a frequency distribution cannot be found with knowledge of only the average and the standard deviation unless the histogram of the data, in which the scale of the data is expressed in standard units, is approximated by the normal curve. The question "What IQ value exceeds 95 percent of IQs?" still makes sense even if the normal curve does not apply. In this case we would have to go back to the frequency distribution itself and to add up the frequencies from the lowest values until 95 percent of the frequencies had been accounted for. The value of IQ at this point would be the 95th percentile. If the histogram were not normal, the 95th percentile might not be 125, even though the average were 100 and the SD, 15. If the histogram looked like Figure 4.18, the 95th percentile would be about 115, even though for this histogram the average is 100 and the SD is 15.

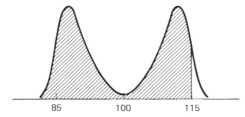

Figure 4.18 Nonnormal histogram with average 100 and SD 15 having a 95th percentile of about 115. For the normal distribution the 95th percentile would be about 125.

A **percentile** is a value that exceeds a certain percentage of the numbers in a list of numbers. The nth percentile of a list of numbers would be naturally defined as a value that just exceeds n percent of the numbers in the list. Because there is more than one value that would satisfy this definition, we have a convention to resolve the ambiguity: we define the nth *percentile* of a list of numbers as the smallest value that equals or exceeds n percent of the numbers in the list. But for practical purposes, we seldom need to be so worried about this ambiguity, and we can usually think loosely of, for example, the 25th percentile as any value that is greater than 25 percent and less than 75 percent of the numbers in the list. Also, it is common to speak of the 25th percentile of a frequency distribution—this just means the 25th percentile of the numbers that were used to form the frequency distribution.

Certain percentiles have special names: the 25th, 50th, and 75th percentiles are called the first, the second, and the third *quartile,* respectively. The 50th percentile is also called the median. The 10th, 20th, 30th, . . . percentiles are called the first, second, third, . . . *decile.*

The percentile idea leads to some additional measures of spread of a distribution. Recall that we have already introduced the SD, the MAD, and the range. A measure of spread that is occasionally used is the *interquartile range* (IQR), which is the difference between the third and the first quartiles. It is the distance spanned by the middle 50 percent of the data. A similar measure of spread could be constructed using the first and ninth deciles, although the term "interdecile range" is not widely used.

4.5.3 Description of Nonnormal Frequency Distributions

The calculation of an average and an SD as a summary of a frequency distribution has two shortcomings when the curve is not normal. Anomalies cannot be detected, and percentiles cannot be calculated. The proper use of average and SD as a summary technique is dependent on the presence of "normality": the histogram in standard units follows the normal curve.

The term "percentile," which was described in the preceding section in the context of a normal distribution, does not depend in any way on normality of the distribution. In fact, percentiles are often used to describe features of nonnormal frequency distributions. For example, the list of numbers {10, 13, 15, 16, 18, 19, 30, 37, 57, 59, 61, 65, 80, 82, 83, 90} does not have a normal distribution. Note that the list has been ordered for easy scanning. Suppose that these numbers represent the expenditures on restaurant meals by 16 residents of a student housing unit. What is the amount of money spent by the bottom 75 percent of the students? Well, 75 percent of 16 is 12, so we should look at the value of the 12th student's expenditures, and this is $65. The value $65 is equal to or greater than 12/16 or 75 percent of the numbers in the list. We call this number $65 the 75th percentile. We have not changed the definition from the discussion of Section 4.5.2. The point is simply that percentiles may be useful summary statistics even when the table of normal percentages does not apply.

Normality should not be interpreted as meaning "the usual situation" or "the proper distribution." Again the word is misleading but widely used in its special statistical context. In Chapter 7 we discuss when distributions can be expected to be normal and when this assumption would be foolish. In many practical situations we simply do not know whether or not the distribution is normal; we must have enough data to draw the histogram to check its shape visually.

We conclude this section with a simple demonstration that the average and the SD provide a poor summary of a distribution unless its shape is known to be approximately normal. The two data sets {0.59, 2, 2, 4.51} and {1, 1, 3, 3} have the same average and SD: namely, average = 2.0 and SD = 1.0. So if we are told these values for the average and SD, there is no way to choose between the distributions. The distributions can be recorded as follows:

Data set A		Data set B	
Value	Frequency	Value	Frequency
0.59	1	1	2
2	2	3	2
3.41	1		

which are obviously very different distributions. There are, in fact, an infinite number of different distributions with average 2.0 and an SD of 1.0: only one of them has the normal shape.

EXERCISES

1. The IQs of five people are 70, 85, 100, 115, and 130. IQs have a normal distribution whose average is 100 and whose SD is 15.
   (a) Express the IQs in standard units.
   (b) Draw a diagram like Figure 4.13 for these five IQ scores.
   (c) A new IQ scale is devised which has an average of 10 and an SD of 3. Assuming that the new scale measures the same underlying mental ability as the original scale, what score would you predict on this new scale for each of the five people?
2. Assume that the distribution of 200 sailing students' error scores follows a normal curve, that the average is 35 errors, and that the SD is 7.0.
   (a) If you were going to fail the 10 students with the greatest number of errors on the test, what would the cutoff score probably be? (*Hint:* What score is exceeded by 5 percent of the scores?)
   (b) What percent of scores are between 28 and 49?
   (c) What percent of scores are less than 20?
   (d) What score is greater than 75 percent of the scores?
   (e) What is the 75th percentile?
   (f) What is the 25th percentile?
   (g) What interval of scores contains the middle 50 percent of the frequency distribution?
3. There is an association called *Mensa* that has an I.Q. test as a test of fitness for membership. To qualify for membership, an applicant must score in the top 2 percent relative to a standard population. The distribution of IQ scores of the standard population is normal with an average IQ of 100 and an SD of 15. What is the IQ needed for membership?
4. On a statistics exam, raw scores average 55 percent and have an SD of 18. If the professor wishes to control the percentage of students receiving a letter grade of A to the top 15 percent of the class, what should his cutoff mark be? Assume normality.
5. When a measurement is expressed in standard units, it is unchanged by changes in the units of the original measurement. Verify this by performing the following steps with the following birth weight data {3700 g, 4100 g, 3300 g}.
   (a) Express the three data values in standard units.
   (b) For each data value, subtract 4000 and divide the result by 1000. What do these new units measure?

(c) Express the result of the calculation of part (b) in its standard units by recalculating the average and the SD.

6. The following data set[10] shows the exchange rate of the monetary unit of several countries expressed in U.S. dollars as of July 25, 1984.

Country	Currency	Exchange value (U.S. dollars)
Algeria	Dinar	0.20
Argentina	Peso	0.018
Australia	Dollar	0.83
Austria	Schilling	0.50
Bahamas	Dollar	1.00
Barbados	Dollar	0.50
Belgium	Franc	0.18
Bermuda	Dollar	1.00
Brazil	Cruzeiro	0.00055
Britain	Pound	1.34
Bulgaria	Lev	0.99
Canada	Dollar	0.76
Chile	Peso	0.011
China	Renminbi	0.50
Cyprus	Pound	1.70
Czechoslovakia	Koruna	0.15
Denmark	Krone	0.099
East Germany	Mark	0.351
Egypt	Pound	1.198
Finland	Markka	0.170
France	Franc	0.11
Greece	Drachma	0.0092
Hong Kong	Dollar	0.13
Hungary	Forint	0.021
Iceland	Krona	0.033
India	Rupee	0.084
Ireland	Punt	1.08
Israel	Shekel	0.0042
Italy	Lira	0.00057
Jamaica	Dollar	0.27
Japan	Yen	0.0041
Jordan	Dinar	2.63
Lebanon	Pound	0.20
Luxembourg	Franc	0.018
Malaysia	Ringgit	0.43
Mexico	Peso	0.0051
Netherlands	Guilder	0.31
New Zealand	Dollar	0.50
Norway	Krone	0.12
Pakistan	Rupee	0.071
Poland	Zloty	0.0092
Portugal	Escudo	0.0067
Romania	Leu	0.21

(continued)

114 Chap. 4 Descriptive Methods for One Variable

(continued)

Country	Currency	Exchange value (U.S. dollars)
Saudi Arabia	Riyal	0.28
Singapore	Dollar	0.47
South Africa	Rand	0.65
Spain	Peseta	0.0063
Sudan	Pound	0.77
Sweden	Krona	0.12
Switzerland	Franc	0.41
Taiwan	Dollar	0.025
Trinidad	Dollar	0.42
USSR	Ruble	1.25
Venezuela	Bolivar	0.084
West Germany	Mark	0.35
Yugoslavia	Dinar	0.0069
Zambia	Kwacha	0.56

(a) What is the percentile of the U.S. dollar among these currencies?
(b) What is the percentile of the Canadian dollar?
(c) What is the interquartile range of the values of the various monetary units?
(d) Is there a difference in the economic style of the countries in the tenth decile compared to those in the first decile?

4.6 ANSWERS TO EXERCISES

Section 4.2

1. (a) and (b)

    Appraised market value: quantitative.

    Number of levels: quantitative.

    Exterior facing material: qualitative.

    Rank based on size of lot: neither. Rank data are sometimes called "ordinal," indicating that such data allow ordering of things but does not describe the size of any particular thing. A possible strategy for summary here would be to retrieve the original data, size of lot, which is clearly quantitative, and use an average.

    Rank based on personal preference: neither. If this is one person's ranking, the data for the whole group will just be the integers 1, 2, 3 . . . . In looking at subgroups, one might try using average ranks, or percentage in the top 10, for example.

    Age of house: quantitative.

    Architectural style: qualitative.

    Number of rooms: quantitative.

    Status of landscaping: neither. The suggested categories are naturally ordered. In this case the three percentages provide a reasonable summary.

    Distance to shopping center: quantitative.

    Type of roofing material: qualitative.

(c)
> *Appraised market value:* real estate records of recent sales
> *Number of levels:* municipal tax records
> *Exterior facing material:* site visit
> *Rank based on size of lot:* tax records
> *Rank based on personal preference:* site visit
> *Age of house:* municipal registry
> *Architectural style:* site visit
> *Number of bedrooms:* municipal records
> *Status of landscaping* (poor, middling, good): site visit
> *Distance to shopping center:* map and ruler
> *Type of roofing material:* site visit

2. (a) Precise measurements
   (b) Accurate measurements
   (c) Unbiased measurements
3. To be sure of answering this question correctly, it would be wise to discuss the matter with the farmer. It is likely that the visual observation is the most reliable since if the volunteer grain is growing well, next year's crops probably will, too. The nitrogen content of the soil is not the only factor that affects its fertility. However, it might be argued that some leaching of the soil between now and next season is predictable and that a measure of nitrogen content would measure "fertility" for the practical purposes of the farmer. Such a consideration might lead to the conclusion that the laboratory measure is more valid. Validity questions are difficult to answer definitively in the absence of expert factual knowledge about the context of the question.

Section 4.3

1. The grouped frequencies are: 40, 80, 160, 150, 60, 10. The anomaly is absent from this distribution. The amount of grouping will depend on the purpose of the display: for data exploration one uses as many intervals as one has time for (or as many as will fit on a page of computer printout)—the ultimate in this case is the ungrouped data histogram. Usually for a data summary, 5 to 15 intervals are used, and the amount of grouping is determined to achieve this number of intervals. The choice among 5 to 15 intervals is determined largely by the prior expectations of the smoothness of the distribution to be displayed.
2. (a) The proportion can be ascertained from the picture without reference to the vertical axis. The proportion is 9/30 whether each □ stands for 1, 10, or 25 employees.
   (b) Your guess should be close to 30 percent.
   (c) 9/30 is 30 percent. 30 percent of the employees have worked with the company for five years or more.
3. (a) The histogram shown in Figure 4.19 is turned sideways from the usual orientation. For categorical data, this is quite usual, although there does not seem to be any compelling reason for this.
   (b) The gaps have no meaning at all. The variable whose histogram is drawn is a qualitative variable, and the position of a subject in the list of categories has no numerical interpretation worth noting.

116   Chap. 4   Descriptive Methods for One Variable

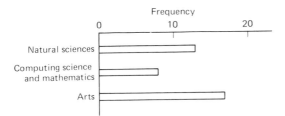

Figure 4.19 Histogram categorizing major fields of study.

4. (a) The distribution has three humps. This can be seen from inspection of the histogram; it is especially clear if grouping intervals of width 5 are used (i.e., 10–14, 15–19, 20–24, . . .). The frequencies are (1, 1, 2, 3, 4, 0, 0, 5, 4, 4, 4, 3, 3, 1, 5, 2, 2). The three humps can be assigned the grades F, B, and A, respectively. A justification of this assignment of grades is that the distribution seems to indicate that there are three distinct categories of performance, and if only three letter grades are being assigned, they should identify these three groups. The argument is not compelling, however. The instructor would have to satisfy himself that the lowest hump really did represent unsatisfactory performance (F), and that the highest group all deserved a grade indicative of excellence (A). This judgment depends on the scores themselves but also on the difficulty of the examination.

   (b) The simplest way is probably to display the one-dimensional scatter diagram of marks and draw vertical lines at the borders between F and B and between B and A. If the scatter diagram is considered a bit arduous, a table of the ranges of marks for A, B, and F would suffice; however, this would not reveal the multiple modes that guided the letter assignment, and this may be of interest to students.

5. • Coins may be taken out of circulation after five years (see the drop from 1978 and later to 1977 and before).
   • All years are plentiful from 1969 on (most recent 15 years); 1968 and earlier are quite rare.
   • Any coin dated 1955 or earlier, and possibly also 1957 and 1958, could be very rare.
   • Either some of the rolls scanned were old rolls (one or two years old), or else the issue of pennies has declined in the last two years. (Perhaps the penny is succumbing to the ravages of inflation and is becoming useless. Or perhaps the geographic dissemination of new pennies takes one or two years to reach the bank of our idle employee.)

6. (a) First one has to make a decision about the two groups of animals, and whether to distinguish them on the histogram. Although it does seem that the African and European animals shown tend to have longer life spans, it seems unlikely that this would be the case if all animals were included. So one may as well include the two groups in one histogram, as shown in Figure 4.20.

   (b) One-year intervals would produce a very "gappy" histogram. Ten-year intervals would produce a histogram similar to the five-year-interval version, although the second hump would be obscured.

   (c) The scatter diagram is shown in Figure 4.21.

   (d) The scatter diagram (Figure 4.21) avoids having to make a decision on interval size (and possibly having to draw more than one diagram to get it right). It shows several possible modes, and sometimes these are meaningful in defining subgroups, but with these data it is probably not useful. The shape of the histogram (Figure 4.20) is easier to describe

Sec. 4.6   Answers to Exercises   117

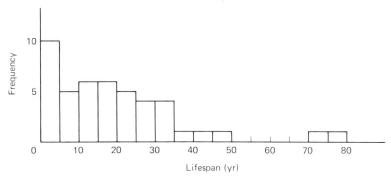

Figure 4.20   Histogram of life-span data.

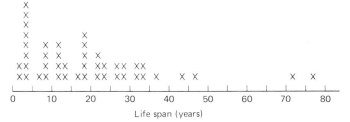

Figure 4.21   Scatter diagram of life-span data.

and to remember. It eliminates the minor modes and this elimination is appropriate in this instance.

7. (a) The histogram using 1-cm intervals is shown in Figure 4.22.
   (b) The histogram based on 4-cm intervals is shown in Figure 4.23.
   (c) The two humps in the histogram of part (a) may well correspond to two successive years' fry. This would presumably be of interest to anyone examining these data. The

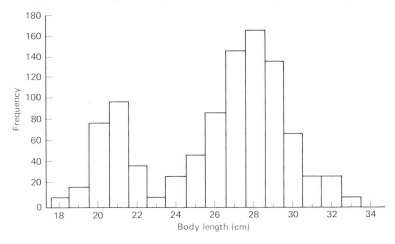

Figure 4.22   Histogram based on 1-cm intervals.

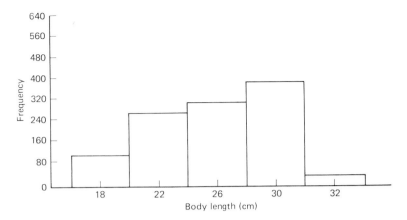

Figure 4.23 Histogram based on 4-cm intervals.

histogram in part (b) fails to show the two humps, and this is an indication that the interval width chosen for the histogram of part (b) is too large.

Section 4.4

1. $(0.40 \times 9.0) + (0.60 \times 9.5) = 9.3$ years. The justification for this method is that if the school had, say, 400 students, 0.4×400 or 160 would be girls and 240 would be boys. Adding up all 400 ages would give $(160 \times 9.0) + (240 \times 9.5)$ and the average age is this number divided by 400. That is, the average age is $[(160 \times 9.0) + (240 \times 9.5)]/400$, which is the same as $[(160/400) \times 9.0) + (240/400) \times 9.5]$, which is the same as $[(0.40 \times 9.0) + (0.6 \times 9.5)]$. Now this same procedure would lead to the same formula no matter what total school size was used. In other words, the procedure for calculating the average of a set of numbers from the averages in subsets depends only on the proportional sizes of the subsets, not the actual total size of the set. This procedure for calculating an average is called a weighted average.

2. **(a)** Average = 111°F; SD = 3.06°F.
   (b) Average = 43.9°C; SD = 1.70°C.
   (c) The Fahrenheit temperatures are more variable as measured by the SD, since $3.06 > 1.70$. Of course, this says nothing about the weather. The SD very much depends on the units of the data.
   (d) Yes, the formula does the job. SD depends on the scale factor, 5/9, only. It ignores the 32. However, we have to use the 32 in converting averages:

$$\text{average C temp.} = (\text{average F temp.} - 32) \times \frac{5}{9}$$

$$43.9 = (111 - 32) \times \frac{5}{9}$$

Also,

$$\text{SD C temp.} = (\text{SD F temp.}) \times \frac{5}{9}$$

$$1.70 = 3.06 \times \frac{5}{9}$$

3. (a) The average is $34,400. 2/10 or 20 percent earn more than average.
   (b) The median is a more typical value.
   (c) The SD is $18,700 approximately, so the $\pm$ 1 SD interval is 15.7, 53.1, which includes all but one salary, so the answer is 90 percent. With the $90,000 omitted, the average is 28.2 and the SD is 6.8, so the interval is now 21.4, 35.0. Thus the answer is now 70 percent, which is closer to the 68 percent predicted by the empirical rule. A reasonable speculation at this point is that outliers can invalidate the empirical rule. (We will see later in this chapter that the empirical rule is actually a property of normal distributions.)
4. Let the two numbers be 1.0 and 2.0. The SD is 0.5, as required.
5.

Data set	Mean	Median	Skew
1	4.13	2.50	Right
2	3.25	3.50	Left
3	4.00	4.00	Not skewed (i.e., symmetrical)

The choice should involve the variability of the list, not simply the average or the median. The lesser variability for newspaper 2 may be more important to the seller than the average number of responses.

6. (a) $(4 \times 0) + (8 \times 1) + (\cdots) + (5 \times 8) = 300$. The average is 300/100 or 3.0.
   (b) $[4 \times (-3)^2] + [8 \times (-2)^2] + \cdots + (1 \times 4^2) + (5 \times 8^2) = 1.62$.
   (c) 3 ± 1.62 is 1.38 to 4.62, which includes $78/100 \times 100$ or 78 percent of the data values. This is not very close to the 68 percent given by the empirical rule.
   (d) The data values ranked 50th and 51st are both 3, and so is their average. So the median is 3 (exactly).

Section 4.5

1. (a) $-2, -1, 0, 1,$ and 2, respectively.
   (b) Figure 4.24 shows the five IQ scores.

Figure 4.24 IQ scores.

120 Chap. 4 Descriptive Methods for One Variable

   (c) $10 + (-2 \times 3) = 4$
   $10 + (-1 \times 3) = 7$
   $10 + (0 \times 3) = 10$
   $10 + (1 \times 3) = 13$
   $10 + (2 \times 3) = 16$
   The new IQ scores would be 4, 7, 10, 13, and 16.

2. (a) $10/200$ is 5 percent. From normal table this is at 1.65 standard units. Thus the score at the cutoff point is $35 + 1.65 \times 7.0$ or about 46 errors.
   (b) 28 is $(28 - 35)/7$ or -1.0 SD; 49 is $(49 - 35)/7$ or $+2.0$ SD. From Table A4.1 the percent below -1 is 16 percent, and below $+2$ is 98 percent, so the percent of values between 28 and 49 is $98 - 16$ or 82 percent.
   (c) 20 is $(20 - 35)/7$ or -2.14. Table A4.1 shows that this value exceeds only 1.6 percent of the scores. Thus 1.6 percent of the scores are less than 20.
   (d) The score that is greater than 75 percent of the scores is 0.67 in standard units (see the table). In original units this is $35 + (0.67 \times 7.0)$ or approximately 40 errors.
   (e) 40 errors
   (f) $35 - (0.67 \times 7.0)$ or approximately 30 errors.
   (g) 30 to 40 errors.

3. We need the 98th percentile. From Table A4.1 this is 2.05 in standard units, or $100 + (2.05 \times 15) = 131$. Thus 131 is the minimum score for membership.

4. The top 15 percent must have scores that exceed the 85th percentile. The 85th percentile is 1.04 in standard units, or $55 + (1.04 \times 18) = 74$ approximately. The cutoff mark should be 74.

5. (a) The average is 3700, SD is 327. The data in standard units are 0.0, 1.22, and -1.22.
   (b) The new values are $-300/1000 = -0.30$, 0.10, and -0.70.
   (c) The average and SD of the new data are -0.30 and 0.33. These new data expressed in standard units are $[-0.30 - (-0.30)]/0.33 = 0.0$, $[0.10 - (-0.30)]/0.33 = 1.21$, and $[-0.70 - (-0.30)]/0.33 = -1.21$. Except for rounding errors, this is the same as in part (a).

6. (a) Of 58 currencies, 49 have less value and two others have equal value to the U.S. dollar. Thus the U.S. dollar is greater than 84 percent of the currencies ($49/58 = 0.84$) and is the 84th percentile. Some statisticians would use $(49 + \frac{1}{2} \times 3)/58$ or $50.5/58 = 0.87$ and call the U.S. dollar the 87th percentile. Either method is acceptable.
   (b) We count that 46 are less than 0.76, so the Canadian dollar would be the $(46/58) \times 100$ or 79th percentile.
   (c) The third quartile is just a little less than 0.76, since the Canadian dollar was the 79th percentile. In fact, if Zambia and South Africa are excluded from the 46, there are 44 left, and since $44/58$ is 0.76, 0.56 is the 75th percentile. $15/59 = 0.25$ and the fifteenth largest value is 0.031. So the interquartile range is $0.56 - 0.031 = 0.53$ approximately.
   (d) The difference is not as marked as one might expect, but the low-value currencies do include those from countries with very high inflation, whereas countries with relatively low rates of inflation tend to have high-value currencies.

4.7 NOTATION AND FORMULAS

4.7.1 About Mathematical Notation in Statistics

First, a general note about the sections on notation that appear here and at the end of each subsequent chapter. These sections may be omitted completely: the material of future sections does not depend on a familiarity with the notation. Why, then, is it introduced at all? Or, some might ask, why has it not been introduced together with the quantities they symbolize?

The use of mathematical notation has been left to the preference of the reader: it has been avoided as much as possible in the description of statistical techniques in this text since it draws one's focus to the symbols and the formulas instead of the more important aspects of the material: the rationale of the methods and the nature of the statistical phenomena being described.

For example, once one has a formula for the standard deviation, there is a tendency never again to think about what it is: a measure of spread based on squared deviations of the numbers from their mean. This understanding has practical value: if one measurement is far from the rest, perhaps due to a gross error, the SD will no longer describe a typical deviation, since the SD is very much inflated by a single large deviation. One would realize this if one has in mind that the SD is based on the *squared* deviations. In such circumstances the SD is a very poor descriptive measure of spread. Many such points of practical importance could be learned rather than spending the time on understanding the definition. It seems simpler to learn the basic definition thoroughly, so that such points become obvious consequences of your mental model of the SD, rather than a list of unrelated facts.

Most formulas in statistics are applied more properly by those who can describe in words what the formula is doing. This verbalization process helps to relate the calculation procedure to the words we use when we are talking to ourselves and others about our calculations. When the procedure is learned as a "plug-into-the-formula" process, we may be unable to explain why we have done the calculations and why they are appropriate to the problem at hand. This can lead to confusion, embarrassment, and subsequent disenchantment with statistics. The verbalization is an important part of statistical expertise.

Notwithstanding the foregoing arguments, mathematical notation has an essential role in statistics. Theoreticians require it to allow the symbolic power of mathematics to simplify technical developments—but that need not concern us here. The value of notation for an elementary statistics course is that it is an aid to memory for some students. Another consideration is this: it is worth a bit of a struggle to master the use of this notation if you intend to continue your study of statistics beyond the scope of this text. A subsequent course in statistics may well assume familiarity with the mathematical notation.

Consequently, the notation is presented, following the introduction of various techniques, for those who find it helpful. Some will find the notation perplexing, others will find it so helpful in remembering methods that they will forget the rationale for the methods and make the right calculations at the wrong time. You and your instructor will have to decide what is best for you. It is stressed that the notation sections are optional, and the remainder of the text does not depend on these sections.

4.7.2 Notation for this Chapter

Let X be a variable whose values are measurements in a data set. $\bar{X}$ denotes the average of the values of a variable X and SD is the standard deviation of the values of X. [Many texts define the standard deviation slightly differently and label it s. The relationship between the two is $\text{SD} = \sqrt{(n-1)/n} \times s$. See Section 4.3 concerning the choice of SD over s.] Given

n = number of measurements in the data set
X_i = ith measurement in the data set

then

$$\bar{X} = \sum \frac{X_i}{n}$$

and

$$\text{SD} = \frac{\sqrt{\sum (X_i - \bar{X})^2}}{n}$$

Note: If a preprogrammed calculator is used to calculate the SD, the answer you get may be slightly greater than the value you would get using the procedure described in this section. The value obtained on the calculator may have to be multiplied by the factor

$$\sqrt{\frac{n-1}{n}}$$

This multiple is very nearly 1, so the difference will be slight except for very small data sets, but you could be confused if you are trying to reproduce calculations in the text.

When data are converted into standard units, the conversion is represented as

$$z = \frac{X - \bar{X}}{\text{SD}}$$

Data values, represented by X, are converted into standard units, represented by z. Standard units are sometimes called z units. Once the data are expressed in z units, they have mean = 0 and SD = 1. The normal distribution with mean 0 and SD 1 is called the *standard normal distribution*. This is the distribution that is tabulated in Table A4.1.

The quantity tabulated in Table A4.1 may be denoted $F(z)$. For example, when

$z = 1$, $F(z)$ becomes $F(1) = 0.84$. We have seen how to use the table to recover z from the equation $F(z) = p$. If $p = 0.84$, $z = 1.0$ can be read from the table. When $F(z) = p$, z is the $(100 \times p)$th percentile. 1.0 is the 84th percentile of the standard normal distribution. What is the 43rd percentile for the standard normal distribution? The answer may be read from Table A4.1. It is -1.48. That is, $F(-1.48) = 0.43$.

When a histogram is drawn using relative frequencies, it does not "grow" as the sample size gets larger. (A histogram of plain frequencies would grow.) It will vary a bit but will stabilize eventually. This histogram, which we can imagine being generated by a huge sample, using relative frequencies, is called a *density function*. (This is not quite correct. When relative frequencies are used for grouped data, the height of the rectangles will depend on the size of the grouping interval. This can be corrected using a "density scale" which expresses the relative frequency per unit of measurement of the variable.) A density function is often written as $f(x)$, or $f(z)$ if the variable is in standard units. It can be shown that $F(z)$ is the area to the left of z under the curve $f(\cdot)$: that is, between the curve, the z axis, and the vertical line drawn at the value z. [For those familiar with the definitions of calculus, $f(z)$ is the derivative of $F(z)$ and $F(z)$ is the definite integral of $f(\cdot)$ between $-\infty$ and z.]

4.8 SUMMARY AND GLOSSARY

1. Measurements are composed of three components:

    measurement = true value + bias + unexplained variation

    Accurate measurements are those that are precise (i.e., small unexplained variation) and with small bias.
2. A frequency distribution describes the relative frequency of various values of a certain measurement. Its graphical form is called a histogram.
3. The location of the center of the frequency distribution is described by the average and the median. The amount of spread in a distribution is described by the standard deviation (SD), the range, and the average absolute deviation (MAD), or the interquartile range (IQR). The combination of average and SD is the most common summary of a frequency distribution, because of its appropriateness for the normal distribution.
4. The average and the SD are sensitive to outliers; the median and interquartile range are not sensitive to outliers.
5. The units of the average, the median, and the SD are the same as the units of the original measurements being summarized.
6. Standard units express the value of a measurement relative to the entire set of values of that measurement by reporting the number of SDs from the average.
7. A frequency distribution of data that has been expressed in standard units may be approximated by the normal distribution described by Table A4.1 of Appen-

dix 4. Alternatively, one can adapt the tabular information of Table A4.1 to describe the data set in its original units.

8. The area under any part of the histogram curve is proportional to the percentage of data values in the interval under that part.
9. A percentile is a particular data value that exceeds a certain percentage of the data values. If the percentage exceeded is p percent, the particular value is called the pth percentile (of the frequency distribution of the data).

Glossary

Measurement may be thought of as the process of assigning a number or numerical code to something. (4.2)

An **instrument** is the device used to assist in the production of a number from the thing measured. (4.2)

Qualitative measurements are those which result in categorical information (without any connotation of size). (4.2)

Quantitative measurements are those which result in numbers whose size indicates the aspect measured. (4.2)

Bias is the amount the measurement tends to be wrong, on average. (4.2)

Imprecision is the amount of variation of measurements from the average measurement. (4.2)

When both bias and imprecision are small, the measurement is said to be **accurate.** (4.2)

A measurement for which the bias is zero may be called **unbiased.** (4.2)

The **validity** of a measurement is the degree to which it measures the quality or quantity it is supposed to measure. (4.2)

A **frequency distribution** of a list of numbers consists of a list of the possible different values that the original list contains and the frequency with which each of these possible values occurs. (4.3)

A **histogram** is a graphical representation of a frequency distribution in which a rectangle represents the possible values and their frequencies: the position of the rectangle indicates the possible values being described while the size of the rectangle represents the frequency of the values being described. (4.3)

A **mode** of a histogram is an interval over which the height of the histogram is greater than for values just outside the interval. It refers to the "humps" of the histogram. A distribution with only one hump is called unimodal. (4.3)

Skewness is the property of a histogram to have one "tail" much longer than the other—a histogram is said to be skewed right when the long tail is to the right. (4.3)

The **average** of a list of numbers is the sum of the numbers in the list divided by the number of numbers in the list. (4.4)

The **median** of a list of numbers is the middle-ranked value in the list. (4.4)

The **range** of a list of numbers is the largest value in the list minus the smallest number in the list. (4.4)

The **standard deviation** (SD) of a list of numbers is the square root of the average squared deviation of the list. "Deviation" in this definition means the difference between a number and the average of the list. (4.4)

The **average absolute deviation** or **mean absolute deviation** (MAD) of a list of numbers is the average absolute value of the deviations of the numbers in the list from the average of the numbers in the list. (4.4)

Invariance means different things in different contexts. The common aspect that justifies the use of the one term for two ideas is that in both cases changing one thing does not change some other thing, in some sense. The "other thing" is said to be invariant with respect to changes in the "first thing." (4.4)

The **normal curve** is a histogram defined by a particular mathematical formula (see Section 4.5.1). (The curve is symmetrical and unimodal but there are other histograms which have this property which are not normal curves.) (4.5)

Standard units for a measurement are the units that have the property that the average measurement in these units is zero and the standard deviation of the measurements in these units is 1. (4.5)

Alternative definiton: **standard units** are the units that describe how many SDs a measurement is from its average. (4.5)

A **normal** frequency distribution is one whose frequencies may be calculated from the normal curve: that is, the relative frequency of any particular interval of values, expressed in standard units, is the area under the normal curve and above the horizontal axis, and marked off on each side by the ends of the interval. (4.5)

A **percentile** of a list of numbers is a value that exceeds a certain percentage of the numbers in the list. The nth percentile of a list of numbers would be naturally defined as a value that exceeds n percent of the numbers in the list. Because there is more than one value that would satisfy this definition, we have a convention to resolve the ambiguity. We define the nth percentile of a list of numbers as the smallest value that equals or exceeds n percent of the numbers in the list. (Certain percentiles have special names: the 25th, 50th, and 75th percentiles are called the first, second, and third quartiles, respectively. The 50th percentile is also called the median. The 10th, 20th, 30th, ... percentiles are called the first, second, third, ... deciles.) (4.5)

The **interquartile range** (IQR) is the difference between the third and the first quartiles. It is the distance spanned by the middle 50 percent of the data. (4.5)

PROBLEMS AND PROJECTS

1. Lumber companies employ specialists called "scalers" to estimate the number of "board feet" of lumber that can be cut from a given log or truckload of logs. As logs tend to be elliptical in cross section and not quite conical in shape, contain knots and other defects,

and contain brittle heartwood, the estimation process involves considerable error. If you were in charge of a company's team of 10 scalers, what techniques would you suggest for the reduction of bias and imprecision in these estimates? What descriptive methods would you use to record the success of your suggestions?

2. Pick 10 books from your bookshelf and without opening any of them, estimate the number of pages in each book. Now record the actual number of pages in each book.
   (a) Record the error for each book. Do you think your estimate was biased, based on the 10 measurements?
   (b) Use a standard deviation calculation to record the degree of imprecision in your estimate. How does this compare with your average error? Explain.

3. (Continuation of Problem 2.) Pick a different group of 10 books, but this time check your estimate immediately with each book. Record the sequence of errors. Have you been able to reduce the bias? Explain your answer.

4. In drawing the histogram of rod diameters, the use of rectangles of width corresponding to 0.001 cm results in rectangles over adjacent intervals touching on a vertical side. This implies that the values actually depicted by the histogram are spread across intervals such as 1.0005 to 1.0015, 1.0015 to 1.0025, and so on. However reasonable this may seem, if the instrument has a digital readout that is limited to displaying to the nearest 0.001 cm, that is, 1.001, 1.002, . . . , this implication is incorrect. Can you think of a way of modifying the histogram so that the viewer is made aware of the fact that the machine records to only three decimal places?

5. For small data sets, the accumulation of frequencies for the various intervals of values is occasionally done by hand instead of on a computer. A useful technique in this situation is the stem-and-leaf plot. For the following marketing survey data a stem-and-leaf plot has been begun below. The data represent the age of head of household in 25 households selected at random from the Riverview area.

$$31, 64, 53, 51, 60, 52, 49, 43, 40, 45, 47, 73, 50$$
$$67, 48, 56, 34, 41, 35, 37, 72, 25, 26, 23, 32$$

```
1 |
2 |
3 | 1
4 | 9
5 | 312
6 | 40
7 |
```

Reproduce this beginning with the first seven data values, and complete the procedure. (*Hint:* Each row of the plot is for recording numbers with a certain first digit. Hence 31 is recorded with a 1 on the third row, 64 with a 4 on the sixth row, and so on.) Do you end up with a histogram-like display? What are the advantages of such a plot relative to a histogram?

6. A family doctor's secretary compiles the weight distribution of all her patients.
   (a) Describe the shape you would expect to see for the histogram of these weights.

(b) If you were to see this same pattern for the histogram of tree heights in a virgin forest, what would you conclude about the forest? (This problem is continued as a computer project in Problem 25.)

7. The times taken for five cyclists to complete a competition circuit are, in minutes: 25.7, 27.1, 31.0, 26.9, 28.3. Assuming this is a random sample of five from a population of 100 competitors, do a calculation to estimate the rank among the 100 of the cyclist whose time is 25.7, assuming that all 100 cyclists compete on the same circuit. (Assume normality.)

8. A biologist measures a sample of maple trees on the local college campus. These trees have trunk circumference (cm) as shown in the following data set:

$$92, 55, 83, 49, 51, 79, 75, 93, 72, 67, 60, 78, 57,$$
$$65, 66, 91, 33, 77, 94, 97, 73, 79, 68, 74, 92$$

Provide a statistical summary of these data that is likely to be useful to the biologist. (*Hint:* Why might a biologist be interested in such data?)

9. Give one reason why the SD would be preferred to the range (largest to smallest) as a measure of spread of incomes, for a comparative study between male and female executives.

10. A data processing company requires applicants for computer programming positions to take a test. The company policy is to reject from further consideration any applicants who fail to obtain at least a score of 70 on this test. Thus far, the average score for applicants has been 77.5 and the SD has been 5.
    (a) About what percentage of the applicants have been rejected by the test?
    (b) State the assumption you need for the validity of your calculation in part (a).

11. Summarize the following data for the president of a small landscaping company.

Month	Number of locations	Total revenues
March	12	$2900
April	23	5000
May	41	6200
June	45	6500
July	24	3100
August	31	5300
September	46	9200

12. In a large statistics class there are 500 students: 300 men and 200 women.[14] The men average 5 ft 10 in. in height with an SD of 3 in. The women average 5 ft 5 in. in height with an SD of 3 in.
    (a) What is the average height of all 500 people?
    (b) Is the SD of heights for all 500 people bigger than 3 in., or smaller than 3 in., or just about 3 in.?
    (c) About how many women in the class will be over 5 ft 10 in. in height?

13. Consider the distribution of examination scores shown in Exercise 4 of Section 4.3. Toss a coin for each score, selecting those scores for which the toss results in a head. Can you still discern the three humps in the distribution? Would you expect to be able to do so most

of the time if you repeated this experiment? Explain. (This problem is continued as a computer project in Problem 26.)

14. The amount of rainfall in millimeters is recorded each day in Seattle for 365 days. Rainfall in Seattle is seasonal, with frequent rain in the winter months and very little rain during the summer. A histogram is constructed of the 365 data values.
   (a) What would be the shape of this distribution? (Draw a rough histogram to illustrate your answer.)
   (b) Is the histogram a useful summary of these data? Discuss.

15. The most commonly used letter of the English language is the letter "e." For example, this paragraph contains more e's than any other letter. Of course, it is impossible to say in a small piece of prose just what proportion of the letters are e's. The number of e's per line will vary and will have a certain frequency distribution. Also, the number of e's in a given line of prose can vary in the sense that we may count them imperfectly—we might call this measurement error. Some variation is due to "error," while some variation is not caused by "error" (in the usual sense of the word).
   (a) *Without* taking the utmost care, count up the number of e's in the entire paragraph above: do *not* note the number of e's in each line. Repeat this count twice more. Your three paragraph counts form the beginning of a frequency distribution. Compute the average and SD based on your three paragraph counts.
   (b) Again, count the number of e's in the same paragraph but this time note the number of e's in each line. Do *not* count the same line twice in order to get it right; just note your first count. Compute the SD of these counts. These counts will be used in part (e).
   (c) For each line in the same paragraph, count up the number of e's. Repeat this count on each line until you obtain the same count twice. Then record these repeated counts (the number of recorded numbers should equal the number of lines). These numbers begin to reveal a frequency distribution of the number of "e"s per line. Compute the average and SD of these numbers.
   (d) Multiply the average in part (c) by the number of lines. Is this product larger or smaller than the average in part (a)? Can you explain why?
   (e) Compare the SDs computed in parts (a), (b), and (c). Comment on, and where possible, explain, the relative sizes of these three SDs.
   (f) For each of the three SDs compared in part (d), predict whether the effect of using a much larger test paragraph (of 100 lines, say) would increase the SD, decrease the SD, leave it about the same, or is it impossible to predict reliably? Explain.

16. Historical studies of eminent people often mention the predominance of eldest sons and daughters among the eminent group. However, most sons (for example) are eldest sons. Show that this is reasonable by assuming that families always have one, two, three, or four children, and that each number of children occurs equally frequently (i.e., 25 percent of the families). Enumerate the possible family patterns (i.e., *BBG*) including birth order (i.e., list *BBG* and *BGB* separately), and for each family size, work out the proportion of boys that are eldest sons. [You should assume that the chance of a male birth is the same as the chance of a female birth, so that all orderings (i.e., *BBG*, *BGB*, *GBB*) occur equally often.] Then combine the results for the four family sizes to verify that about 70 percent of sons are eldest sons.

   State in a simple sentence why this percentage is as large as 70 percent. (At first, this seems surprising since the property of being an eldest son seems a fairly specific attribute, and we naively expect a small proportion of people to be eldest sons.)

17. Table 4.5 presents the final examination scores for a class of 400 students of a statistics course. Construct three histograms of these data based on intervals of width 2, 10, and 20. (A freehand drawing with attention to the frequency levels should suffice for this problem.) Comment on the relative utility of the three histograms to
    (a) students, to assess their relative performance on the exam
    (b) the statistics department chairman, who is interested in monitoring grading standards in his department
    (This problem is continued as a computer project in Problem 27.)

TABLE 4.5 FREQUENCY DISTRIBUTION FOR FINAL EXAMINATION SCORES IN A STATISTICS COURSE

Score	Frequency	Score	Frequency	Score	Frequency
0–1	0	34–35	2	68–69	12
2–3	0	36–37	3	70–71	17
4–5	0	38–39	8	72–73	19
6–7	1	40–41	11	74–75	14
8–9	0	42–43	14	76–77	8
10–11	0	44–45	15	78–79	5
12–13	1	46–47	19	80–81	6
14–15	0	48–49	17	82–83	7
16–17	2	50–51	18	84–85	6
18–19	0	52–53	23	86–87	3
20–21	0	54–55	27	88–89	0
22–23	2	56–57	19	90–91	2
24–25	3	58–59	20	92–93	1
26–27	5	60–61	21	94–95	0
28–29	10	62–63	13	96–97	0
30–31	9	64–65	14	98–99	0
32–33	6	66–67	17	100	0

18. An ecologist knows that the numbers of trees of each species in a certain west coast forest are in the following proportion: 35 percent fir, 40 percent cedar, and 25 percent hemlock. (Ignore other species.) She wishes to estimate the average volume of wood per tree over the whole forest. This volume is measured for three trees of each species selected at random from the forest. These volumes are:

    Fir: 15.0, 13.5, 18.0 m^3

    Cedar: 18.5, 20.5, 18.0 m^3

    Hemlock: 12.5, 15.0, 14.5 m^3

    (a) Estimate the average volume of wood per tree in the forest.
    (b) If the forest contains 1,000,000 trees of the species described, estimate the total volume of wood in the forest.
    (c) Comment on the estimation procedures you used in parts (a) and (b) in contrast to the multiplication of the average of the nine volumes by 1,000,000. That is, which method is better?

19. An alternative measure of the center of a distribution is called the "trimmed" mean. If 10

data values are to be summarized, the 10 percent trimmed mean is the average of the eight central values—one orders the 10 observations from smallest to largest and eliminates 1 (10 percent of 10) from each end of this list. For example, the 20 percent trimmed mean of {1, 10, 12, 17, 19, 28, 30} would be the average of {10, 12, 17, 19, 28} or 17.2, whereas the ordinary mean (or average) is 16.7. Twenty percent of the seven data points is 1.4, but since we can only trim an integral number of values, we trim 1 from each end of the ordered list.

(a) Compute the 15 percent trimmed mean, the average, and the median of the values {13, 7, 28, 47, 29, 13, 19, 24, 8, 16, 13, 9, 3, 34, 24}.

(b) Describe the relationship of the trimmed mean to the average and the median (in general).

(c) Describe a situation in which the trimmed mean would be a better summary of the center of a distribution than either the average or the median.

20. Transformations of data sets can make a distribution that cannot be approximated by a normal distribution into one that can be so approximated. Consider the logarithmic transformation, for example: the numbers 1, 10, 100, and 1000 become, after taking the logarithms to the base 10, the numbers 0, 1, 2, and 3, respectively.

Now suppose that your sample is 1.5, 8.7, 105.3, and 935. These data would not be modeled by a normal approximation, but the logarithms of these numbers might be. Taking logs, we have 0.18, 0.94, 2.02, and 2.97. These numbers are more reasonably represented by a normal distribution. Try out this transformation on the following subset of the animal-life-span data (years; the complete data set is given in Exercise 6 of Section 4.3).

Domestic dog	20
Domestic cat	28
Canadian beaver	20
Gray squirrel	10
Brown rat	3.3
Raccoon	14
Human being	75

(a) Is the distribution of the transformed data more normal?

(b) When would this logarithmic transformation be of some practical use? (That is, what advantage is there that a distribution be normal when the distribution is to be described?) (This problem is continued as a computer project in problem 28.)

21. Refer to the exchange-rate data of Exercise 6 of Section 4.5. The comparable list for exchange rates in any country's currency can be obtained from this list (applicable to July 25, 1984).

(a) Obtain current exchange rates from a local paper. (Use whatever countries are listed.) Convert the two lists to the same currency.

Obtain the differences of the exchange rates (in some currency) between July 25, 1984, and a current date. Express this difference as a percentage of the July 25, 1984, figure, and be sure to include the sign of the difference.

(b) Summarize the distribution of the percentage changes in exchange rates [the differences obtained in part (a)] using a graphical technique.

(c) Summarize this distribution using a measure of center and a measure of spread.

(d) Attempt a verbal summary such as might constitute the lead paragraph in a newspaper article on exchange-rate changes since July 25, 1984. (*Hint:* It may be helpful to consider lumping together countries that are geographically, economically, or politically close.)

22. One thousand guinea pigs arrive at a large medical laboratory. The supplier claims that at least 50 percent of these guinea pigs are female, and the researcher is anxious that this be verified. Rather than check all of them, the researcher selects a simple random sample of 100 guinea pigs and finds that only 41 of them are female. Is the evidence strong enough that she would be justified in asking the supplier for compensation or additional animals, or would it be necessary for her to inspect more animals to decide whether the supplier's claim is valid? Or is it impossible to make any judgment based on this sample?

23. In a statistics test, your test paper is returned with the information that your score was in the 80th percentile. The average mark on the test was 55 and the SD was 15. There are 100 students who take the test.
    (a) What additional information would you need to be able to estimate a good approximation to your actual score on the test?
    (b) Assuming that this information is provided, demonstrate the method of estimation.

24. Draw a freehand sketch of the histogram (a smooth-curve histogram will suffice) of the age distribution of students currently enrolled in this course. Based on your sketch, estimate the average age, the median age, and the SD of age. (Your solution to this problem will be judged on the consistency of your answers, on the general shape of the histogram, and on the use of reasonable scales for your histogram.) Include a brief explanation of your choice.

Computer Projects

25. (Continuation of Problem 6.) Generate 250 standard normal deviates. To the first 100, add 1.0. To the next 50, add 2.0. Then display the histogram of the resulting 250 data values.
    (a) Histograms that show more than one hump suggest that more than one population is being observed. Discuss this claim with reference to your computer-generated histogram.
    (b) Do small subpopulations show up in a histogram as readily as large subpopulations? Explain.

26. (Continuation of Problem 13.) Refer to the data set in Exercise 4 of Section 4.3. Use a random number generator to select, with replacement, a random sample of size 44 from the population consisting of the original 44 numbers. (This sample will contain many of the original values more than once, and many not at all.) Plot the histogram of this new sample. Observe whether the three modes (the same ones, not just any three modes) are apparent. Repeat this process a few times to help to decide whether the three humps would usually be discernible or not.

27. (Continuation of Problem 17.) Construct histograms from the data of Table 4.5 using intervals of size 2, 4, 6, . . . , 22, 24, 26. Comment on the appearance or disappearance of features of the histogram (such as modes, smoothness, isolated frequencies, skewness) as the interval size increases from 2 to 26. Does this change your answer to Problem 17?

28. (Continuation of Problem 20.) Do Problem 20 but on the entire data set of Exercise 6 of Section 4.3. Try the square-root transformation as well. [Do parts (a) and (b).]
    (c) Which transformation does the best job of converting the distribution to normality?

Descriptive Methods for Two Variables

*You must lie upon the daisies
And discourse in novel phrases
Of your complicated state of mind,
The meaning doesn't matter
If it's only idle chatter
Of a transcendental kind.*

W. S. Gilbert (1836–1911)

Education is what you have left over after you have forgotten everything you have learned.

Anonymous

Many data sets consist of more than one kind of measurement on each item. A variety of models of cars may be assessed for fuel economy, weight, engine displacement, and retail price, and any one of these measurements is more useful in conjunction with the others than it would be by itself. We cannot analyze the four variables simply as four one-variable data sets without missing useful information. Many of the complications in the description of multiple-variable data sets can be explained using two-variable examples, and this is the overall goal of this chapter.

Key words are variable, case, group, scatter diagram, prediction, predictor variable, predicted variable, association, agreement, positive association, negative association, correlation, oval diagram, outlier, linearity, correlation of averages, five-number summary, regression, point of averages, SD line, regression line, regression method, SD_{pred}, homoscedasticity, heteroscedasticity, residual plot, regression effect, regression fallacy, augmented scatter diagram, Chernoff's faces, star plots, cluster analysis, multiple regression, discriminant analysis, factor analysis, principal components, contingency table, and dependent variables.

5.1 SOME PRELIMINARIES

5.1.1 Terminology

In Chapter 4 we discussed strategies for the description of very simple data sets: one measurement on each item (or individual) in one group. We need an unambiguous way to refer to more complex data sets. Statistics is a discipline that demands an uncommon precision in the meaning of words. To avoid ambiguity it is necessary to take ordinary words and give them meanings which are more limited than they would have in general usage. Lewis Carroll, a mathematician as well as an author, expressed this point through Humpty Dumpty in *Through the Looking Glass*[1]:

> "But 'glory' doesn't mean 'a nice knockdown argument,'" Alice objected.
> "When I use a word," Humpty Dumpty said, in a rather scornful tone, "it means just what I choose it to mean—neither more nor less."
> "The question is," said Alice, "whether you can make words mean so many different things."
> "The question is," said Humpty Dumpty, "which is to be master—that's all."

We have already encountered the words "distribution," "random," "deviation," "normal," and even "statistics" itself, which we have anointed with meanings more specific than the words have in general usage. We must agree with Humpty Dumpty about the mastery of our vocabulary, that it do what we want it to do, but we have to sympathize with Alice that too many meanings for a word can be confusing. In this text we stick to a single definition for each of these slippery words.

We will use an example to introduce some further terminology. Suppose that we have the final exam scores in English and mathematics for a high school graduating class, and we are interested in whether the oft-claimed male–female difference in performance in these subjects is confirmed by these data. There are 80 students in the class, of which 45 are female. The data set consists of $80 \times 2 = 160$ measurements. There are two **groups,** males and females, and two **variables,** mathematics score and English score. The female group has 45 cases, and the male group has 35 cases. Thus the 160 measurements should be thought of as $45 + 45 + 35 + 35$. The individuals in each group are called the **cases.** Sometimes we will refer to the measurements as **values of a variable;** that is, a variable, mathematics score, would have 80 values in this example.

So "cases" are the number of items examined, "variables" are the different kinds of measurement performed on each case, and "groups" are the collections of cases that we may wish to compare with each other.

When confronted with a data set, an important step in absorbing the structure of it is to identify how many cases, groups, and variables the data set contains.

Many statistical techniques are designed to be applicable to particular combinations of the number of variables and the number of groups; occasionally, the techniques depend on the number of cases as well. Note the structure of the data set in the

134 Chap. 5 Descriptive Methods for Two Variables

example. This structure is best thought of as a rectangular table of numbers (sometimes called a *matrix*) in which each student's scores appear on a particular horizontal row and each variable's values appear in a vertical column. The rows corresponding to the females form a rectangular table within the larger table. Thus the data set has the structure shown in Figure 5.1.

```
              Variables
            X X X X X
            X X X X X
            X X X X X
     Cases  X X X X X
            X X X X X
            X X X X X
            X X X X X
```

Figure 5.1 Common structure for data sets. Each row is one item or individual—a "case." The cases are usually sorted so that the groups appear as separate rectangular blocks of data, one above the other. Each column contains the values of a single variable. The example shown here has five variables and seven cases.

Almost all the techniques in this chapter are for data sets that have two variables, several cases, and one or more groups.

5.1.2 Scatter Diagrams

Any trainer knows the importance of feedback in improving the performance of the trainee (whether athlete, child, or circus bear). The following simple guessing game demonstrates the futility of training in the absence of feedback. This unsurprising result will not distract the reader from the main purpose of this example—to introduce scatter diagrams.

Table 5.1 records the author's attempt at guessing the number of pages in 10 books, without feedback. (See Problem 2 of Chapter 4.) Every single estimate was low, so there does appear to be a bias here. The average of the errors is about −88 pages and the average percentage error is about −18 percent. Is the bias more like a constant shortfall of 88 pages or a constant percentage shortfall of 18 percent? We

TABLE 5.1 DATA FOR ESTIMATION OF NUMBER OF PAGES IN TEXTBOOKS, WITHOUT CORRECTIVE FEEDBACK

Book	Estimate	Actual	Error	Percent Error
1	325	380	−55	−14.5
2	360	426	−66	−15.5
3	350	432	−82	−19.0
4	425	540	−115	−21.3
5	375	386	−11	−2.8
6	430	526	−96	−18.3
7	400	546	−146	−26.7
8	340	446	−106	−23.8
9	485	560	−75	−13.4
10	470	594	−124	−20.9

could do some calculations, but there is a graphical way to examine these options. The graphical display in Figure 5.2 is called a **scatter diagram** or **scattergram.** A scatter diagram is a figure depicting data values as points in a scaled graph; it has one axis for one variable and two axes for two variables. (See Figure 4.9 for one-variable example.)

What features can be "read" from this scatter diagram? There is certainly a tendency for books with more pages than average to be estimated to have more pages than average. This is expected. But note that the data fall well above the line of perfect guesses, that is, the line through (0, 0) and (400, 400). (For each point on this line, both coordinates have the same value, and since the two variables plotted are the guess and the actual, a data value on the line would mean that the guess for that book was perfect.) How would you describe the bias?

There are two simple possibilities for describing the bias. The line of perfect guesses goes through (0, 0) and (400, 400), and if a bias exists, it may well result in a line, but different from the line of perfect guesses. If our data points appear to follow a line through (0, 0) but with a slope different from 1.0, the bias would be described as being a constant percentage. If the points follow a line that has slope 1.0 but does not go through (0, 0), the bias would be described as a constant.

To see whether the error is more like a constant or a constant percentage, we simply see whether the data points are closer to a line not through (0, 0) but with slope 1.0, or to a line through (0, 0) but with slope different from 1.0. With these data, the line through (0, 0) with slope less than 1.0 seems a better fit, although the choice is by no means clear cut. The practical result of reading this scatter diagram is that one

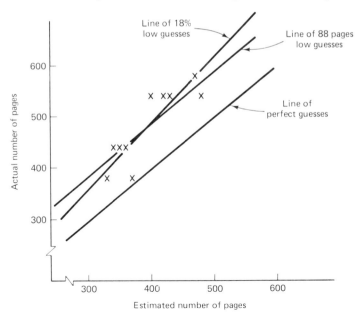

Figure 5.2 Scatter diagram for pages estimation data. (See Table 5.1.)

would summarize the data as exhibiting an 18 percent shortfall in the guesses (rather than an 88-page shortfall). (See Exercise 5 at the end of this section.)

Scatter diagrams have proved useful to data analysts for decades in spite of being tedious to draw. With the ease of computer programs to do the work, they are becoming even more popular. In this book they are often used to describe certain statistical phenomena of the relationships between two variables.

What kinds of data sets can be displayed with a two-axis scatter diagram? There must be two variables (or two versions of the same variable as in the pages example above) and several cases. Note that the two variables must be measured on the same cases. Ten-page estimates and 10-page actuals could not usefully be plotted if they were measurements for 20 books. When two variables are measured on each case in a group, the data are sometimes called **paired data.** All the techniques of this chapter apply only to paired data.

It would be possible to represent more than one group of books on a single scatter diagram: say, paperbacks and hardcovers. One could plot a P for the paperbacks and an H for the hardcovers instead of the X shown.

Here is a checklist of tips for drawing effective scatter diagrams.

1. Label the axes, including the units of measurement.
2. A brief title should indicate the context of the data. This may appear on the scatter diagram if it is very short, or just underneath or above if more lengthy. The source of the data may be referenced in a footnote, if appropriate.
3. When a causal or predictive relationship is being examined, the outcome variable (effect or predicted variable) should be scaled on the vertical axis, and the cause or predictor variable should be scaled on the horizontal axis. This is just a convention that many people are used to, but it is helpful.
4. When more than one point occupies the same position on the diagram, the multiplicity is plotted at that position instead of the X or whatever symbol is used for single points. For example, three coincident points are indicated by plotting a "3" instead of an X.
5. When the data consist of natural subgroups, the points in each subgroup should be plotted with a different symbol (M for males, F for females, for example).
6. The scales of the two axes should be chosen so that the vertical spread of the scatter is roughly the same as the horizontal spread of the scatter. (Of course, if the main point of the diagram is to compare vertical and horizontal spread, this rule should be ignored.)
7. The points should occupy a significant portion of the diagram. It is often necessary to omit a portion of the axis between the origin and the range of the data values so that the data are not squeezed into one or the other corner of the diagram. When this is done, be sure to indicate it with a break (or a zigzag) in the axis itself to alert the viewer to the omission.
8. The scale numbering should be simple and uncluttered: usually, three or four numbers on an axis are adequate, although for some purposes more detail may be needed.

5.1.3 Prediction, Association, and Causation

The concepts of prediction, association, and causation involve two or more variables in their application. These concepts motivate many statistical techniques; correlation, regression, and contingency table analysis are the most widely used of these. Before discussing details of calculation, the concepts themselves should be clearly understood.

It is common to speak of predicting the weather, interest rates, growth rates for livestock, or World Series outcomes. In all these examples there is an implication that the predicted event is in the future. In statistics we are interested in past and present events as well, the key being that these events are unkown. For example, if we know the amount of money spent annually to run and maintain a car, but not the separate expense of gasoline, we may wish to "predict" this gasoline expense based on something we do know—the miles traveled over the period of interest and the approximate number of miles per gallon consumed by the car. We could then subtract this from the total running expense to estimate the nongasoline expenses. This example is presented simply to emphasize that "prediction" need not relate to the future to be useful.

A typical **prediction** problem in statistics is of the form: Predict the value of variable A given the value of variable B. Predict the food costs for a family of four. (Variable B is the size of the family.) The technique for making such a prediction is the following: first, collect measurements of food costs and number in family for a sample of families in the target population. Then examine the relationship between food costs and family size for the sample data. Finally, use this relationship to predict the food costs for families of any particular size for which this information is desired. This information may well be of interest to a social worker or a welfare agency—either might require that estimates be based on only very basic information about a particular family.

In statistical prediction problems, there are usually at least two variables: one variable whose value is known, called the **predictor variable,** and one variable whose value is not known, called the **predicted variable.** Actually, there are many synonyms for these names:

Predictor variable	Predicted variable
Independent variable	Dependent variable
Control variable	Outcome variable
Input	Output
Criterion	Performance
Given	Unknown
Stimulus	Response

The variety of names results from the breadth of disciplines which have customized the prediction vocabulary.

The important thing to have firmly in mind in dealing with statistical prediction techniques is that it matters which variable is which. As has been pointed out, the relationship between two variables is studied with a sample of data for which both predictor and predicted variables are measured, so it is possible to be confused unless the different roles of the two variables are clearly recognized.

We proceed now to a related concept, **association.** Two variables are said to be associated with each other if measurement of one variable can help to predict the other variable. The word "association" is usually used to describe a relationship among two or more variables, without addressing the question of which variable is to be considered the predictor. It is used in situations where the strength of the association is of more interest than the actual predictive relationship itself.

One research use of a measure of association is in the identification of pairs of variables for which one variable merely mimics the other; both variables appear to measure the same thing. For example, suppose that we have 25 variables which describe the socioeconomic portrait of counties over a large geographic area; variables such as average income, farming income per capita, population density, years of education per family head, and so on. A first analysis of these data may include an assessment of which of these variables are redundant: that is, which variables are so strongly associated that we can simplify matters by deleting those that do not provide differentiating information about the counties. For example, the average number of years of formal education for adults in a county may convey the same information as the proportion of adults with some university education. A high degree of "association" between two variables is an indication that both variables are not needed in a description. [There are exceptions to this. The lengths of a person's two legs are very highly correlated, yet one may need both variables (left-leg length and right-leg length) to predict the presence of back pain from unequal leg length.]

Another research use of associations between variables is in the detection of causal links. The observation that eye inflammations are prevalent in areas where public swimming pools are intensively used was an association that led to the recognition of these public swimming pools as a source of infection. But there is no direct logical implication in such associations. For example, it is conceivable that the wearing of eye goggles to protect the eyes from swimming-pool chlorine physically irritates the eye and leaves the eye more vulnerable to infection. This causal path would also have caused the observed association, but the wearing of goggles would have been the culprit, rather than the infectious pools themselves. The difference would have practical implications: ocean swimmers would be warned of the dangers of goggles. However, in this instance, follow-up studies showed that the link between pool use and eye infection was a causal one; these laboratory studies were motivated by the hint of causation provided by the observed association. The conditions necessary to demonstrate causal links were discussed in Section 2.5, and are relevant to this discussion.

Associations can be misleading indicators of causation. This important point deserves one more example. One cannot fail to notice that fat people drink a lot of diet pop—ergo, diet pop makes people fat. Fallacies such as this can have serious implications in other contexts.

The description of two variables naturally raises questions of prediction, association, and causation. The proper analysis of two-variabale data is more complex than simply performing two one-variable analyses, because of these additional issues.

EXERCISES

1. The following data[2] are based on a historical study of the numbers of German submarines actually sunk each month by the U.S. Navy in World War II, together with the Navy's reports on the numbers sunk. Draw a scatter diagram of these data. What does the scatter diagram reveal about the errors in the original U.S. Navy reports, assuming that the historical study provides accurate numbers?

    Month: 1, 2, 3, 4, 5, 6, 7, 8, 9, 10, 11, 12, 13, 14, 15, 16
    Actual: 3, 2, 6, 3, 4, 3, 11, 9, 10, 16, 13, 5, 6, 19, 15, 15
    Report: 3, 2, 4, 2, 5, 5, 9, 12, 8, 13, 14, 3, 10, 13, 10, 16

2. Overweight people have a much higher incidence of heart attacks than have people of the same age who are of normal weight. Does this fact imply that weight reduction in overweight people will reduce the incidence of heart attacks in this group? Explain.

3. A trucking firm retrieves the following information from its records:

Truck	Annual distance traveled (km)	Maintenance costs
1	50,000	$ 9,000
2	35,000	5,000
3	70,000	14,000
4	43,000	4,500
5	61,000	10,500

    How might these data be used by the firm's management? (*Hint:* First decide which variable management would best be able to control.)

4. Sociological surveys have shown that people who have experienced many "life changes," such as marriage, divorce, children, moving, buying a home, losing a job, and so on, tend to have more medical problems. Discuss the implications for further research suggested by this finding.

5. For the game involving the guessing of the number of pages in various textbooks (see Section 5.1.1), calculate for each book listed in Table 5.1 the values of guesses that would have been exactly 88 pages too few. Calculate the guesses that would have resulted in exactly 18 percent too few pages. Now plot the following points on one scatter diagram:
    (a) The original 10 guesses (just as shown in Figure 5.2)
    (b) The 88-page-short guesses
    (c) The 18-percent-short guesses
    Which of the two descriptions of the shortfall in the guesses best describes the author's errors? Explain.

6. Infants who are born deaf are often born by mothers who have had rubella during their

pregnancy. That is, the number of deaf babies born is greater if the mothers have had rubella during the pregnancy than if they have not had rubella during the pregnancy. This association suggested that rubella was the cause of the subsequent deafness of the infant. Can you suggest any alternative explanations of the observed association? Would these alternative explanations suggest different strategies for the attempt to reduce deafness in infants?

5.2 CORRELATION

When two variables tend to differ from their averages in the same direction, they may be said to **agree**. The "agreement" between two variables is the degree to which the variables tend to be on the same side of their respective averages. Variables that agree are said to be **positively associated** and when they "disagree" they are said to be **negatively associated**. (The latter terms are widely used, but the "agreement" terminology will be slightly better for the exposition here.)

On the whole, tall people tend to be heavier than short people, and light people tend to be short, although there are many exceptions. We may therefore say that height and weight "agree," although the agreement is not very strong. The strength of agreement among quantitative variables is most commonly gauged by the calculation of a number called the *correlation coefficient*. (The correlation coefficient will be carefully defined shortly; let us concentrate for the present on the possible uses of a measure of agreement, using the correlation coefficient as an example.) The strength of agreement among the variables of a data set is sometimes the most important aspect to describe.

For example, consider the ratings of six movies by six film critics as displayed in Table 5.2. The ratings are constructed using a point system, with 0 indicating very poor and 10 indicating excellent; in other words, the ratings may be thought of as quantitative measurements, not simply rankings by the critics. (Ranks are a bit like a quantitative measurement and a bit like a qualitative measurement. Rank data require special procedures for summary and analysis. We touch on the point in Section 7.6 on nonparametric tests.)

TABLE 5.2 CRITICS' RATINGS OF SIX FILMS

Film	Critic					
	1	2	3	4	5	6
A	2	5	5	6	3	6
B	5	6	3	6	5	4
C	8	7	6	9	7	6
D	4	5	3	6	4	5
E	5	3	2	5	5	2
F	7	4	5	4	5	7

How well does a table of averages and SDs summarize these data? See Table 5.3. To get some sort of consensus about the films, we might compare the average scores for each movie. Movie C seems most acclaimed. The SD for each movie would tell us how representative of the individual ratings this consensus was—the SD of ratings for movie C was one of the smaller ones, so the agreement among critics was fairly good on this movie. Or we might possibly be interested in comparing the critics. The averages and SDs for critics reveal that critic 4 is relatively generous and critic 3 is harsh; and also that critics 2 and 5 give all the movies about the same score, while critic 1 assigns a wider range of ratings. But there is another aspect to these data that will be of interest if we want to compare the critics: Do they agree on the relative merits of the various films? Actually, we will concentrate on a slightly simpler question: How can we measure the degree of agreement of any two critics' ratings?

TABLE 5.3 AVERAGE AND SD FOR EACH MOVIE AND EACH CRITIC[a]

Critic	1	2	3	4	5	6
Average	5.2	5.0	4.0	6.0	4.8	5.0
SD	2.0	1.3	1.4	1.5	1.2	1.6
Movie	A	B	C	D	E	F
Average	4.5	4.8	7.2	4.5	3.7	5.3
SD	1.5	1.0	1.1	1.0	1.4	1.3

[a] Raw data are displayed in Table 5.2.

Look at the ratings of critics 1 and 5—they are quite close. The correlation coefficient gives a numerical index of how close they are. In this case the value of the correlation coefficient is 0.93. The value of a correlation coefficient ranges from -1.0, indicating perfect disagreement, to 1.0, indicating perfect agreement. Table 5.4 provides all the correlation coefficients among the six critics. The correlation coefficient will henceforth be abbreviated r. None of the r values is negative, but the agreement is quite weak for certain pairs, such as for critics 5 and 6. Let us compare the scatter diagrams for pairs 1 and 5 ($r = 0.93$) and 5 and 6 ($r = 0.00$) (see Figure 5.3).

TABLE 5.4 CORRELATIONS AMONG THE FILM RATINGS OF SIX CRITICS

		Critic					
		1	2	3	4	5	6
Critic	1	1.00	0.27	0.36	0.28	0.93	0.21
	2		1.00	0.55	0.85	0.43	0.40
	3			1.00	0.46	0.29	0.87
	4				1.00	0.54	0.13
	5					1.00	0.00
	6						1.00

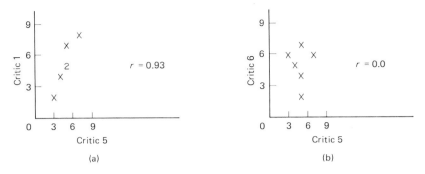

Figure 5.3 Scatter diagrams for data having (a) high correlation ($r = 0.93$) and (b) zero correlation ($r = 0.00$).

The **correlation coefficient,** r, is quite simple to calculate. After the two variables have been expressed in standard units, it is the average product of the two variables. The calculation for the r value between critic 1 and critic 5 is shown in Table 5.5. The average of the products of the scores is $5.52/6 = 0.92$, which agrees with the figure quoted earlier, except for rounding error. Why does the number calculated in this way measure the degree of agreement between the two critics?

TABLE 5.5 CALCULATION OF THE CORRELATION COEFFICIENT AS THE AVERAGE PRODUCT OF THE STANDARDIZED VARIABLES

Critic 1 (raw score)	2	5	8	4	5	7
Subtract mean (5.2)	−3.2	−0.2	2.8	−1.2	0.2	1.8
Divide by SD (2.0)	−1.6	−0.1	1.4	−0.6	0.1	0.9
Critic 5 (raw score)	3	5	7	4	5	5
Subtract mean (4.8)	−1.8	0.2	2.2	−0.8	0.2	0.2
Divide by SD (1.2)	−1.5	0.2	1.8	−0.7	0.2	0.2
Product of scores in standard units	2.40	−0.02	2.52	0.42	0.02	0.18

First it should be stressed that there are many other respectable measures of agreement. Since critics 1 and 5 agree exactly on their scores for three of the six films, we could say they agree 50 percent of the time; or we could say that the average difference ignoring signs is 0.7. Nevertheless, the correlation coefficient is by far the most widely used measure. Let us examine this calculation more closely.

The first step of expressing the scores in standard units implies that we are interested only in how the critics rate films relative to each other, not the overall rating of the six films by a particular critic. (We saw that this overall rating can be examined by using averages.) Now the next step is to calculate the products of the standardized scores; these products will be positive when the two standardized scores, which are multiplied together, have the same sign. That is, if critic 1 rates film C as above average, and so does critic 5, the product of the standardized scores will be positive, just as it is for film A, which both critics rate as worse than average. So we have agreement of the critics, resulting in a positive contribution to the average product.

Disagreement occurs when one critic rates a film "above average" and the other rates it "below average": numerically, this results in a negative product and reduces the average of the product. This is why the correlation coefficient can be thought of as measuring agreement between two variables: when the variables are "large" together or "small" together, the correlation coefficient will be positive. When one of the variables measures an object as greater than average and the other measures the same object as less than average, and this happens for most objects, the correlation coefficient will tend to be negative.

It is not quite obvious that this average product will be 1.0 when the agreement is perfect, but you can verify this numerically by seeing how much critic 1 agrees with himself.

The ultimate in disagreement occurs when the correlation coefficient is -1.0. Let us suppose that a new critic-in-training decides to make a name for herself by rating films as contrary as possible to the established critics. She accomplishes this by reporting a rating for a film as a number between 0 and 10, as do the others, but this critic determines her rating by subtracting the first critic's rating from 10. The scattergram in this situation is shown in Figure 5.4. In this case $r = -1.0$, which represents as much disagreement as is possible.

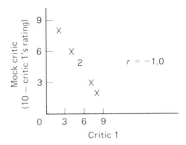

Figure 5.4 Scatter diagram for perfectly disagreeing critics.

Why is r preferred as an indicator of the degree of agreement to the many other possible indicators? One reason is that r has the convenient property that its value does not depend on the particular units used. If critic 1 recorded his preferences as 1.2, 1.5, 1.8, 1.4, 1.5, 1.7 instead of 2, 5, 8, 4, 5, 7, all the correlations would remain unchanged. The reason for this is that the first step in calculating r was to express the ratings in standard units; you can check that the transformation indicated (divide by 10 and add 1) does not change the ratings once they are expressed in standard units. Similarly, the correlation between height and weight for a group of people is the same number whether these heights and weights are expressed in feet and pounds, or in centimeters and kilograms. This unit invariance allows us to measure agreement in a way that is quite separate from what averages and SDs can tell us. We might say that since critics 2 and 6 have the same average rating for the six films, their ratings "agree"; but this kind of agreement is not what is measured by the correlation coefficient.

The description of exactly what *is* measured by the correlation coefficient is best given in terms of the scatter diagram. The correlation coefficient may be interpreted in terms of the familiar concept of distance. To describe this interpretation of the correlation coefficient, let us begin with the scatter diagram for the film-rating data expressed in standard units. Furthermore, we will need to think of the r as the product of a sign, plus or minus, and a positive value, which we will abbreviate $|r|$. That is, $r = $ (sign of r) times $|r|$. (The usual verbalization of $|r|$ is "the absolute value of r".) In the same way that the SD measures the distance of points from their average value, $\sqrt{(1 - |r|)}$ measures the distance of points in a scatter from a line: namely, the line through the origin with slope $=1$ or -1, depending on the sign of r.

That is, $\sqrt{(1 - |r|)}$ measures the typical size of the distance of the points, expressed in standard units, to a certain straight line. (This line would be the SD line based on standardized data. The SD line based on standardized data has slope $+1$ or -1, depending on the sign of r. The SD line is defined in Section 5.3.) Note that $|r|$ itself does not depend on the units of the variables correlated. When one is trying to assess the correlation coefficient for two variables based on a scatter diagram, it is advisable to draw the scatter diagram using the standard units so that the actual units do not mislead the eye. If the scatter diagram is drawn using the original units, and the SDs of the two variables are not the same, the scatter will appear less tightly clustered around a line and give the false impression of a lower correlation.

The fact that the correlation coefficient has an interpretation in terms of widely understood concepts (standardization and distance) is another reason for its popularity.

Figure 5.5 shows the scatter diagram for the ratings of critics 2 and 3. The ratings have been expressed in standard units for Figure 5.5. In terms of the units shown on the axes, what would you say was a typical distance of the points from the SD line? Since the correlation between the two critics' ratings is $+0.55$, your answer should be close to $(1 - |0.55|)^{1/2}$, which equals 0.67.

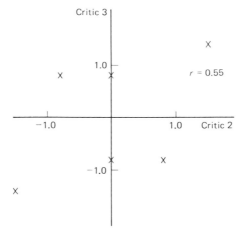

Figure 5.5 Scatter diagram for two variables that are expressed in standard units. The typical distance of the points from the line is 0.67, and the units of this distance are the same as shown on either axis of the diagram (i.e., standard units).

What are the relationships among the concepts "association," "agreement," and "correlation"? **Association** is a general term which describes relationships of all kinds between two variables, as long as this relationship allows the prediction of one variable from the other. **Agreement** is association with the additional property that the two variables tend to be on the same side of their averages for the items measured. **Correlation** is a word in general use that is synonymous with association, but our use of the word will be exclusively in reference to the correlation coefficient—as such, correlation is a particular measure of agreement and is defined unambiguously as the average product of the two variables expressed in standard units. So agreement is a special kind of association and correlation is a special kind of agreement.

Note that the correlation coefficient, r, measures agreement, but that $|r|$ measures the strength of association between the variables. Perfect disagreement ($r = -1.0$) can occur when the variables are perfectly associated ($|r| = 1$). We reiterate that the correlation coefficient is not the only possible measure of agreement or association, but that it is the one most commonly used. (See Problem 2.)

When a scatter diagram has enough points on it, say 20 or more, one can often discern an oval shape to the scatter. The points tend to be concentrated at the center of the oval and the scatter becomes sparse at its circumference. In describing properties of data scatters, it helps to be able to do schematic drawings which simply display the oval, rather than reproduce the entire scatter itself. We will refer to such a diagram as an **oval diagram**: a schematic diagram depicting by an oval the approximate shape of the boundary of a two-variable scatter. The oval diagram associated with Figure 5.5 is drawn in Figure 5.6. This is introduced simply as a device to assist in learning; it is not being proposed as a descriptive technique. We will be referring to this oval diagram in the section on regression.

To fix the position of an oval diagram, we need five numbers: the average and SD for each variable, which locates the center of the oval and its vertical and horizontal extent, and the correlation coefficient, which indicates how narrow the oval

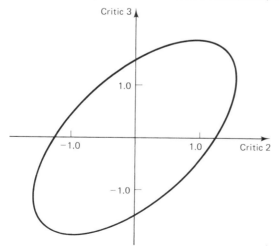

Figure 5.6 Oval diagram for data of Figure 5.5.

is and also whether it is positively or negatively oriented. These five numbers are called the **five-number summary** of the two-variable scatter. This summary is appropriate only for two-variable scatters that really are oval-shaped. The same is true of the oval diagram itself.

In certain circumstances, the correlation coefficient has some shortcomings as a measure of agreement. A problem called **nonlinearity** occurs when the two variables are associated but not linearly. A **linear** relationship between two variables is one which can be represented by a straight line: a unit increase in one variable causes a fixed increase in the other variable, determined by a certain slope constant. For example, let us generate some scores for an imaginary critic whose habit is to copy critic 1 in the following way. If the score is 5 or less, he uses that score; if it is 6, 7, 8, 9, or 10, he uses 4, 3, 2, 1, or 0, respectively. This mimic's score is perfectly associated with critic 1's scores in a sense, but the correlation coefficient in this case turns out to be only -0.1. The scatter diagram for this case, Figure 5.7, shows clearly that the relationship is not linear, and thus the correlation coefficient should not be used to measure association in this case. (The fact that the relationship is fairly well described as two straight lines does not make it linear; for linearity it must be describable as one straight line.)

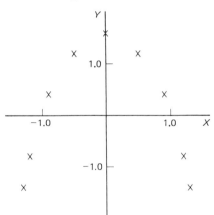

Figure 5.7 Example of a nonlinear association. The association is perfect but the correlation is zero.

The other main problem with the correlation coefficient as a measure of agreement is indicated in Figure 5.8. It shows one observation that should not be included with the rest of the data for the usual descriptive procedures; inclusion would cause the description to be misleading about the typical data scatter. Such an observation is called an **outlier**. In general, an outlier is an observation among a group of observations that is suspected, by virtue of its distinctive size, of not belonging to the group. Although $r = 0.0$ here, there is clearly an association between the variables. The outlier distorts the usual meaning of r. With the outlier set aside, the correlation is 1.0.

Nonlinearity and outliers can confuse and mislead the naive user of correlation coefficients. It is a good rule to look at the scattergrams of every pair of variables for which a correlation coefficient is being used as a descriptive measure. Treatment of

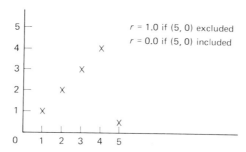

Figure 5.8 Effect of an outlier on the correlation coefficient.

the outlier will depend on the particular situation, but it will usually be separated from the rest of the data when describing the original data. The problem of nonlinearity is usually solved using transformations of the data to linearity, by means of a logarithmic or power transformation, but a discussion of this technique would lead us too far into the mathematical side of statistics, so we omit this topic.

There is one more pitfall that users of correlation descriptions will encounter. When correlations are calculated for groups of items or individuals, there are two possible correlations to calculate: the correlation of the two variables for each group separately (i.e., one correlation coefficient for each group), or one correlation coefficient based on the averages from each group. These correlations can be very different, as the following example shows.

Suppose that we have three groups of three observations each, as shown in Figure 5.9. Within each group the correlation is $+1.0$, but if we calculate the group averages $\{(5, 25), (15, 15), (25, 5)\}$, the correlation among these points is -1.0. Do we conclude that the variables agree perfectly, or disagree perfectly? Careful interpretation of the correlation coefficient will resolve this problem. The variables disagree across groups but agree within groups.

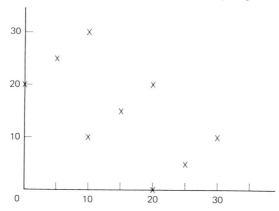

Figure 5.9 Scatter diagram showing the dangers of correlating averages. Correlation within groups is 1.0, but correlation of averages is -1.0. Correlation ignoring groups is -0.60.

But one must be aware of the groups. If we ignore the grouping, the correlation turns out to be -0.60 and we would be led to think that the variables "disagree," which is not the whole story. It is a good practice to think of how homogeneous a data set is with respect to the variables of interest, so obvious clusters will be analyzed separately. A data-based approach is to look for clustering in the scatter diagram. An example in which we might expect clusters would be in height–weight data for groups of athletes in various sports. Basketball players should be treated as a separate cluster from power lifters.

The problem of sorting out correlations for clusters will be referred to as the *correlation-of-averages* problem. When two variables are first averaged by subgroups, and then these averages are correlated, there is a possibility that the correlation calculated will not describe the correlation between the variables themselves; in this case the correlation is a descriptor of the relative position of the subgroups.

Quite a few details about the use of correlation coefficients have been covered. We should not lose sight of the role of this technique in descriptive statistics; it is used as a supplement to averages and standard deviations when more than one variable is being described. To describe the height and weight data of a group of basketball players, we would record the average height, the average weight, the SD of the heights, the SD of the weights, and the correlation coefficient: five numbers to summarize 200 measurements from a group of 100 basketball players. These five statistics allow us to recreate in our minds the scatter diagram, which really depicts the data itself. In other words, except for anomalous cases such as we have described, this five-number summary contains most of the information we would usually need for the description of two variables measured on a single group.

EXERCISES

1. Describe in words, as concisely as possible, the similarities and differences among the six film critics as evidenced by their ratings and the correlations shown (see Tables 5.2 and 5.3).
2. Calculate the correlation coefficient between the GPAs (grade-point averages) in computing science courses and other courses, based on the data shown below. Interpret your result.

Student	GPA in computing	GPA in all other courses
1	2.7	3.1
2	3.3	3.2
3	1.9	2.3
4	3.9	2.0
5	2.3	2.2
6	3.6	3.7
7	3.1	3.5

3. The following data have been collected for a random sample of North American model cars. Calculate the correlation coefficient between the two measures of gasoline consumption, and comment on its appropriateness as a summary measure for this data.

Model	Miles per gallon	Liters per 100 km
1	23	10.3
2	35	6.8
3	19	12.4
4	27	8.8
5	41	5.8
6	18	13.1
7	12	19.7

4. In Exercise 3, the gallons referred to are U.S. gallons. If the data were recalculated in miles per imperial gallon, would the correlation increase, decrease, or be unchanged? What if the miles per gallon were changed to feet per ounce?

5. Ten tenth-grade students were given tests of memorization ability and of creativity.[3] Describe the results recorded below.

Student	Creativity	Memorization
1	11	13
2	96	85
3	15	27
4	88	69
5	92	76
6	34	30
7	44	39
8	67	32
9	37	13
10	38	58

6. How much correlation is $r = 0.7$? To gain an intuitive feeling for this, examine scattergrams for variables with various correlations. One way to do this is to construct two variables (which are to be correlated) from two uncorrelated variables ($r = 0.0$) as shown below.

V_1	V_2	$V_3 = V_1 + 2 \times V_2$	$V_4 = V_1 + V_2$	$V_5 = V_1 + 0.5 \times V_2$
1	1	3.0	2.0	1.5
0	1	2.0	1.0	0.5
1	0	1.0	1.0	1.0
2	1	4.0	3.0	2.5
1	2	5.0	3.0	2.0

(a) Draw scattergrams of
(i) V_1 and V_2, (ii) V_1 and V_3, (iii) V_1 and V_4, and (iv) V_1 and V_5, and calculate the

correlation of each pair. (It helps to have a statistical calculator or a computer for this exercise. Try to use the actual scale of the data for the scattergram. Computer plotting programs may adjust your plot so that differences in the apparent scales are eliminated.)
(b) Standardize each variable and redo part (a). What is the effect of standardization on the scatter diagrams and on the correlation?
(c) What is the five-number summary for V_1 and V_4? Use it to draw an oval diagram for the V_1–V_4 scatter. Also plot the original data points on the oval diagram, and comment on the adequacy of the oval diagram as a summary of the scatter in this instance.

5.3 REGRESSION

5.3.1 Regression as Prediction

Regression is a method for determining predictive relationships among variables. In this text we discuss only the **regression** of one predicted variable on one predictor variable (i.e., the prediction of one variable by another variable). Notice that the two variables involved play different roles in this prediction.

One example of a study that would use regression analysis concerns the relationship between the duration of a treadmill exercise and heart rate. The duration of exercise is the predictor and the heart rate is the variable to be predicted. It is unlikely that we would be interested in reversing the role of the variables. However, there are some situations where the choice is not so obvious. For example, if we had data on the daily caloric intake of a group of obese people, and also the average daily weight gain of these same people, one investigator might wish to predict weight gain as a result of caloric intake, while another might wish to study caloric intake as a response to weight gain. (Does weight gain slow caloric intake?) In this case the statistical analysis cannot proceed until the investigator's question is clearly posed. As we shall see, the relationship between X and Y that is used to predict Y from X is not the same as the one that is used to predict X from Y.

There are two reasons why a predictive relationship may be desired. One is that the investigator wants to determine what the relationship is between the two variables. The other is that the investigator may wish to do some predicting. Occasionally, both reasons may be important. The one constant in regression problems is the idea of **prediction.** We may summarize the uses of regression:

1. To make predictions
2. To determine predictive relationships

Table 5.6 displays data indicating the relationship between the pulse rate of an animal and the animal's consumption rate of oxygen per gram of body weight. The readily apparent fact that oxygen consumption rate per gram is higher for animals having a lower pulse rate can be quantified. Quantification is desirable because the relationship can then be communicated without ambiguity, and the quantitative re-

TABLE 5.6 PULSE RATE (BEATS PER MINUTE) AND OXYGEN CONSUMPTION RATE FOR A VARIETY OF MAMMALS[4]

Species	Pulse rate	Oxygen consumption rate ($m\ell$/g body weight/hr)
Camel	55	0.12
Guinea pig	270	0.78
Elephant	30	0.07
Horse	40	0.14
House mouse	625	1.59
Little brown bat	588	2.90
Harbor seal	34	0.47
Chimpanzee	44	0.25
Alpine marmot	200	0.50
Muskrat	225	0.98

lationship can be used to predict the oxygen requirements of an animal for which only the pulse rate is known. Regression techniques allow us to do these things in a way designed to extract as much information from the data as possible. The next few subsections introduce the details of the regression technique.

5.3.2 Use of Averages for Prediction

The following conversation between a professional handicapper (a gentleman found more often making bets at the race track than anywhere else) and a statistician (who wishes to try his expertise in the real world) contains a point about the use of averages in regression (see Figure 5.10).

Pro: I think I'll bet on Sure Thing to "show" in the fifth.

Statistician: How long do you think she will take to get around the track?

Pro: I would guess about 1 minute and 5 seconds.

Statistician: But 1:05 minutes is just the average time for Sure Thing published in the racing form. I was hoping that you might have some inside information about this horse.

Pro: Well, I noticed that Sure Thing's owner was in a good mood this morning—sometimes this is a good sign.

Statistician: Sounds a bit unreliable; guess I'll assess the horse on the basis of its average time of 1:05. The horse may well run slower or faster than this average time, but without any additional information, 1:05 has to be my best guess. This would place Sure Thing about fourth in the field. Are you sure you want to waste your bet on Sure Thing?

Pro: If I bet on the basis of the average all the time, I would do no better than the amateurs who also read the racing form. Besides, if you're so smart, why aren't you rich?

152 Chap. 5 Descriptive Methods for Two Variables

Figure 5.10 Comment ça va?

Statistician: If I knew the other sources of information that you have, I could do better than bet on the basis of average time. In fact, I could probably make a fortune. But the average is the best I can do in the absence of more detailed information.

Pro: Would you like me to fix you up with a job in the stables?

The point is that an average can be a good predictor in the sense of coming as close as possible, on the average, to a new outcome value. If we obtain more detailed information which suggests that the outcome will be above or below average, then of course we would modify our prediction. Otherwise, we use the average for prediction.

Let us apply this idea to a more detailed example. An entrepreneur is examining various locations in a large metropolitan area for a new fitness center. For attracting customers, a downtown location would be desirable, but rental costs of space there is very high. Lower rental rates are available in the suburbs and beyond, but of course these locations are less desirable for business. The first step in choosing an appropriate balance is to study the relationship between rental costs and distance of the location from the downtown core.

The entrepreneur obtains the data in Table 5.7 from a trust company specializing in the rental of commercial space. These data are derived from the trust company's records of space that has been rented recently. The distributor must find other space

TABLE 5.7 DATA ON THE RENTAL FOR COMMERICAL SPACE IN THE ENVIRONS OF A LARGE METROPOLITAN AREA[a]

Site	Monthly rental (ft^2/month)	Distance (km)
1	$0.97	4
2	0.35	10
3	0.88	17
4	0.70	19
5	0.63	31
6	0.36	34
7	0.42	36
8	0.62	41
9	0.35	49
10	0.41	58

[a] The distance measure is the distance from the site to a central downtown location.

since these sites are no longer available. A site that is rumored to be coming available in a few weeks is 40 km from downtown. What would you guess that the rental cost would be? The average rental is 57 cents per square foot, and this might be a reasonable first guess. We do not really expect our guess to be correct, but we use the average for our prediction because we think that this will make our error small. After all, it is not so important that the guessed value be exactly correct, but we hope that it is close to the truth.

Although it is not important that our guess be exactly correct, we are nevertheless concerned that the error be small. How can we assess the probable size of the error in our guess? One idea is to pretend that we were using this guess of 57 cents for the rental amounts of each of the 10 sites in the original data set. Our errors would then be, subtracting 0.57 from each value: 0.40, −0.22, 0.31, 0.13, 0.06, −0.21, −0.15, 0.05, −0.22, −0.16. (This method of assessing the errors is not quite right: we have used the data to determine the average in the first place and then reused it to see how big the errors are. Nevertheless, we proceed as if these were errors typical of what we would experience in estimation of the rental for a new site.) Since we used the average 57 cents as a prediction of the rental, it would be natural to attempt to use the average error as a prediction of error. This average error is 0.0, as it always is with this method, so this is not telling us anything. Another possibility is to calculate the **average absolute error:** the average size of the absolute error (this is the same as the MAD introduced in Chapter 4). This is more helpful. In this case it is 0.19. We could say that our guess of 57 cents is likely to be in error by about 19 cents, because this is the average error that would have been incurred with the original 10 sites.

The average absolute error is not used in statistics very often, in spite of its apparent simplicity. As discussed in Chapter 4, we are led to accept the standard deviation of the errors as a summary measure of spread rather than the average absolute error.

In the example the standard deviation turns out to be 0.22. Because the standard deviation can be considered to be a "typical" deviation, we can express our guess as

$$57 \text{ cents} \pm 22 \text{ cents}$$

We have suggested using the average and SD for the prediction of a particular unobserved value. However, the accuracy indicated by the SD (22 cents in the example) might not be good enough. To improve on it we must record some information that is associated with the predicted variable, and more easily available than the predicted variable itself. In the example we suspect that the location of the site (kilometers from downtown) would help. The scatter diagram of rental versus distance is useful in examining this possibility. See Figure 5.11.

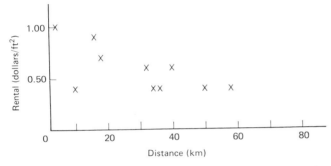

Figure 5.11 Scatter diagram of monthly rental and distance from downtown.

There is a hint in the scatter diagram that the average rental declines as the distance from downtown increases. If we wish to know the rental for a site that is at a distance of 40 km, the distance itself should be used in our our prediction strategy. Certainly it should if we believe that the average monthly rental declines as the distance from downtown increases. How is this decline taken into account in our estimation scheme?

We would not want to use the 41-km site (the one in the data set that is at a distance closest to 40 km) as our estimate since the rest of the data indicate that a single data value can be quite far from the average. Futhermore, we may predict the 40-km rental to be, on average, slightly lower than the average rental over all the rentals in the data set, since the scattergram suggests a decline in the average rental with increasing distance. How can we use all the data to estimate the rate of this decline? Let us do some averaging over sites that are close to each other; that is, calculate the averages of the monthly rentals for 20-km vertical strips whose width corresponds to 20 km in the scatter diagram. The average for the 1- to 20-km strip is 72.5 cents, for 21 to 40 km is 46.7 cents, and for 41 to 60 km is 46.0 cents. Thus our "smoothed" scattergram looks like Figure 5.12. We will attempt to use a line to represent this smoothed scattergram, and we can then read off the rental for the 40-km distance.

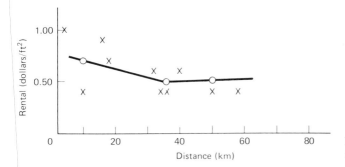

Figure 5.12 Smoothed scattergram of monthly rental and distance.

5.3.3 The SD Line and the Regression Line

Our objective now is to fit a line through the scatter of data points that is as close to the points as possible, so that this line will take the place of the smoothed scatter diagram. It will indicate the rental prediction pertaining to a 40-km distance. But we must be more precise about the words "fit close."

If we mean the line such that the SD of the distances of the points to the line (i.e., perpendicular distances) is as small as possible, the line will be the **SD line.** The SD line will be defined by specifying one point that it passes through, called the **point of averages,** and the slope of the line. The point of averages is the point whose coordinates are (average of X, average of Y), where X is the horizontal coordinate and Y is the vertical coordinate. The SD line is the line through the point of averages that has slope

$$\text{sign of } r \times \frac{\text{SD of } Y}{\text{SD of } X}$$

It is the line that indicates changes in Y that are the same size as changes in X when both X and Y are expressed in standard units (see Figure 5.13).

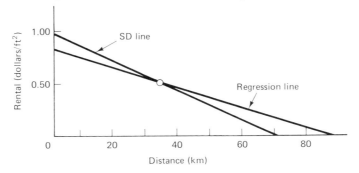

Figure 5.13 SD line and regression line. Both lines pass through the point of averages. The slope of the regression line is $|r|$ times the slope of the SD line.

On the other hand, if we want our predictions to be close to the true Y values (rental costs), the line that will give us the best predictions is the line that will minimize the prediction errors. Graphically, this amounts to choosing the line such that the *vertical* distances from the points to the line are made small. The line having this property is called the **regression line;** it is the line through the point of averages whose slope is $|r|$ times the slope of the SD line. [The SD line has a slope whose sign is $+$ or $-$ depending on the sign of r. The slope of the regression line is defined in terms of the slope of the SD line: that is, $|r|$ times (sign of r) times $\text{SD } Y/\text{SD } X$. But (sign of r) times r is just r, so indeed the slope of the regression line is r times $\text{SD } Y/\text{SD } X$, as claimed elsewhere. The point of defining the regression line in terms of the SD line is to emphasize that it is "flatter" than the SD line, and the flattening factor is $|r|$, a positive number.] The regression line also has the property that the standard deviation of the vertical distances of the points to this line are as small as for any line. The regression line is the line that makes the prediction errors small.

For prediction, the regression line is the one to use. For description in a context in which prediction is irrelevant, the SD line is more suitable.

Let us calculate the SD line and the regression line for the fitness center example. The point of averages is (30, 57); the correlation coefficient, r, is -0.6; the SD of monthly rental per square foot is 22 cents; and the SD of distance is 16 km. The SD line and the regression line both pass through (30, 57) on the scatter diagram, but the SD line has slope $-22/16 = -1.4$ while the regression line has slope $-0.6 \times 1.4 = -0.8$. Note that the regression line is "flatter" than the SD line, and this will always be so when $|r|$ is less than 1 (i.e., virtually all the time).

Note that we have not yet displayed the formula for the regression line. We next show that the description of the regression line that has been given already is complete enough to use the line for prediction. The distance 40 km is 10 km further than the average, or in other words, $10/16 = 0.63$ SD above average. The corresponding monthly rental, based on the regression line, must be $r = -0.6 \times 0.63$ or -0.38 SD or 0.38 SD below average. But the average and SD of monthly rental are 57 and 22, respectively. So -0.38 SD is $57 - (0.38 \times 22) = 57 - 8$ approximately, which is 49 cents. Thus the predicted monthly rental for a site 40 km from the downtown area would be 49 cents. This method of prediction is called the **regression method.**

The regression method for predicting one variable from another uses the regression line to indicate the predicted value of the predicted variable that corresponds to a specified value of the predictor variable. Graphically, this process is as shown in Figure 5.14.

It is possible to make the regression method more algebraic, and this may yield some additional insight into the regression method. The regression line has the equation

$$\frac{Y - 57}{X - 30} = r \times \frac{\text{SD } Y}{\text{SD } X}$$

which specifies the relationship that coordinates of points (X, Y) on the regression line

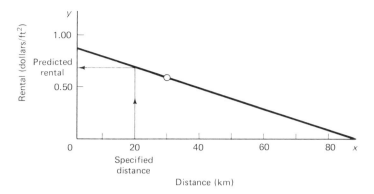

Figure 5.14 The regression method of prediction of Y from X uses the regression line to do the prediction. This graphical explanation of the method is equivalent to the method described elsewhere in this section.

must satisfy. Thus putting $r = -0.6$, SD $X = 16$, and SD $Y = 22$, the equation becomes

$$\frac{Y - 57}{X - 30} = -0.6 \times \frac{22}{16}$$

The regression method amounts to finding the value of Y that satisfies this equation, using the specified value for X. (It is not difficult to reexpress this relationship in the form of an equation for a line, that is, for the regression line:

$$Y = \text{intercept} + \text{slope} \times X$$

The intercept turns out to be

$$\bar{Y} - r \times \frac{\text{SD } Y}{\text{SD } X} \times \bar{X}$$

and the slope turns out to be

$$r \times \frac{\text{SD } Y}{\text{SD } X}$$

Here $\bar{X}$ is the average value of X and $\bar{Y}$ is the average value of Y.)

5.3.4 Prediction Error with the Regression Method

Regression is used for prediction or in establishing predictive relationships. In either case it is useful to be able to say how accurate the predictions are likely to be. The calculation and use of this prediction error are demonstrated in this subsection.

We have claimed that the regression equation does minimize the prediction error. What this means is that the regression line shown is the one whose vertical distances from the points in the data set are as small as possible, in a sense to be explained shortly. The actual size of these errors is clearly still relevant, even if they are as small as possible—we need to know if the errors are small enough to make predictions that are useful in practice.

Here is a definition of the error we want to estimate, which we will label SD_{pred}. It is SD of the prediction error incurred while using the regression equation to predict the values of Y in the data set. The first thing to clarify is why we want to estimate the SD of the prediction error rather than to estimate the prediction error itself. In fact, these two things are about the same size. If the errors in the commercial space rental example are about 17 cents, then the typical squared error is 17×17, and the SD is therefore 17. The SD can be thought of as the size of a typical data value (ignoring the sign) when the average of the data values is zero. The prediction errors do average to zero when the regression line is used for prediction.

We will proceed with our calculation of the typical prediction error: more specifically, of SD_{pred} defined above. There is both a hard way and an easy way to calculate SD_{pred}.

First, the hard way. For each pair of values (X, Y) in the data set we want to calculate the error in using the prediction obtained for Y from the regression equation: that is, the Y on the left-hand side of the above equation that results from inserting the X value on the right-hand side.

After we substitute values for $SD X$, $SD Y$, r, $\bar{X}$, and $\bar{Y}$, the equation simplifies to, approximately, $Y = 83 - 0.82X$. This is the formula for the predicted value of Y. The calculation of the prediction errors is shown in Table 5.8. Now the SD of the errors is 17.5. That is, $SD_{pred} = 17.5$. When we make a prediction using the regression method, the typical error in this process is about 17.5 cents. How does this compare with the typical error of 22 when we just used the average? The regression

TABLE 5.8 WAREHOUSE RENTAL DATA AND REGRESSION METHOD PREDICTIONS OF THE RENTAL RATES, BASED ON THE DISTANCES OF THE SITES IN THE DATA SET[a]

X (distance)	Y_{pred} (cents) (predicted value)	Y (cents) (actual value)	Error
4	79	97	18
10	75	35	−40
17	69	88	19
19	67	70	3
31	58	63	5
34	55	36	−19
36	53	42	−11
41	49	62	13
49	43	35	−8
58	35	41	6

[a] The errors are represented graphically by vertical distances from the points to the regression line, since the predicted variable is labeled on the vertical axis.

method is better (although in this instance it is not much better). Of course, reduction of error in our prediction is a good thing. Whether the task of using regression is worthwhile depends on whether the prediction error is reduced enough for practical purposes. Are we going to establish what the rental should be based on its location? Will we be able to accept the error in the estimate that our regression provides?

The key to how much improvement to expect in the error is the correlation, r. In fact, $\sqrt{(1-r^2)}$, which in the example was $\sqrt{(1-0.6^2)}$ or 0.8, will always equal the ratio of "typical" prediction errors, $17.5/22 = 0.8$.

This relationship between the correlation and the ratio of the prediction errors is useful. It tells us that to get a 50 percent reduction in the prediction error, we would have to find a predictor variable correlated 0.87 with the result variable.

This relationship hints at the easy way of calculating SD_{pred}. If we have $SD\,Y$, the standard deviation of the result variable, and r, then SD_{pred} may be computed from

$$SD_{pred} = \sqrt{(1-r^2)} \times SD\,Y$$

In the example,

$$SD_{pred} = \sqrt{(1-0.6^2)} \times 22 = 17.6$$

which would be exactly the same as we calculated the hard way except for rounding errors. This easier method is less subject to rounding errors.

The quantities SD_{pred} and r^2 are each used to indicate the usefulness of a regression. SD_{pred} is best for indicating if the predictive ability of the regression is good enough, whereas r^2 measures the proportion of the variability in the result (in squared units) that is eliminated by making use of the regression on the predictor. r^2 signals whether or not the predictor improved the prediction. It says nothing about how good the resulting prediction is. Exercise 3 at the end of this section explores this distinction.

5.3.5 Assumptions of the Regression Method

Now it is time to look critically at the regression method and to see when it works best and when it fails. We have already entertained the idea that the relationship between Y (or predicted variable) and X (or predictor) might not be a straight line (i.e., might not be *linear*). The regression method assumes linearity. You may wish to try out Exercise 5 of this section to see why it is important to check that your data are reasonably linear before proceeding with the calculation of a regression line.

When we choose the line that best fits the data in the minimum SD_{pred} sense, as we do when we calculate the regression line, there is an implicit assumption that all the points should have an equal vote in this process. There are some situations when this is not desirable, such as when some points have been determined more carefully than others. The standard regression procedure assumes equal weights. Note that it does not make sense to talk about the typical value of the prediction error if these typical values change over the range of X in the data set. The typical error for predictions based on small X would be different than the typical errors for predictions based on large X. There would not be one typical value. The calculation of SD_{pred} would not then apply to regression method predictions for all X, as it does when the prediction errors tend to be about the same size over the whole range of X. This property is called **homoscedasticity,** and its opposite is **heteroscedasticity.** There is little danger of these technical words being confused with their everyday usage! The regression method assumes homoscedasticity of errors (i.e., that the errors tend to have the same size over the whole range of X).

When we use SD_{pred} as a measure of a typical deviation, there is a hidden assumption. SD_{pred} is calculated by an operation that is a bit like an averaging of the size of the errors. (The process is called the root-mean-square operation.) If the frequency distribution has a histogram that has more than one hump, say one at 0.5 SD and one at 1.5 SDs, a 1.0 SD error may not be "typical" at all. A similar problem of interpretation of SD_{pred} arises when the error distribution is skewed—a long tail on one side and a short tail on the other. Symmetric unimodal error distributions are the ones that are best summarized by the calculation of SD_{pred}.

When we examine the errors of prediction for the various X values in the data set, it is assumed that these errors are independent of each other. This would not be the case if the points seemed to move from left to right in "waves," as, for example, they might if X were time and Y were temperature. If we wanted to predict the noon temperature at a certain location, we would be foolish to ignore the temperature at 9 A.M. But we would also be foolish to try to use regression to describe the increase of temperature during the morning. One problem with such a regression is that it assumes that the relationship is a straight line: if so, the increase from 9 A.M. to 10 A.M. would be about the same as the increases from 10 A.M. to 11 A.M. and 11 A.M. to noon, and this is not likely to be true. The other problem relates to the assessment of SD_{pred} when the independence assumption is violated. We will elaborate on this difficulty in Chapter 6. Regression assumes independence of errors.

Another feature of data sets that can cause problems with the regression technique is the presence of outliers. Look at Figure 5.11, which again shows the commercial space rental data. If you feel a little uncomfortable about inclusion of the data value (10, 35) in the regression analysis, your instincts are right. This observation appears to be an outlier. Outliers can ruin an otherwise useful regression prediction. In the example it will have greatly increased the estimate of SD_{pred}. In fact, when this point is removed from the data set, r becomes -0.86, so SD_{pred} decreases to about 11 instead of 17. The regression method works best in the absence of outliers. Of course, there is a scientific difficulty with throwing out awkward observations, and this does not disappear with the rejected outlier. (In fact, outliers can occasionally be the most informative of data values. A test score of 165 percent should be corrected before being included with the other elements of a student's final grade. A test score of 0 on an objective exam may expose a diabolical genius.) The point is that the regression method should only be used to describe the part of the data that excludes outliers. Some other method, such as a report of exceptional findings, must be used for the outliers themselves.

The assumptions that we have stated so far can conveniently be summarized as statements about the errors calculated from the regression line (i.e., the distribution of the errors as calculated in Table 5.8). In Figure 5.15 each narrow vertical strip could conceivably give rise to a different distribution of errors. (That is, the spread of the points around the regression line could be different for each X.) *A usual assumption for regression is that these error distributions, for each vertical strip, are all identical distributions.* To check the assumptions of regression, we should look at a plot of these errors and decide whether the vertical errors from each vertical strip confirm the assumption or not.

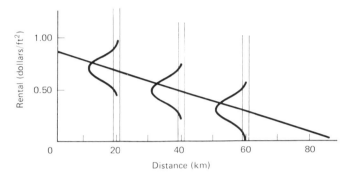

Figure 5.15 Assumptions for regression may be expressed in terms of the distributions in vertical strips: that they are normal, all with the same variance, and such that the averages lie on a straight line. The independence assumption is not portrayed but should be kept in mind (see Section 5.3.6).

If the errors are plotted against the predictor variable, one can assess the constancy of the error distribution visually. This error plot is used to examine the validity of the assumptions. For the rental example, this plot, sometimes called a **residual plot,** is shown in Figure 5.16. Do you think the assumptions were valid for these data?

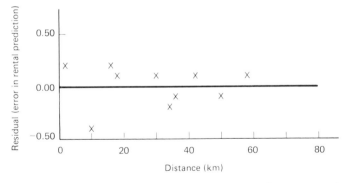

Figure 5.16 Residual plot. This scatter diagram plots the prediction errors versus the predictor, that is, error in predicting rental rates versus warehouse distance.

Here is a checklist of regression assumptions:

1. Linearity of predicted variable in terms of the predictor
2. Homoscedasticity of error distribution
3. Symmetric unimodal error distribution
4. Absence of outliers
5. Independence of errors

There are ways to get around all these restrictions. However, the basic technique of regression works only when these assumptions are satisfied.

Could we use the regression method to predict the rental rate for space 100 km from the downtown core? The arithmetic could be done as before, but the prediction might be very poor. We might be in a different city. The regression method is valid only for X values in the range of the X values of the data set. Not only do we not know if the relationship is a straight line outside this range, but even if the straight-line relationship does hold, we have to have the right line. A small change of the slope can make a big difference in the estimate for X values that are far from the average of the X values. Using the regression line from our data set of 10 sites between 4 and 58 km from downtown to predict the rental price for a site 100 km distance would be a foolish application of the technique.

5.3.6 Causality in Regression

In Section 5.2 the distinction between correlation and causation was emphasized. In Section 5.3.4 we demonstrated the close connections between regression predictability and the correlation coefficient. The point to be made here is that a predictive relationship between two variables can exist without the relationship being causal. Whether or not a regression is indicative of a causal relationship depends on the experimental design. More explicitly, a causal relationship can be demonstrated unequivocally, in a regression setting, only if the values of the predictor variable are assigned to the experimental units by the experimenter in a randomized way. The principle is exactly the same as the one introduced in Chapter 2. But in the regression setting the comparison groups are defined by the different values of the predictor variable.

The example involving the rental of commercial space was one in which causality could not be established, no matter how high the $|r|$ value. The causality referred to in this example would be that the rental rates were set on the basis of their distance from the downtown core. This is plausible, but to prove it with data would require an impossible experiment: prospective sites cannot be "assigned" a distance from downtown—the downtown is already in place and the lots cannot be moved into their assigned locations.

Even if distance were perfectly correlated with rental rate, the "cause" of the rental rates being what they are could not unequivocally be ascribed to the distance factor. An unlikely but possible alternative causal path might be: the commercial spaces serve an active boat-building industry. For these sites there is an advantage to being near the water. The core of the city is built beside a body of water, so this explains the differential rental rates. It would be wrong to infer that distance to the city core is the crucial factor. The highest rentals would be for sites on the water, which perhaps would not need to be very close to the downtown core.

Another example of the relationship between regression and causality is the following. Suppose that a study is done of the value of landscaping to the resale price of a house. The study could be done two ways:

1. A number of houses could be selected, perhaps at random, from a list of houses in a community. These houses could be appraised by appraisers unaware of the study, and these houses could also be evaluated as to the value of the landscaping done, preferably assessed by someone who did not know the appraisal prices of the houses. The appraisal price could then be regressed on landscaping assessment to gauge the extent to which the price was dependent on the value of the landscaping.
2. Alternatively, we could assume that houses in a new subdivision are under construction and will soon be sold by the builder. Buyers are told that they must accept landscaping paid for by the builder and not alter it for a three-month period. The builder agrees to allocate randomly to the houses dollar amounts for landscaping, based on a given list of such proposed amounts, as long as the total for the whole group of houses matches the forecast expense to the builder. After the houses have been lived in for a month or two, the appraisals are done and the regression analysis proceeds as with the first study design.

Let us assume that the ratio of appraised value to builder's cost is the result variable, and that the ratio of landscaping expense to builder's cost is the predictor variable. If the regression confirms the expected relationship, with the two variables positively correlated, method 2 could be said to establish the causal link. Method 1 has a less clear interpretation because it is not a true experiment. People who choose to spend more money on landscaping may well spend more money on other value-increasing additions to their house, such as expensive fixtures, carpets, and decor; swimming pools; greenhouses; and garden sheds. So the causal link would not be demonstrated by the regression in this case.

The scatter diagram for two variables usually has an oval shape. An exception is when the predictor variable has assigned values rather than observed values. In this case, if, for example, an equal number of data points are obtained for each assigned value of the predictor, the scatter would look more like a parallellogram than an oval. The range of values of Y for each value of X should be the same for the usual regression assumptions (homoscedasticity), and the only difference X would make is to shift the distribution upward (or downward for a negative r) for the larger X values.

Does an oval scatter indicate heteroscedasticity? Not usually. It might seem that the SD of the errors is small for small X, then increases as we move to the center of the oval, then decrease again as we move to larger X values. However, this is usually an illusion. The phenomenon that causes this illusion is the increase in the range of a group of numbers as we enlarge the group. For two people, salaries may differ by only $5000, but for 100 people the maximum difference will surely be larger. But this does not mean that the SD is greater for the 100 than for the two. The SD is the square root of the average deviation, not of the maximum deviation. The eye sees range, but the variation is better measured by SD, which does not grow with the size of the data set. Similarly, the reason for the oval shape of many scatters where regression is applicable is that there are fewer data points with low or high X values than with intermediate X values, so at X extremes the range of errors is smaller than at middle X values. The error SD may nevertheless be constant for all X.

So an oval shape would usually not indicate heteroscedasticity. (A rare exception would be in the situation where equal numbers of Y values are given for each X value. In such a situation the oval shape would indicate that the errors are smaller for large and small X than for intermediate values of X.)

5.3.7 Regression Pitfalls

The term "regression" was first used in its statistical context by Sir Francis Galton about 1870. He used it to describe the following phenomenon. Sons of tall fathers tend to be shorter, on average, than their fathers, and sons of short fathers tend to be taller, on average, than their fathers. A naive conclusion from this is that in a few generations, all males will be the same height. Actually, the phenomenon would be observed even if the distribution of heights of males remained unchanged from one generation to the next. Nor is a difference between fathers and men in general a key factor in this phenomenon. The conclusion that something is changing in the population as a result of the "regression" toward the average is called the **regression fallacy.**

To see what is happening in this example, look at Figure 5.17, which shows the heights of the 1078 father–son pairs. The line equal heights for fathers and sons is almost identical to the SD line, because the average and SD of fathers' heights are almost identical to the average and SD of sons' heights. The point of averages is on the line of equal heights, the ratio of SDs, which is the slope of the SD line, is close to 1. We will proceed as if these two lines (the SD line and the line of equal heights) were, in fact, identical.

We wish to examine the distribution of sons' heights for fathers who have a particular height. Graphically, this amounts to focusing on a narrow vertical strip of the scatter diagram, the position of the strip being determined by the particular fathers on which we wish to focus. In a vertical strip that is to the right of the average fathers' height, the average sons' height is below the SD line. Even though the SD line bisects the scatter symmetrically, the vertical strips cut the oval at an angle to the SD line, so the bottom half of the strip has more points than the top half. Now, keeping in mind that the SD line in Figure 5.17 is also a line of equal heights of fathers and sons, to say that more sons' heights are below the line for a particular tall father's height means that sons of tall fathers tend to be shorter than their fathers. Similarly, a vertical strip on the left side of the scatter would verify that sons of short fathers tend to be taller than their fathers. Yet the distribution of heights of fathers and sons is the same. There is, in fact, no tendency demonstrated by these data for subsequent male generations to converge to one particular height.

The tendency for a son's height to be closer to average height than his father's was is called the **regression effect.** In general, the regression effect is the tendency for the averages of vertical strips to be closer to the overall average than is indicated by the SD line. The **regression fallacy** is to assume the regression effect implies something about the entire data set—when really the effect occurs in vertical strips but not over the whole range of values of X.

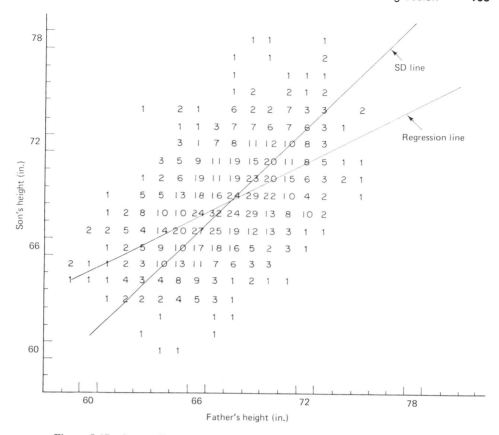

Figure 5.17 Scatter diagram of heights of 1978 father–son pairs, collected under the direction of Karl Pearson in 1903.[5] The numbers in the graph refer to the number of pairs at that location. (Heights are rounded to the nearest inch.) The regression line for predicting son's height from father's height is shown for comparison with the SD line.

Note that the same diagram and argument could also show that fathers of tall sons tend to be shorter than the sons. The argument is not wrong—only the naive implication that may be drawn from it. It is true that fathers of tall sons tend to be shorter than their sons, and also that the fathers of short sons tend to be taller than their sons, but this does not imply that the spread of all fathers' heights is less than the spread of all sons' heights.

We have stated earlier that the slope of the regression line is $|r|$ times the slope of the SD line. Since $|r|$ is less than 1 usually, this implies that we will predict sons to be shorter than their fathers when the fathers are taller than average. Thus the regression effect concurs with predictions made according to the regression method. This explains the origin of the word "regression." Another way of describing the regression effect is to say that it is the result of the regression line being flatter than the SD line.

Historically, the regression effect was noticed for the situation in which the same measurement is measured for two groups, where there was a natural pairing between the groups (as there was between fathers and sons). The regression effect is more general than this. It can be described as the effect of the regression line having a slope closer to zero (flatter) than the SD line (see Figure 5.18). The SD line is a line of symmetry through the oval scatter, but the regression line is not. In fact, there are two regression lines associated with each scatter: that of the regression of Y on X (to predict Y) and that of X on Y (to predict X). The two regression lines cross the SD line at the point of averages, but are otherwise on opposite sides of the SD line. (It is not possible to get the line for regression of X on Y by solving the equation for the regression of Y on X for X. This procedure gives the same Y on X line again, expressed differently.)

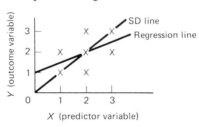

Figure 5.18 Regression toward the mean. The regression line is a line of averages of the vertical strips, but this is not a line of symmetry of the oval scatter.

The problem of correlations of averages discussed in Section 5.2 is also troublesome for regressions. Refer to that discusssion again to relate it to regression problems. The same is true of the discussion of outliers. One outlying observation can give a completely erroneous impression of the relationship between variables when regression methods are used carelessly. Examination of scatter diagrams before attempting to interpret a regression equation will protect you from these pitfalls.

EXERCISES

1. For the following data set:

X	Y
−1	−1
0	−1
1	2

   (a) Calculate the averages, SDs, and r.
   (b) Draw a scatter diagram of the data, with the scale in standard units (i.e., plot the standardized data).
   (c) On the scatter diagram, identify the point of averages and draw in the regression line and the SD line.

(d) Verify numerically, using the scatter diagram, a ruler, and a calculator, that the SD of the signed distances of the points to the regression line, in a vertical direction, is
$$\sqrt{1 - r^2}$$
(e) Verify that the SD of the signed shortest distances of the points to the SD line is $\sqrt{1 - |r|}$.

2. A university psychology professor wants to predict the effect of university education on IQ. One hundred students newly admitted to university are given IQ tests. When they graduate they take another IQ test. The formula for the regression line which predicts "after" IQ (AIQ) from "before" IQ (BIQ) is

$$(AIQ - 120) = 0.8 \times (BIQ - 120)$$

Does this imply that these graduating students tended to have IQs which were closer to 120 after college than they were before college? Explain why or why not.

3. A statistics professor records the amount of gasoline required at each fill-up of his car, and the number of miles since the last fill-up. These data may be summarized as follows:

Liters: average = 33.5, SD = 2.0

Kilometers: average = 450, SD = 30, r = 0.4

The professor notes that $r^2 = 0.16$ and $SD_{pred} = 1.8$ liters. On the one hand, SD_{pred} seems reasonably small, as one would expect. The distance traveled does allow one to predict the gasoline required fairly well. On the other hand, r^2 is only 0.16, indicating that the prediction of gasoline required is not helped much by knowing the distance traveled. Is the predictive relationship a good one?

4. Use the rental data of Table 5.7 to predict the monthly rental per square foot of commercial space that is 20 km from downtown.

5. Determine the equation of the regression line for predicting Y from X, based on the following data.

X	Y
0.7	0.7
1.0	0.0
0.7	-0.7
0.0	-1.0

Draw the scatter diagram of the data and superimpose the regression line on the same diagram. What does the diagram tell you about your regression line?

6. Perform a regression analysis of the data shown in Table 5.6. Predict the oxygen consumption of a human being, whose pulse rate is about 65 (according to the same source as the rest of the data).

7. The Seychelles are a group of islands in the Indian Ocean. The diets on the islands consist mainly of marine fish. Marine fish are known to contain relatively high concentrations of mercury even in unpolluted waters, and it was expected that the Seychelle residents would

have high levels of absorbed mercury, at least in their hair, where it is usually measured. Investigators wondered whether these high concentrations in mothers were passed on to the fetuses during pregnancy. The following data have been extracted from this study.[6]

	Mercury levels (parts per million)	
Mother	Mother	Newborn
1	14	12
2	27	48
3	7	11
4	8	62
5	6	6
6	17	18
7	15	19
8	16	21
9	7	7
10	17	14

(a) Perform regression calculations related to the research objectives.
(b) Discuss the implications of the regression analysis for the research objectives.

5.4 OTHER METHODS FOR QUANTITATIVE VARIABLES

5.4.1 Graphical Display of Three or More Variables

In discussing correlation and regression, graphical methods have already been used to explain many points that would have been very difficult to explain with just words and numbers. Graphical modes of description are ideal for revealing and portraying complex quantitative relationships. Until recently, when computers have made the construction of graphs much less time consuming, there has been little interest in the development of new graphical methods. This is changing.

Techniques for graphical display are hard to subject to purely mathematical analysis. The invention of graphical techniques to suit particular data sets will probably be left to nonmathematical data analysts who are expert in the field of application pertaining to the data.

Real data sets usually have many variables, not merely one or two. The methods we have discussed so far often have to be adapted or extended to deal with practical statistical problems. In this section we mention three simple ways that have been proposed for the graphical display of many-variable data sets. This field is in its infancy and you may be able to suggest better ways. This is an area that has few experts at the moment.

When a scientist wishes to study the relationship between two variables, he usually varies the values of one of the variables and observes the response of the other variable. The investigator tries to keep all other variables constant during the experiment, so the response can be attributed to the one source of change. For example, to compare the quality of various brands of photographic film, one should examine pictures taken by the same camera, of the same subject, under the same lighting conditions, for the various brands of film. Thus the investigator can examine the relationship between quality and brand (or price) without having to make allowances for confounding factors (see also Section 2.5). When it is not possible to study two variables without changing other relevant variables, one is stuck with the necessity of dealing with the description of the relationships of several variables at once. Examining the variables two at a time is usually not good enough.

As an example, suppose that we have two kinds of film that we want to compare; one costs $1 and the other $2. We have one picture taken with each film on a cloudy day (lighting conditions coded "1"), and one picture with each film on a sunny day (lighting conditions coded "2"). The picture quality is rated a "1" if it is acceptable, "0" if not. The results might be as follows:

Picture quality	Price of film	Lighting conditions
0	1	1
1	2	1
1	1	2
0	2	2

It is easy to see that there is a perfect dependence of picture quality on price of film and lighting conditions. With poor light we need the expensive film, but the expensive film is too sensitive for good light. Arithmetically, this predicted variable, picture quality, may be expressed as the absolute value of the difference between the two predictor variables, but examination of the variables two at a time would indicate that absolutely no relationship existed. (Draw the three scatter diagrams to confirm this. Of course, we would need a bigger data set to convince us that this relationship was not just a fluke, but let us ignore this point for now; we are only concerned with the use of graphs to sort out the information in several variables.)

The general rule is that if many variables are interrelated, the analysis of the data must examine all the variables simultaneously; important links can be missed by only looking at the variables two at a time. How can we do this graphically? The most common method for three or four variables is the **augmented scatter diagram.** Instead of points at each position indicated by two variables, an indication of the size of the third and fourth variables is plotted at this same location. The photograph data, which have three variables, would be plotted as shown in Figure 5.19. A fourth variable, say shutter speed, could be added to the diagram by using it to specify the size of the figure plotted. All this still could be done if each variable had several possible values, as is usual for quantitative data. See Figure 5.20.

With more than three or four variables, even augmented scatter diagrams become too confusing to "read." Two other methods, Chernoff's faces and star plots,

170 Chap. 5 Descriptive Methods for Two Variables

Figure 5.19 Augmented scatter diagram for three variables.

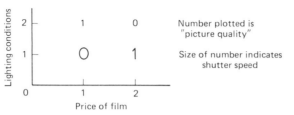

Figure 5.20 Augmented scatter diagram displaying four variables: picture quality, price of film, lighting conditions, and shutter speed.

are outlined as examples of a technique that can handle up to about 15 variables. To illustrate these techniques, we will return to the film critic data introduced in Section 5.2. The previous analysis used correlations to study the similarities and differences among critics: "critics" were the variables and "films" were the cases. For our present purposes we will reverse these roles, the critics being the cases and the films being the variables. That is, each film can be thought of as measuring a different aspect of the critics. Our objective is to graph the data so that we have a different "point" for each critic. This will again enable us to see which critics are alike in their ratings.

We have ratings on six films to characterize each critic: that is, a graph depicting the data of a critic will have to use six numbers. In **Chernoff's faces**,[7] these six numbers are used to draw a schematic face. The integrative ability of the eye is used to great advantage in comparing the "faces" of the six critics. The rating of each film (i.e., each variable) defines a different aspect of the schematic face. We arbitrarily choose the assignment as follows:

Film A: curvature of mouth
Film B: length of nose
Film C: width of nose
Film D: size of eye on right
Film E: size of eye on left
Film F: size of ears

The result is displayed in Figure 5.21.

The similarity between critics 1 and 5, between 3 and 6, and between 2 and 4 is immediately apparent. Critics 1 and 6 are about as unlike as any pair. These observations are in agreement with the correlation analysis of Section 5.2. An advantage with this graphical method over the correlation analysis is that nonlinear associ-

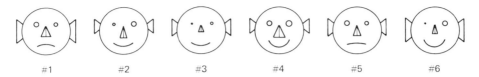

Figure 5.21 Chernoff's faces for the film critic data.

ations could possibly be detected from the faces, whereas the correlation analysis would miss this.

Another method for graphically portraying a data set with several variables is called a **star plot.** This method marks off the size of each variable along the spokes of a wheel, with one spoke for each variable; these marks are then joined to form a star. This produces one star for each case, just as there was one face for each case. It is usual to suppress the "wheel" and the "spokes" so that the diagram is less confusing. The critics would be portrayed as shown in Figure 5.22. Again, one can quickly spot the like pairs of critics. This method has one advantage over the faces: the choice of features to represent certain variables is less arbitrary, since it is unlikely that the apparent similarity of stars would depend on the particular spokes chosen of the six variables, whereas it may well be that a viewer would give more weight to eyes than to ears in noticing similarities and differences.

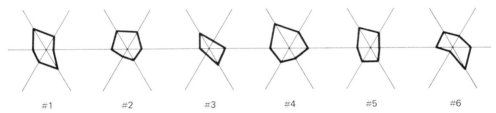

Figure 5.22 Star plot representation of critics' ratings.

The use of stars and faces for the critics data is not a practical suggestion. The data set is small enough that the information discernible from a graph could be more easily inferred from the numbers themselves, or perhaps with the aid of correlation analysis. If the data set had included 15 critics, the graphs would be more useful, but then the graphing itself would be tedious. Except, of course, that the graphing can easily be done by computer. We reiterate the importance of considering graphical innovations that have been ignored in the past because of the time required to construct the graphs. The only deterrent to using novel techniques for graphical analysis is the need to program the method the first time it is used. For complex data sets, it is often worth the effort, even for the first use.

This discussion of graphical techniques is intended primarily to be illustrative rather than dogmatic. Each application will suggest different requirements, and the creation of appropriate graphical strategies for particular applications may take a considerable investment of time. For serious studies, the time may be well spent in revealing hidden patterns or anomalies, and for less serious studies, the graphical approach may be more fun!

5.4.2 Multivariable Strategies

The purpose of this section is to drop some names of famous statistical techniques. There is not time in a first course in statistics to discuss all the useful statistical procedures. This subsection introduces you to the kinds of things you might discuss in a subsequent course in statistics, or if another course in statistics is not in your plans, you will at least have some idea of what you did not learn about statistics. To "know what you do not know" is of practical use in statistics as in other fields of study.

In statistical reports of research studies, you will frequently run into techniques with such names as "cluster analysis," "multiple regression," "discriminant analysis," "factor analysis," and "principal components." These are techniques that are used to explore and describe data sets with many variables. The techniques themselves are beyond the scope of this text, but it is useful to be familiar with the approaches taken even if the detailed analysis is to be left to someone more advanced in statistics. One can at least learn how to ask statistical questions relating to these multivariable strategies.

Cluster analysis is the process of grouping similar cases into homogeneous units called clusters. If you sort the books in your bookcase into, say, scientific books, novels, reference books, and so on, you are doing a cluster analysis. In statistics, the term refers to the use of numerical information to sort items. Usually, the numbers are quantitative measurements, although codes for qualitative data may also be used for this purpose.

In biology, cluster analysis is called taxonomy or classification. The identification of new species or subspecies is sometimes done using cluster analysis. Cluster analysis has been used in paleontology to study the relationships between various primate fossil teeth.[8] Teeth of modern man and modern apes were measured in various ways, and fossil teeth were similarly measured. The profile of measurements of the fossil teeth were compared to the teeth of human and ape data in an attempt to decide which fossil teeth were more like human teeth and which were more like ape teeth. This sorting of all the teeth into groups on the basis of the data alone is an example of cluster analysis.

Another example where a cluster analysis would be useful would be in the detection of groups of students who, although they have not declared a major, seem to be heading toward one or other major based on the courses they are taking. The college administration may well find this information helpful for planning purposes. The variables used might be:

1. Number of science courses taken in first semester
2. Grade-point average in math courses
3. Grade-point average in humanities courses divided by overall grade-point average

4. Age of student
5. Number of courses taken per semester

Many other variables could be tried. The end result of the cluster analysis is the identification of subgroups of students having similar values on all, or most, of the variables. It might then be hypothesized that these subgroups would be likely to end up in the same major programs.

Eventually, the predictive value of these categorizations would have to be tested. This could be done using a kind of regression which has one predicted variable (major eventually chosen) and several predictor variables (such as the ones listed above). This technique is called **multiple regression.** (The example is a bit different than is usual: the predicted variable in regression is usually a quantitative variable, just as it was in our discussion of regression in Section 5.3. To make the predicted variable, major eventually chosen, a quantitative variable, we would have to concentrate on one major at a time, and code the variable as 0 or 1 indicating presence or absence of that major.)

If we had the historical record of graduates who had received degrees in certain majors, say psychology and commerce, we could use the record of courses taken in their early years at college to differentiate these groups. In other words, we have several variables, each describing some aspect of the students' early course selections, and two groups, psychology and commerce, and the aim is to describe the difference between the two groups, using the variables. This is called **discriminant analysis.** The difference between discriminant analysis and cluster analysis is that in discriminant analysis one starts with known groupings of cases and tries to describe this in terms of the variables, whereas in cluster analysis one has no prior groupings and the aim is to form groups based on the variables.

In **factor analysis** and **principal components,** the aim is to develop indices combining several variables so that the data can be described using a few indices instead of many variables.

In Section 7.5.1, another multivariable technique, **analysis of variance,** is introduced. It is particularly useful in studying relationships between variables when some variables are qualitative and some are quantitative.

The mathematical and logical difficulties in making wise use of these multivariable techniques are many. Expert help should be sought in using them.

5.4.3 Adjustment of Averages

In this section we give a thumbnail sketch of a topic that, at its most advanced level, goes by the name **analysis of covariance.** But the basic idea is quite simple and also quite useful. We will illustrate the idea with an example.

It is well known that blood pressure increases with age. Presumably, blood vessels become constricted and resist blood flow. The difference between the blood pressures of men and women is also well known, but let us suppose that we want to learn about this difference from the following small data set for four men and four women. (SBP stands for "systolic blood pressure"):

	Men		Women	
	Age	SBP	Age	SBP
	23	120	19	110
	30	135	24	120
	50	140	35	120
	65	140	55	130
Average	42	134	33	120

From the averages shown it is clear that the men have higher blood pressures than the women. (Of course, we would need a larger data set to make any firm conclusion, but this is not the issue here.) In this data set, the men are older than the women, and since blood pressure is thought to increase with age, the higher blood pressures of men might be entirely attributable to age. In fact, it is not certain whether the men, if they were the same age, would have lower blood pressures. We really need a method to adjust for age before we compare the two blood pressure averages.

The method for this adjustment is to use regression in each group to determine the typical blood pressures for the group at a specified age, and then compare these typical blood pressures for the two groups at the specified age. Let us arbitrarily pick 35 years of age as the age at which we would like to make our comparison. (Is 35 midlife?) First we perform the two regressions.

For the men we predict, using the regression method, a 35-year-old to have a SBP of

$$134 + 0.81 \times \frac{35 - 42}{16.6} \times 8.2 = 137 \quad \text{(approximately)}$$

For the women

$$120 + 0.92 \times \frac{33 - 42}{13.8} \times 7.1 = 124 \quad \text{(approximately)}$$

The data indicate that a 35-year-old man will have a SBP of 137, on the average, while a 35-year-old woman will have a SBP of 124. The difference cannot be explained away by age differences in the groups. We say that the SBP values of 137 and 124 are "averages that have been adjusted for age."

Regression can be used to adjust group averages to allow for differences in nuisance variables (such as age).

Note that the difference in the group averages, adjusted for age, can depend on the specific age chosen to make the comparison. Only if the two regression lines were to have the same slope would this difference be the same value at all specified ages.

EXERCISES

1. Construct star plots and Chernoff's faces for the picture-quality data discussed in Section 5.4.1. Compare these plots with the augmented scatter diagram for the same data in Figure 5.19.
2. Discuss the likelihood of success of the proposed method in Section 5.4.2 to predict the major that would be declared by a student at the beginning of his or her third year, given information such as described from the student's first-year studies.
3. A data set describing economic characteristics of 100 counties has the following characteristics:

    Output: dollar value of goods and services produced
    New: number of building permits issued
    Taxes: taxes paid to the federal government
    Geog: a 0–1 variable indicating whether coastal or inland

    Among the techniques listed in Section 5.4.2, which would be appropriate to determine the following?
    (a) Unforeseen relationships among the four variables
    (b) The difference between the characteristics of coastal counties from inland counties
    (c) Whether some description of the counties (using these four variables) is a more useful grouping of counties than the Geog variable alone
4. Confirm and discuss the claim of the last paragraph of the text of this section.

5.5 METHODS FOR QUALITATIVE VARIABLES

In Section 4.2.2 the difference between a quantitative variable and a qualitative variable was discussed. In this section we outline some methods for describing data sets that consist of two qualitative variables. (The summary of several qualitative variables can be attempted two at a time, but this strategy has the same limitations as with quantitative variables. Better strategies for the description of several qualitative variables exist; they are not overly complex, but are beyond the scope of this text.[9,10])

When a data set contains one qualitative variable and one quantitative variable, descriptive methods for quantitative variables will often suffice. For example, in a survey of the price paid for bicycles owned by a class of eighth-grade children, the new price and the type of bike might be recorded. The "type" variable might be one of the categories "conventional bike" or "BMX bike," the latter being the type with the small wide wheels and the stronger frames for jumping curbs and climbing steps. The summary of this data set would be simply a display of the distribution of prices for each of the bike categories, using the methods outlined in Chapter 4.

5.5.1 Contingency Tables

But when we have two qualitative variables, and when the relationship between the two is of primary interest, we have a new problem. Consider, for example, the question of relating the type of bike (conventional versus BMX) to the presence or absence of gears. The data consist of a list, as indicated in Table 5.9. Clearly, averages and standard deviations of the categories do not make sense. For this type of data, the usual summary is a table called a **contingency table.** See Table 5.10. Note that using the frequencies to summarize the qualitative data has given us some numbers to work with, and furthermore, these numbers really do have all the relevant information about the relationship between the two variables. How would you summarize the findings of the bicycle survey based on Table 5.10?

TABLE 5.9 QUALITATIVE DATA

Student	Bike type	Gears?
1	BMX	Yes
2	Conventional	Yes
3	Conventional	Yes
4	BMX	No
5	BMX	No
⋮	⋮	⋮

TABLE 5.10 CONTINGENCY TABLE[a]

		Type of bike	
		Conv.	BMX
Gears	Yes	33	29
	No	7	37

[a] The entries in the table are frequencies. The 106 bikes are all bikes from a certain neighborhood and are classified according to type and the presence of gears.

Contingency tables are summarized using percentages rather than averages. For example, 33/40 or 83 percent of the conventional bikes have gears, but only 29/66 or 44 percent of the BMX bikes have gears. A conscious effort must be made to decide which way to calculate the percentages. One could have concluded that 29/62 or 47 percent of bikes with gears were BMX bikes, and 37/44 or 84 percent of the bikes without gears were BMX bikes. It depends on who is interested: the manufacturer of gears might well want the latter summary.

Table 5.10 is called a two-by-two contingency table. Table 5.11 is a three-by-four contingency table. An *n*-**by-***m* **contingency table** is one with n horizontal rows and m vertical columns. How many variables are involved? Only two! Two-variable data sets, for qualitative variables, are usually summarized by contingency tables which include percentages by rows (horizontal percentages add to 100 percent) or

TABLE 5.11 CONTINGENCY TABLE WITH SEVERAL CATEGORIES FOR EACH VARIABLE[a]

		Type of fruit			
		Apple	Peach	Cherry	Other
	Okanagan	45	10	40	5
Region	Niagara	30	30	10	30
	Annapolis	70	5	0	25

[a] The table entries are frequencies: they are the number of fruit farms, among the 100 from each region that have been selected, that concentrate on a particular fruit (indicated by the column category). The data are hypothetical.

percentages by columns (vertical percentages add to 100 percent) or, very occasionally, both row and column percentages. If one wished to compare rows, percentages should be calculated so that the column percentages should add to 100 percent; if one wishes to compare columns, the percentages should be calculated so that the row percentages add to 100%.

To decide which way to do the comparison, by rows or by columns, you should be guided by the context of the data. Once you have decided what it is you want to compare, the rules above may help you to compute the percentages appropriately.

In Table 5.11, 100 fruit farms have been randomly sampled from each region. So the numbers in this table can be thought to represent row percentages as well as absolute frequencies. Does it make sense to conclude, based on these row percentages, that the Annapolis region has more fruit farms that concentrate on apples than has the Okanagan region? No. The only sensible regional comparisons would be based on the row percentages, which reveal the relative frequency of orchards of various kinds in each region, not the absolute frequencies. That is, the row percentages indicate that a greater percentage of farms in the Annapolis region concentrate on apples than in the Okanagan region. The column percentages in this table are meaningless, since they are sensitive to the arbitrary choice of row totals used in this example. If we had decided to collect data from 200 farms in Niagara, and 100 in Annapolis and Okanagan, the column percentages would be different without there being any difference in the distributions that are being described. These data cannot be used to describe the relative prominence of the various regions with respect to any particular fruit.

What is it about contingency tables that makes them different from other tables? It is the fact that the entries in the table are frequencies—not averages, not proportions, and not ratios. It is true that the interpretation of contingency tables certainly involves proportions (or percentages), as we saw in the discussion of the bicycle data (Table 5.10). The contingency tables we have described so far are "two-way" or two-variable contingency tables. We often examine them to see if the two variables are related in some way. We do this by comparing proportions. In Table 5.10, we determine whether the proportion of conventional bikes with gears is the same as the proportion of BMX bikes with gears. If it is not, the characteristic "bike type" is related to the characteristic "gears." We say that the two categorizations depend on each other or that they are **dependent.** Two categorical variables are said to be dependent if the relative frequency distribution of one of the variables is not the same for all the categories of the other variable.

There are such things as three-way tables: that is, contingency tables where the categories for which we are tabulating the frequencies are defined by the values of three variables. In fact, we used such a table to describe Simpson's paradox in Section 2.5. The three variables in the example we used there were "pass–fail," "short course (Prob or Read)" and "GPA (High or Low)." The conditions in which Simpson's paradox have a chance of occurring are when the three variables are dependent. That is, the proportion of students in the Prob course who pass depends on their GPA. This idea of dependence is an important one in statistics and is not limited to qualitative variables. However, the description of dependence among three or more variables is beyond the scope of this book. The techniques used to unravel the complexities of dependence among several variables are those alluded to in Section 5.4.2. However, we have discussed Simpson's paradox in a contingency table, and you are reminded of this by Figure 5.23.

Common sense will help to sort out what summaries are appropriate. But one should be aware that there are several wrong ways to calculate percentages and to draw conclusions, and take heed.

5.5.2 The Histogram

The histogram technique can be adapted to qualitative data. The orchard-type distribution for the Niagara region (the data are presented in Table 5.11) can be graphed as shown in Figure 5.24.

If the order of the fruit types is changed to, say, peach, cherry, apple, and other, the graph would look different, yet the information portrayed would, of course, be the same. This is in contrast to the histogram of a quantitative variable such as price. Determining the shape of the distribution is one of the principal aims of drawing the histogram for a quantitative variable, but the shape is irrelevant for a qualitative variable. In fact, the order of the categories is sometimes chosen to coincide with the order of the frequencies, so one can tell at a glance which categories have the highest and lowest frequencies. Note the separation of the rectangles; it helps to remind us that the frequency shown by the height of the rectangle is not an accumulation of frequencies for grouped data, as is usually the case for quantitative data.

Data sets are seldom entirely qualitative or entirely quantitative. Figure 5.25 displays a data set containing one quantitative variable, age, and one qualitative variable, sex. The "population" depicted by each bar is really a frequency, so this really is a two-variable display, not a three-variable display, using our usual terminology (that is, reserving the word "variable" for a data measurement). This diagram

Sec. 5.5 Methods for Qualitative Variables

Miss Lonelyhearts

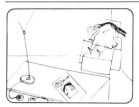 Miss Lonelyhearts, a statistician, is tired of sitting home alone.
Miss Lonelyhearts: I wish I knew some men who weren't married. I think I'll join a group for single people.

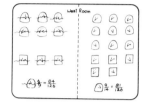 Her statistics for the West group were similar. The proportion of mustached swingers was 84/126. This was greater than 81/126 for the cleanshaven swingers.

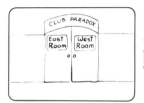 Miss Lonelyhearts joined *two* such groups. One evening both groups had parties at Club Paradox. One group met in the East Room, the other in the West Room.

 **Miss Lonelyhearts:** How simple! At *both* parties, I'll have a better chance to meet a swinger if I look for men with mustaches.

 **Miss Lonelyhearts:** Some men have mustaches and some don't. Some men are swingers and some are squares. I'd like to meet a swinger tonight. Should I look for a man with a mustache?

 By the time Miss Lonelyhearts got to Club Paradox, the two groups had decided to combine. Everybody had moved to the North Room.

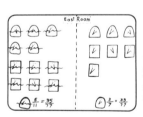 Miss Lonelyhearts made a statistical study of the men in the East group. She found that the proportion of mustached swingers was 5/11 or 35/77. The proportion of cleanshaven swingers was smaller. It was 3/7 or 33/77.

 **Miss Lonelyhearts:** What shall I do now? If a mustached man is my best bet in each group, he should still be the best bet. But I'd better check out the combined party to make sure.

 **Miss Lonelyhearts:** So— when I attend the East Room party, I'll go after the men with mustaches!

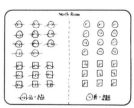 When she finished her new chart, she was flabbergasted. The proportions had changed places! Now her best bet was a man *without* a mustache!

Figure 5.23 Simpson's paradox.[11]

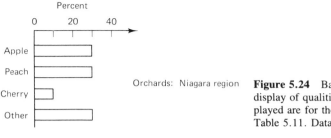

Figure 5.24 Bar chart for frequency display of qualititive data. (Data displayed are for the Niagara category of Table 5.11. Data are hypothetical.)

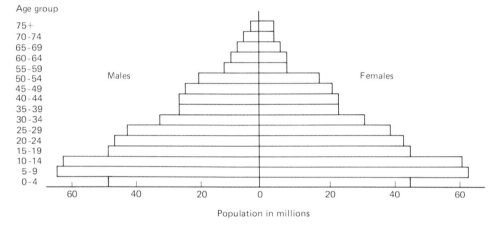

Figure 5.25 Population pyramid for China's population in 1978. The horizontal bars indicate the number of people of a particular sex in each age group.[12]

makes quite clear the recent anomaly in the age distribution of China's population. The diagram is called a *population pyramid* by demographers.

EXERCISE

1. Refer to Figure 5.24. Draw a freehand diagram indicating the shape you think the population pyramid will have in the year 2000.

5.6 ANSWERS TO EXERCISES

Section 5.1

1. The scatter diagram reveals that the U.S. Navy reports tended to report fewer sinkings than actually occurred. Note also that the errors tended to be largest in months in which the largest numbers of submarines sunk. Big numbers often have big errors.

2. Not necessarily. If the presence of extra fat is causing some heart attacks, this would be one explanation for the higher incidence of heart attacks in the overweight. But there are other possibilities that cannot be excluded; some environmental or genetic factor may predispose a person to both overweight and heart attacks.
3. Presumably, the management controls the distance a truck is allowed to cover in a year, and this policy then determines, with some error of estimation, the maintenance costs. A first step might be to look at a scatter diagram to see if there is a simple relationship that will help to predict maintenance costs from the planned distance for each truck. (It would be possible to argue that the expenditure on preventive maintenance would determine the distance the truck could cover in a year. If this is thought to be plausible, the role of the variables in the answer would be interchanged.)
4. A central problem in this kind of research is what is causing what. Does a person's failing health lead him or her to seek a change in life-style? Or do life changes exert stress that predisposes the body to medical problems? Or do people who do not look after themselves tend to have disruptive social lives and poor health? Further research should be aimed at assessing the validity of these different hypotheses.
5. See Figure 5.2. The points are closer to the line of 18 percent-short guesses than to the line of 88-page-short guesses. Hence the 18 percent shortfall seems the better description.
6. The association in this case was later shown to be causal. However, there are some logical possibilities that had to be considered. If the discomfort of rubella induced victims to take analgesic medications (such as aspirin or alcohol), the association would have been consistent with the theory that these analgesics caused the deafness. Another possibility would be loss of sleep caused by the rubella at a crucial stage of development of the fetus. (In this case it could still be argued that rubella was the cause, although the suggestion is that it is not the immediate cause.)

   The practical implications of these alternative explanations would be that instead of vaccination of potential mothers against rubella, the elimination of analgesics (or alcohol) during pregnacy would be the strategy for the first alternative, and for the second one it would be to induce sleep during discomfort by some method (alcohol?).

Section 5.2

1. Critics 1 and 5 rank the films similarly, as do 2 and 4, and also 3 and 6. (See the correlation coefficients for these pairs.) Critic 4 gives ratings about 20 percent higher than the rest, while critic 3 gives ratings 20 percent lower. Critic 1 is most variable in his ratings and critics 2 and 5 are least variable. (The latter comments are based on averages and SDs.)
2. The correlation is 0.31, which usually would indicate a very mild correlation between the computing science GPAs and the other GPAs. However, in this case there is one outlier, student 4, that appears to be atypical. The other students appear to have a very high correlation between these scores. The sample is very small, but we may tentatively conclude that the two GPA ratings are highly correlated in most students in the same academic environment as those yielding this data set.
3. The correlation is -0.93. However, the calculation is inappropriate in this example because either variable can be calculated exactly from the other. There is a perfect relationship between the variables, but it happens not to be linear; this nonlinearity is another reason, although a secondary one in this instance, because the nonlinearity is not extreme.

182 Chap. 5 Descriptive Methods for Two Variables

4. It would remain unchanged.
5. The five summary statistics are:

$$\begin{array}{ll}\text{Averages} & 52.2,\ 44.2\\ \text{SDs} & 29.9,\ 24.2\\ r & 0.85\end{array}$$

6. (a) The correlations are
   (i) 0.00; (ii) 0.45; (iii) 0.71; (iv) 0.89.
   (b) Standardization does not change the correlation, but it does change the look of the scatter plot. To guess correlations visually, the variables should have the same SD.
   (c) Average of $V_1 = 1.0$, SD of $V_1 = 0.63$
   Average of $V_4 = 2.0$, SD of $V_4 = 0.89$
   $r = 0.71$
   To draw the oval diagram, first draw a rectangle with center at (1.0, 2.0) and width 2×0.63 and height 2×0.89. An oval consistent with a positive association and inside (just touching) this rectangle would be a reasonable, but rough summary of the data.

Section 5.3

1. (a)

	X	Y
Average	0	0
SD	0.82	1.41
r		0.87

   (b) The standardized data plotted in Figure 5.26 is

Z_x	Z_y
-1.22	-0.71
0	-0.71
1.22	1.42

   (c) The regression line has slope 0.87 and passes through the point of averages (0, 0) ($Z_y = 0.87 Z_x$). The SD line has slope 1.00 and passes through the point of averages (0, 0) ($Z_y = Z_x$).
   See Figure 5.26.
   (d) The vertical distances are 0.35, 0.71, and 0.36. To indicate the direction, we record the signed distances as 0.35, −0.71, and 0.36. The SD of these is 0.50. The formula agrees $\sqrt{1 - (0.87)^2} = 0.49$, except for rounding errors.
   (e) Perpendicular signed distances are 0.36, −0.50, 0.14, and the SD of these is 0.36. The formula agrees $\sqrt{1 - |0.87|} = 0.36$.

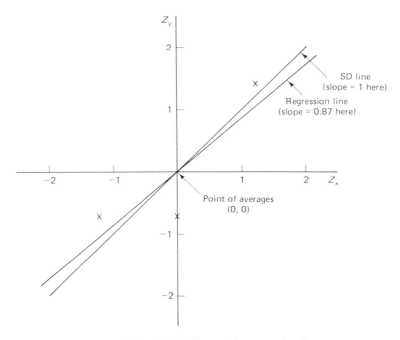

Figure 5.26 The SD line and the regression line.

2. The answer to the question is "no." The suggested conclusion would be an example of the regression fallacy.

3. The problem here is that the car was refilled after about 450 km each fill-up. So there is not much variation in the X variable in this regression. The prediction in this data set of the liters required is quite good without knowing the kilometers traveled: The SD for liters is 2.0, and this is only reduced to 1.8 using the knowledge of the kilometers. This does not mean that liters do not depend on kilometers. If one fill-up had occurred at a distance traveled of, say, 50 km, r^2 would have been close to 1.0.

   Yes the predictive relationship is good. (But it was not improved much by knowing the kilometers in the range typical of this data set.)

4. The summary statistics required to use the regression method are:

   Rental: average = \$0.57, SD = \$0.22

   Distance: average = 30 km, SD = 17 km

   $$r = -0.6$$

   Now, 20 km is $(20 - 30)/17 = -0.59$ SD or 0.59 SD below average. The regression prediction of rental rate is -0.59 SD $\times$ -0.6 or $+0.36$ SD above average; that is $0.57 + (0.36 \times 0.22)$ or \$0.65 (65 cents) per square foot per month.

5. Average of X = 0.60   average of Y = 0.00
   SD X = 0.37   SD Y = 0.86
            $r = 0.78$

The regression line is

$$\frac{Y - 0.00}{X - 0.60} = 0.78 \times \frac{0.86}{0.37}$$

which simplifies to $Y = 1.81 \times X - 1.09$.

A scatter diagram shows that points trace a circular path, whereas the regression line cuts across the circle. Clearly, the linear fit is a poor summary.

6. Pulse rate: average = 211.10, SD = 215.15
   Oxygen rate: average = 0.78, SD = 0.84
                 $r = 0.90$

A pulse of 65 is $(65 - 210.10)/215.15 = -0.67$ (i.e., 0.67 standard unit below the average). The regression method says to predict an oxygen rate for this pulse as

$$0.78 - 0.90 \times 0.67 \times 0.84 = 0.78 - 0.51 = 0.27$$

The prediction is that human beings would use 0.27 mℓ of oxygen per pound of body weight per hour.

(A 150-lb man would use 40.5 liters per hour, or about 70 mℓ of oxygen used from each breath. This volume of 70 mℓ is about the size of a 2-oz shot glass. The actual amount of oxygen consumed, according to the data referenced, is 0.21, so the regression prediction is pretty good considering the type of animals in the data set).

7. (a) A scatter diagram (which is advisable to draw for regressions) would show that the point for mother 4 is an outlier. [In the original paper, the investigators redid this point and got (11, 12); even without knowing this, the point should be removed from the regression analysis.] The averages are 14.0 and 17.3, the SDs 6.3 and 11.9, and the correlation 0.90. (The corresponding values before separating the outlier are 13.4, 21.8, 6.2, 17.5, and 0.34.). The most useful regression for the research described would be the regression of newborns' mercury (Y) on mothers' mercury (X). The regression line is

$$\frac{Y - 17.3}{X - 14.0} = 0.9 \times \frac{11.9}{6.3}$$

which may be simplified to

$$Y = 1.7 \times X - 6.5$$

SD$_{pred}$ is 5.2.

(b) The mother's mercury level is certainly a good predictor of the newborn's mercury level (compare 5.2 and 11.9, or else note that $r = 0.9$), and the slope of the regression line is positive as expected. It is interesting that the slope coefficient is greater than 1.0, suggesting that mercury is more concentrated in the fetus. The intercept is close to 0 (especially when we see that SD$_{pred}$ is 5.2), suggesting that the fetal concentration of mercury may be proportional to the mother's concentration. Of course, the causal link is only suggested and not proven. (Storage of mercury could well be explained by genetic factors.)

Section 5.4

3. (a) Principal components analysis or factor analysis
   (b) Discriminant analysis
   (c) Cluster analysis
4. Just work through the calculation for another age, say 50 years. If the regression lines have different slopes, the distance from one regression line to the other will vary as the value of the nuisance variable is varied. If the two lines are parallel, this difference stays constant.

Section 5.5

1. It depends on what birthrates are for the period 1978–2000. If the drastically reduced birthrate, indicated by the populations for 0–4 age category in 1978, is indicative of a new trend, the population pyramid would look like that shown in Figure 5.27.

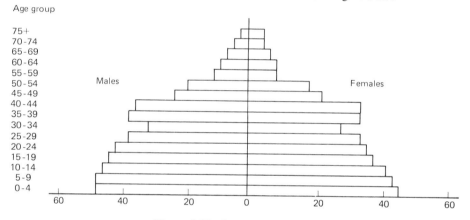

Figure 5.27 Population pyramid.

5.7 NOTATION AND FORMULAS

Let X and Y be two variables, and their averages and SDs are denoted $\bar{X}$, $\bar{Y}$, SDX, and SDY. Then X and Y in standardized units are

$$Z_X = \frac{X - \bar{X}}{SD\ X} \quad \text{and} \quad Z_Y = \frac{Y - \bar{Y}}{SD\ Y}$$

The procedure described in the text for the calculation of r becomes, in this notation,

$$r = \left(\sum Z_X \times Z_Y\right)\Big/n$$

$$= \left(\frac{\sum (X - \bar{X}) \times (Y - \bar{Y})}{SD\ X \times SD\ Y}\right)\Big/n$$

186 Chap. 5 Descriptive Methods for Two Variables

In these formulas the Greek capital sigma (Σ) is used to indicate summation over the cases in the data set.

The notation for regression lines that is most commonly used in elementary texts is
$$Y = A + BX + E$$
where A is the Y intercept, B the slope, and E the error (the amount by which the regression line errs in predicting Y), and BX means B times X. This expression is used to indicate the model, and
$$Y_i = a + bX_i + e_i \quad i = 1, 2, \ldots, n$$
is used to indicate the actual relationship between the data values for X and Y. a, b, and e_i are estimates of A, B, and E. The e_i are defined to make the equation true.

The SD of the e_i is the quantity we have denoted SD_{pred}. Note that the calculation of SD_{pred} from r and SD Y as suggested in Section 5.3.2 is more convenient than starting with a, b, and the e_i's. (In this text we do not discuss the relationship between the A and B and their estimates a and b.)

The regression method is a consequence of the form of the regression line, which may be written
$$Y = \bar{Y} + \left(r \times \frac{X - \bar{X}}{SD\ X} \right) \times SD\ Y$$

Alternatively, the regression line may be written
$$Y = \bar{Y} - \left(r \times \frac{SD\ Y}{SD\ X} \times \bar{X} \right) + \left(r \times \frac{SD\ Y}{SD\ X} \right) \times X$$

which may be compared with $Y = a + b \times X$ to give formulas for a and b.

The SD line may be written
$$Y = \bar{Y} + \text{sign of } r \times \frac{SD\ Y}{SD\ X} \times (X - \bar{X})$$

That is, points (X, Y) on the SD line satisfy this equation exactly. Note that if the data are expressed in standard units, the SD line is
$$Y = \text{sign of } r \times X$$

Note that the SD line does not depend on the actual value of r, only on its sign. The SD line has slope $+1$ or -1 for standardized data. (Sometimes "sign of r" is written "sgn r.")

5.8 SUMMARY AND GLOSSARY

1. Regression and correlation are techniques for use with data sets with more than one variable. Correlation is used to describe the strength of the relationship between variables, and regression is used to estimate a predictive relationship

between variables. Both correlation and regression measure linear relationships. They are most applicable when the scatter diagram is oval-shaped.
2. When two variables are large together, and small together, they are said to be positively associated. If one is small when the other is large, they are said to be negatively associated.
3. Variables that are highly correlated *might* be causally related. Correlation evidence is only suggestive of causality, no matter how high the correlation.
4. The correlation between two variables is the average product of the variables after they have been expressed in standard units.
5. Regression is a predictive technique. The two variables in a regression have different roles: one is used to do the predicting of the other.
6. The regression method of prediction of Y from X consists of adjusting the average value of Y by an amount equal to r times the difference between the value of X and the average value of X expressed in multiples of the SD of X.
7. The regression line has a lesser absolute slope than the SD line. The SD line is a line of symmetry through the oval scatter; it is the longest diameter of the oval. The regression line is a line of averages of the scattergram after it has been divided up into vertical strips.
8. The relationship between two variables should be studied in homogeneous groups of subjects if possible, to avoid confusing subgroup-to-subgroup variation with subject-to-subject variation.
9. Practical studies often use several variables. The techniques available for these multivariable situations are more complex than are covered in detail in this text. However, a partial description of such data sets can be achieved using the two-variable techniques, regression and correlation, as described in this chapter. The multivariable graphical techniques are useful for exploratory purposes.
10. Scatter diagrams are useful adjuncts to any correlation or regression analysis. They assist in spotting anomalies such as outliers and nonlinearity. Residual plots are a special kind of scatter diagram suited to checking the assumptions of regression analysis.
11. Qualitative data are described using contingency tables and some associated percentages. The regression and correlation methods described here do not apply to such data.

Glossary

Cases are the number of items examined, **variables** are the different kinds of measurement performed on each case, and **groups** are the collections of cases that we may wish to compare with each other. (5.1)

A **scatter diagram** (or scattergram) is a figure depicting data values as points in a scaled graph; it has one axis for one variable and two axes for two variables. (5.1)

A typical **prediction** problem in statistics is of the form: predict the value of variable A given the value of variable B. (5.1)

In statistical prediction problems, there are usually at least two variables: one variable whose value is known, called the **predictor variable,** and one variable whose value is not known, called the **predicted variable.** (5.1)

Two variables are said to be **associated** with each other if measurement of one variable can help to predict the other variable. (5.1)

The **agreement** between two variables is the degree to which the variables tend to be on the same side of their respective averages. (5.2)

Variables that agree are said to be **positively associated;** when they "disagree" they are said to be **negatively associated.** (5.2)

The **corrrelation coefficient,** r, is quite simple to calculate: after the two variables have been expressed in standard units, it is the average product of the two variables. (5.2)

The **oval diagram** is a schematic diagram depicted by an oval the typical approximate shape of the boundary of a two-variable scatter. (5.2)

In general, an **outlier** is an observation among a group of observations that is suspected, by virtue of its distinctive size, of not belonging to the group. (5.2)

A **linear** relationship between two variables is one representable by a straight line: a unit increase in one variable causes a fixed increase in the other variable, determined by a certain slope constant. (5.2)

The problem of sorting out correlations for clusters will be referred to as the **correlation-of-averages** problem. To fix the position of an oval diagram, we need five numbers: the average and SD for each variable, which locates the center of the oval and its vertical and horizontal extent, and the correlation coefficient, which indicates how narrow the oval is and also whether it is positively or negatively oriented. These five numbers are called the **five-number summary** of the two-variable scatter. (5.2)

Regression is a method for determining predictive relationships among variables. In this text we discuss only the regression of one predicted variable on one predictor variable. (5.3)

The **point of averages** is the point whose coordinates are (average of X, average of Y), where X is the horizontal coordinate and Y is the vertical coordinate. (5.3)

The **SD line** is the line through the point of averages that has slope (sign of r) × (SD of Y/SD of X). (5.3)

The **regression line** is the line through the point of averages whose slope is $|r|$ times the slope of the SD line. This method of prediction is called the **regression method.** (5.3)

SD_{pred} is the SD of the prediction error incurred while using the regression equation to predict the values of Y in the data set. (5.3)

A **residual plot** is a plot of the errors from a regression fit against the predicted values of the result variable. (5.3)

The property of the prediction errors to tend to be about the same size over the whole range of *X* is called **homoscedasticity,** and its opposite is **heteroscedasticity.** (5.3)

The **regression effect** is the tendency for the averages of vertical strips to be closer to the overall average than is indicated by the SD line. (5.3)

The **regression fallacy** is to assume that the regression effect implies something about the entire data set when the effect occurs in vertical strips but not over the whole range of values of *X*. (5.3)

A **contingency table** records the frequencies of the various possible contingencies defined by categorical variables. When used for two variables, the rows correspond to the categories of one variable and the columns correspond to the categories of the other variable. (5.5)

An *n*-**by**-*m* **contingency table** is one with *n* horizontal rows and *m* vertical columns. (5.5)

Two categorical variables are said to be **dependent** if the relative frequency distribution of one of the variables is not the same for all the categories of the other variable. (5.5)

PROBLEMS AND PROJECTS

1. Draw the scatter diagram for the data below. Use scales for your axes that are the same as the original units of the data. Calculate the correlation coefficient. Demonstrate the effect of changing the last value of "weeks worked" from 10 to 20. Comment on the adequacy of $|r|$ as a measure of the degree of association between the two variables.

Days absent from work	Weeks worked
4	6
6	13
14	9
19	14
20	10

2. Two wine connoisseurs provide the following assessment of 10 wines:

Wine	Connoisseur 1	Connoisseur 2
1	Good	Good
2	Mediocre	Poor
3	Great	Good
4	Mediocre	Mediocre
5	Mediocre	Good
6	Poor	Poor
7	Good	Good
8	Poor	Good
9	Good	Great
10	Mediocre	Poor

Invent a measure of association for these data, to describe the similarity of the assessments. Comment on the strengths and shortcomings of your invention.

3. A student majoring in physical education undertakes a project for her statistics class. She obtains heights and weights from 10 members of her family and finds that the correlation of height and weight is 0.55. Deciding that she needed a larger data set, she then collects the same data for 50 of her fellow physical eduction students. This data group has a height–weight correlation of 0.92. Explain this apparent disparity in the two correlations.

4. Guess the correlation between income and education of "heads of household" in New York City. Would you expect the correlation between these same variables in the combined data for the cities of New York and Los Angeles to be higher, lower, or about the same as for New York alone? Explain your reasoning. (*Note:* Your assumptions need not be true to fact, but your reasoning should be sound.)

5. Use the data of Problem 1 to predict the number of days absent from work for an employee who had worked 12 weeks. Provide an estimate of the size of the error in this prediction. Check the assumptions of regression, and describe what you have done to perform this check.

6. Two conservationists, X and Y, are independently estimating the size of flocks of Canada geese passing near a particular observation point. Each flock is identified by the time of passing so that both observers are estimating the sizes of the same flocks. Assess the extent to which the observers agree, based on the following data.

Flock	X	Y
1	23	29
2	14	17
3	32	37
4	98	81
5	17	18
6	10	10
7	28	35
8	30	30

How well does observer Y's estimate predict observer X's estimate? Comment on the validity of the regression assumptions in this instance.

7. An hydrologist has collected the following data to study the relationship between the flow rate of streams and their depth.[13]

Depth	Flow rate
0.34	0.636
0.29	0.319
0.28	0.734
0.42	1.327
0.29	0.487
0.41	0.924
0.76	7.350
0.73	5.890
0.46	1.979
0.40	1.124

(a) Compute the r^2 value for the prediction of stream flow rate from stream depth. Do the same calculation for the prediction of the square root of stream flow rate from stream depth. Explain the difference in the two values of r^2. (*Hint:* Draw a rough scatter diagram.)

(b) Could the two regressions in part (a) be usefully compared using SD_{pred}? Explain why or why not.

8. Based on the following data, make a recommendation about the film to buy for high-quality pictures:

Lighting conditions	Film price	Picture quality
1	$1	0
2	1	0
2	2	1
2	1	1
1	1	1
2	1	0
1	2	0
1	2	1
1	1	1
2	2	1
2	2	0
1	2	0

9. The data set below is part of a larger data set describing characteristics of cars currently available in North America. What summary of this small data set would illustrate the value of a statistical description to a business manager interested in adding a new model to his company's line of cars? Attempt such a summary, and add a paragraph illustrating to the businessman the possible use of this kind of data summary.

Model	List price (1000 dollars)	Weight (1000 kg)	Interior volume (m³)	Max. number of passengers	Country of origin	Gas cons. (km/liter)	N.A. sales volume (1000s)
(1)	13.5	1.3	3.4	5	United States	8.5	350
(2)	9.6	1.2	3.4	6	United States	9.0	170
(3)	15.3	1.0	2.9	5	Japan	7.4	85
(4)	24.5	1.4	1.9	2	Great Britain	10.0	10
(5)	35.5	1.3	3.0	5	Germany	8.0	25
(6)	6.9	1.0	2.7	4	United States	7.0	140
(7)	14.5	1.2	3.0	6	United States	9.5	210
(8)	7.5	1.1	2.8	5	Japan	7.2	160
(9)	9.5	1.2	3.3	6	United States	8.5	100
(10)	17.4	1.5	3.5	6	United States	9.9	55

10. Using data such as those shown for Problem 9, what type of analysis would produce several lists of cars such that a car in a particular list is like the other cars in that list, but different

Chap. 5 Descriptive Methods for Two Variables

from the cars in other lists? What type of analysis would show how to estimate price based on the other information given?

11. A health clinic advertisement claims to be able to assist customers to attain an ideal weight. As evidence for this, they present data from past customers which show that the SD of weights for customers when they join the "weight adjustment" program is greater than the SD weight of these customers on completion of the program.
    (a) Does this fact support the clinic's claim? Discuss.
    (b) A second advertisement claims that in a special group of customers, some overweight and some underweight, whose ideal weights were all about 130 lb., 75 percent of these customers experienced weight changes toward the ideal. Does this fact support the clinic's claim? Discuss.

12. To predict the gasoline consumption of a proposed vehicle (yet to be built) that would weigh 1.75 metric tons, data showing the relationship between vehicle weight and gas consumption are collected. (We choose a very small sample here to simplify calculations.)

    Vehicle weight (metric tons):   1.60, 1.65, 1.70, 1.75, 1.80
    Gas consumption (km/liter):   5.0, 4.9, 4.8, 4.1, 4.0

    (a) Use the regression method to predict the gas consumption for the proposed 1.75-metric-ton vehicle.
    (b) Is the the prediction of 4.1 km/liter, based on the fourth pair of values, inferior to the prediction based on all five pairs of values? Explain.

13. A study is done to assess the delivery time for first-class letters mailed between cities in Canada. (Canada's postal service was very unpopular around 1980: it was said that increases in postage were used to pay for storage costs.) Pairs of cities are chosen at random from a list of 50 cities in Canada, and arrangements are made to record the sending time and the receiving time for letters mailed from the first city of the pair to the second city of the pair. The resulting data set is summarized as follows:

    Length of time: average = 70 hr, SD = 25 hr
    Distance: average = 1500 km, SD = 700 km
    Correlation between length of time and distance = 0.6

    Assume that the data have the familiar oval scatter.
    (a) How long would you predict that it would take for a letter mailed in one city to be received in another 1000 km away? Include an indication of the precision of your estimate.
    (b) What percentage of letters mailed between cities that are 100 km apart take 48 hr or less?

14. A sequence of 10 tests is given in a certain course. The scoring of the tests is such that the average mark on each test is 30, and the SD on each test is 10. The correlation between scores on any pair of successive tests turns out to be close to 0.8 for each such pair. A person receiving 90 on the first test is predicted (using the regression method) to obtain approximately 78 on the second test. A person receiving 78 on the second test is predicted to obtain 68 on the third test. If one repeats this calculation, it can be shown that a score of about 37 is predicted for the tenth test.

    Does this calculation show that a student who performs exceptionally well on the first test is doomed to mediocrity by the tenth test? Give reasons for your answer. No calculations are necessary.

15. A doctor is in the habit of measuring blood pressures twice.[14] He notices that patients whose readings are unusually high on the first reading tend to have somewhat lower second readings. He concludes that patients are more relaxed on the second reading. A colleague disagrees, pointing out that the patients who are unusually low on the first reading tend to have somewhat higher second readings, suggesting they are more nervous. Which doctor is right? Or perhaps both are wrong? Explain in detail.

16. An experiment was conducted in a supermarket to observe the relationship between the amount of display space allotted to a brand of coffee (brand A) and its weekly sales.[15] The amount of space allotted to brand A was varied over 3-, 6-, and 9-ft^2 displays over 12 weeks, while the space allotted to competing brands was maintained at a constant 3 ft^2 for each. The following data were observed:

Week	Weekly Sales	Space allotted (ft^2)
1	526	6
2	421	3
3	581	6
4	630	9
5	412	3
6	560	9
7	434	6
8	443	3
9	590	9

What are the additional weekly sales for an additional square foot of display space? (This problem is continued as a computer project in Problem 32.)

17. Guess the correlation coefficient for the scatters shown in each of the three diagrams of Figure 5.28. Explain why this is possible when the scales for the axes of the scatter diagram are not given. Suggest, but do not carry out, a method that would produce the answer almost exactly. (This problem is continued as a computer project in Problem 33.)

18. Monthly ticket sales for a movie theater are recorded over a 48-month period. How would you summarize these data for the owner of the theater?

19. Ten tenth-grade students were given tests of memorization ability and creativity.[3] The results were as follows:

Student	Creativity	Memorization
1	11	13
2	96	85
3	15	27
4	88	69
5	92	76
6	34	30
7	44	39
8	67	32
9	37	13
10	38	58

194 Chap. 5 Descriptive Methods for Two Variables

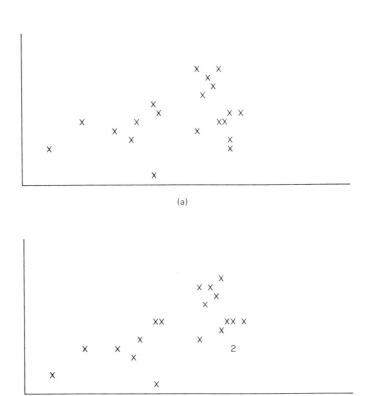

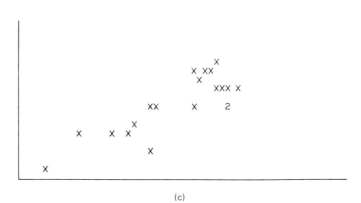

Figure 5.28 Scatter diagrams for Problem 17.

(a) Summarize the results.
(b) An eleventh student had a memorization score of 80. Predict his creativity score, indicating the precision of your prediction.

20. A college basketball team and its football team have height and weight measurements recorded. The correlation between height and weight is 0.8 among the basketball players, 0.7 among the football players, and −0.3 in the combined group of football and basketball players. Explain how this could be so. (*Suggestion:* Use a rough scatter diagram in your explanation.)

21. Ten students keep records of the time they spend, in addition to attendance at lectures, on their statistics course. At the end of the semester, these study times are compared with the grade-point average received for their overall mark. The data are as follows:

Student	Study time (hr/week)	Grade points (letter)
1	4.5	3.0 (B)
2	7.2	3.3 (B+)
3	1.5	1.0 (D)
4	5.0	4.0 (A)
5	3.8	2.7 (B−)
6	6.1	3.3 (B+)
7	5.5	3.0 (B)
8	2.0	2.0 (C)
9	4.5	4.0 (A)
10	3.5	2.3 (C+)

Student 3 visits his professor at the beginning of the next semester suggesting that the amount of studying he did, 1.5 hours per week, should have been enough to earn him a B grade, worth 3.0 grade points. The professor proposes that the student work out answers to the following exercises, which you should try, too! [For parts (a) and (b), you may assume that the regression method is applicable.]
(a) Estimate the grade points that a student typical of this group of 10 students would receive if he or she spent 1.5 hr per week of study time on the course. Indicate the precision of your estimate. (Precision is usually quantified by a measure of spread.)
(b) What proportion of students who receive a grade of B spent 1.5 hr per week or less of study time?
(c) How would you check the assumptions underlying the regressions in parts (a) and (b)? (Only words are required here; no calculations are necessary.)

22. The proud owner of a new car records the gasoline consumption for her first few weeks of driving. Each time she fills up the gas tank, she notes not only the gallons purchased but the number of miles covered since the last fill-up. As the car owner is also the proud owner of a new statistical calculator, she calculates the correlation coefficient between gallons for a fill-up and miles driven. The calculator produces the correlation of 0.2. Knowing that there is a very good relationship between gallons used and miles driven in most cars, the calculator is returned to its place of purchase for repairs or a refund. The salesman checks the correlation program and assures the purchaser that the program is correct. The purchaser wonders if she should return the car instead. It seems that the

gasoline consumption does not depend on the number of miles the car has been driven. Can you offer this distressed soul some explanatory words?

23. A real estate company wishes to study the relationship between the selling price of houses and the number of square feet of living space that the houses contain, in a certain town. Over the past year, 100 houses have been sold in the town, and their prices have averaged $90,000 and have an SD of $10,000. These houses have averaged 1800 ft^2 with an SD of 100 square feet. The correlation between price and square feet is 0.9.
   (a) Which is more typical of these houses: a 1900-ft^2 house that costs $100,000 or an 1800-ft^2 house that costs $100,000? Explain your answer with a diagram.
   (b) What is the average price of all the houses with 1900 ft^2 of living space?
   (c) Of the houses that cost $100,000, what proportion have more than 1900 ft^2 of living space?
   (d) If a house has 1900 ft^2 of living space, estimate the price of the house. Indicate the accuracy of your estimate.

24. A medical school wishes to evaluate the usefulness of its applicants' MAT (a medical-school admission test) scores in predicting success at medical school. From a file of 1000 graduates of the medical school, the students' medical-school grade-point average (GPA) and MAT score are recorded. The regression of GPA on MAT score has an r^2 value of 0.2. Is the MAT a poor indicator of GPA in medical school?

25. Do words that have the vowels e or u tend not to have the vowels a, i, or o?
   (a) Take a passage of 100 words or more from this book (provide a reference of your passage with your solution) and for each word in it record both X and Y, where

   $X:$ the number of vowels that are e or u

   $Y:$ the number of vowels that are a, i, or o

   Draw a scatter diagram of your results, indicating multiple points by plotting a multiplicity number instead of an × at the points' locations. Examine your scatter diagram visually to arrive at an answer to the original question.
   (b) Does your answer to part (a) indicate a special linguistic property of the English language, or is there another explanation?

26. It is proposed to study the relationship between the area of quadrilaterals and their perimeter. (Ns are closed four-sided figures, like a rectangle but possibly with unequal opposite sides.) Regression analysis of the area on the perimeter for 100 different quadrilaterals is proposed as a first step in unraveling this relationship. Comment on the usefulness of the regression approach to this problem.

27. If a two-variable scatter has an oval shape, is it possible that the homoscedasticity assumption is satisfied—or would the SD in vertical strips necessarily be greater in the center of the oval than at the left and right edges? Explain.

28. At a county fair, a weight guesser has set up a booth with a sign as follows: "For a mere 25 cents, I will guess your weight; if I fail to guess your actual weight to within 6 lb, you will win a $5.00 prize!" As a bystander, you record the guess and the actual weight for 100 adult male customers, and calculate the following summary statistics:

   Average guess: 150.3 lb
   Average actual: 149.5 lb
   SD of guesses: 15.0 lb
   SD of actuals: 14.0 lb

SD of (guess − actual): 3.0 lb

Correlation between guesses and actuals: 0.98

(a) What assumptions would be necessary to make valid the prediction of the actual weight from the guessed weight using the regression method? (These assumptions may be considered to be justified for the remainder of this question.)
(b) What proportion of the weight guesser's guesses are within 6 lb of the actual weight?
(c) What proportion of the bystander's guesses (using the regression method) are within 6 lb of the actual weight?
(d) Does the weight guesser make any money from the 100 customers? Estimate how much she won or lost.
(e) Explain the difference between the answers to parts (b) and (c).
(f) Suppose that the weight guesser guesses a weight of 180 lb for Mr. Big. Show that the regression method predicts an actual weight for this same man that is less than 180 lb.

29. Two supermarket employees repack oranges from a large crate into approximately 2-kg packages. Since the packages are to be sold at a fixed price per package, rather than by weight, the manager asks that the packages vary in weight as little as possible. To check this the manager samples five packages from each of the two employees, and the resulting weights of the packages are:

Employee 1: 2.05, 2.13, 1.95, 2.02, 2.20

Employee 2: 2.03, 2.10, 2.07, 1.99, 2.18

(a) Compute the range, standard deviation, and average absolute deviation (average deviation neglecting signs) for employee 1.
(b) For each measure of dispersion in part (a), state the advantage(s) and disadvantage(s) of each in the proposed application.
(c) Is a measure of dispersion an adequate summary for the supermarket manager, given his concern?
(d) Use regression, if possible, to predict the employee 1 weights from the employee 2 weights. If not possible, explain why not.
(e) What should be the manager's aim to maximize his profits on the oranges? How should he instruct the employees to do their job better?

30. A commuter records the time it takes her to drive to work, and the the time it takes to reach the halfway point (i.e., to traverse the first half of the distance to work). The data for one week is 30.0, 27.5, 29.0, 28.0, and 32.5 min for the whole trip and 10.5, 10.0, 10.0, 9.5, and 11.0 min for the half-trip respectively.

(a) What use could the commuter, a punctual employee, possibly make of this data?
(b) Provide some calculation details to support your suggestion in part (a).

31. A random sample of 20 houses from a municipality provides data on appraised value and square footage of each house. The data are summarized as follows:

	Appraisal price	Square feet
Average	$100,000	2000
SD	15,000	100
Correlation	$r = 0.8$	

(a) Predict the appraisal price of a 2500-ft² house.
(b) Provide an estimate of the precision of the estimate in part (a).
(c) Comment on the validity of your answers to parts (a) and (b); that is, if the survey were actually done as described, and the data resulted as shown, what reservation might you have about the regression-method procedures used in parts (a) and (b)?

Computer Projects

32. (Continuation of Problem 16.) Select five random samples of size 9, with replacement, from the 9-week data of Problem 16. For each of the five samples, compute the slope of the regression line for estimation of weekly sales from the space allotted. Compute the SD of this slope. Comment on the relevance of this SD to the estimation problem for the original sample.

33. (Continuation of Problem 17.) Generate 20 cases of two variables that each have a normal distribution and have a correlation of -0.5. Repeat for correlations 0.0, 0.1, 0.2, ..., 1.0. Plot scatter diagrams in each case. Use these plots to assist you in answering Problem 17. Add the point (3, 3) to each plot and recalculate the correlation coefficient. Summarize in your own words the effect that this one additional point has on the calculation of the correlation coefficient. What practical conclusions can you glean from these examples?

Descriptive Methods for Time Series

6

Time makes monkeys of us all.

Anonymous

Time reveals all things.

Erasmus (1465–1536)

Descriptive methods have been introduced for one-, two-, and even many-variable data sets. But even with one variable, there are data sets outside the scope we have covered so far. Time-series data sets are in this category. The crucial difference of time from many other variables is that observations close in time tend to be good predictors of each other, so two observations only have a little more information than one observation. This is in contrast to random samples in which knowledge of one observation of a variable tells almost nothing about any other observations of that variable.

Time-series methods are very widely used. A **time series** is an ordered sequence of values of a single variable in which the values are observed at successive times. World population in each of the years 1900–1985 forms a time series. The motor vehicle accident rate for drivers under 25 years of age, for each month during the period 1971–1980, is a time series of length 120.

Time series are studied for many reasons, but the commonest reasons are for forecasting and for control. We may wish to forecast world population or control accident rates. Anyone who could confidently and correctly forecast interest rates or exchange rates could make themselves fabulously wealthy. Controlling interest rates and exchange rates is an even tougher task. Both forecasting and control are multi-disciplinary tasks requiring a wide range of skills. The approach to time series in this text will be very modest indeed, and will only touch these central issues. The exposition here will be descriptive, and the principal goal will be to emphasize to the reader that time series are a category of data set that must be treated differently from other data sets.

Key words are seasonal effects, seasonal adjustment, smoothing, moving average, forecasting, trend, fit, residuals, pattern of residuals, and serial correlation.

6.1 EXTRACTION OF TRENDS AND SEASONAL EFFECTS

We will begin our discussion of time series with a look at monthly sales of new motorized passenger vehicles during the boom and bust years of 1981–1984. Table 6.1 shows these data in the second column. This column of numbers is not very revealing by itself—a scatter diagram is very helpful for exploratory study of a time series (see Figure 6.1). Note that the points are joined by line segments in chronological order. This clarifies the pattern of month-to-month changes. The annual cycle of car sales should be apparent: a dramatic rise for the first few months of the year, followed by a steep drop, then a smaller rise in the fall. Furthermore, when one looks at the years as a whole, one can clearly see a drop from 1981 to 1982, followed by rises in 1983 and 1984. It is not hard to imagine a bowl-shaped curve (perhaps a parabola) representing the underlying trend. If we were going to forecast the figures for the rest of 1984 and beyond, we would be foolish not to make use of these apparent trends. The technical problem we have is that the annual pattern and the four-year trend are mixed up together.

There are several ways to forecast time series like this one. Let us attempt to forecast the 1984 sales figures assuming only that we know the values from the 1981–1983 data. (This information would have been very useful to auto-industry planners in December 1983.)

How we can separate these trends? One simple way to extract the seasonal component is:

1. To reexpress each month's sales as a percentage of the year's average monthly sales

The total sales figure for 1981 is $8270 million, so the monthly average for 1981 was $689 million. January's sales were $514 million, so we can reexpress January 1981 sales as (514/689) × 100 percent = 75 percent. January's figure for 1982 is (388/578) × 100 percent = 67 percent and for 1983 is 56 percent. So we have 75, 67, and 56 percent representing the percentage of annual average monthly sales experienced in January. If we define the "seasonal effect" as a multiplier of the annual average monthly sales, these three percentages are trying to tell us what the seasonal effect is. But to get a single figure, we need a compromise between the three percentages. To keep things simple, let's just average the three numbers, yielding 66 percent as the January "seasonal" percentage.

Sec. 6.1 Extraction of Trends and Seasonal Effects

TABLE 6.1 MONTHLY SALES (MILLIONS OF DOLLARS) OF NEW PASSENGER VEHICLES, JANUARY 1981–APRIL 1984[1]

Month	Total passenger vehicles	Seasonally adjusted	Smoothed seasonally adjusted	Fit	Residual
Jan. 1981	514	779	—	—	—
Feb.	601	742	749	607	6
Mar.	834	725	730	839	5
Apr.	912	724	709	893	−19
May	828	679	691	843	−15
June	813	670	694	881	68
July	681	732	684	636	−45
Aug.	604	649	682	634	30
Sept.	612	665	649	597	−15
Oct.	626	632	667	660	34
Nov.	780	703	634	704	−76
Dec.	465	567	619	508	43
Jan. 1982	388	588	592	391	3
Feb.	504	622	591	479	−25
Mar.	646	562	584	672	26
Apr.	716	568	575	725	9
May	726	595	596	727	1
June	792	624	585	743	−49
July	500	537	580	539	39
Aug.	559	601	586	545	−14
Sept.	576	619	598	550	−26
Oct.	528	574	569	563	35
Nov.	572	515	579	643	71
Dec.	532	648	600	492	−40
Jan. 1983	421	638	638	382	−39
Feb.	508	627	667	540	32
Mar.	848	737	698	803	−45
Apr.	921	731	739	857	−64
May	913	748	740	903	−10
June	941	741	749	951	10
July	706	759	759	706	0
Aug.	723	777	747	695	−28
Sept.	655	704	783	720	65
Oct.	860	869	796	788	−72
Nov.	903	814	822	912	9
Dec.	642	783	832[a]	682[b]	50[c]
Jan. 1984	650		848[a]	560[b]	−90[c]
Feb.	807		863[a]	699[b]	−108[c]
Mar.	1085		879[a]	1010[b]	−75[c]
Apr.	1069		894[a]	1126[b]	57[c]

[a] Forecast based on trend of seasonally adjusted series.
[b] Forecast based on trend and seasonal effect.
[c] Forecast residuals (based on linear projection of trend).

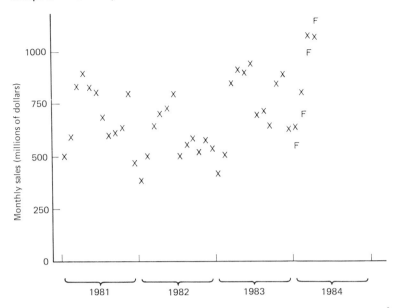

Figure 6.1 Dollar value of passenger vehicle sales, January 1981–April 1984.[2] X, actual; F, forecast.

The same method produces the percentages for the other months:

January	66%
February	81%
March	115%
April	126%
May	122%
June	127%
July	93%
August	93%
September	92%
October	99%
November	111%
December	82%

Now if January tends to be 66 percent of the monthly average, we can recover the seasonally adjusted time series by dividing the actual sales by 0.66. So the second step is:

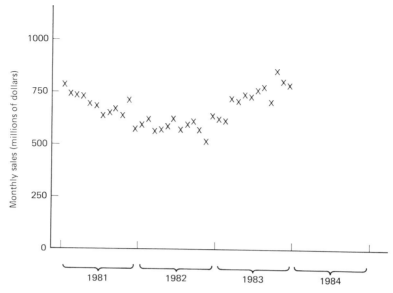

Figure 6.2 Dollar value of passenger vehicle sales, January 1981–April 1984, seasonally adjusted.

2. To divide all the "actuals" by the appropriate monthly proportion

January of 1981 becomes $514/0.66 = 779$, February of 1981 is $601/0.81 = 742$, and so on. These seasonally adjusted sales figures are displayed in Figure 6.2. The trend that you may have vaguely discerned in the raw data is quite clear in the seasonally adjusted series. Moreover, it has been produced by an objective method. This means that the method can be computerized, and this is an obvious benefit.

But we should return to our main purpose: forecasting the 1984 sales. It would be possible to draw, freehand, a smooth curve through the points in Figure 6.2, extend the curve into 1984, and then reapply the seasonal percentages to get our forecast. However, an objective method of producing the smoothed curve for Figure 6.2 is desirable. The method we shall use is called a **moving average.** To begin this process, we average the values, which are recorded in Table 6.1, for January, February, and March of 1981; this is 749; 749 is the smoothed value of the series for February. Next we average the figures for February, March, and April, and record the result, 730, as the March value of the smoothed series. This process is repeated until all the values are used, the last one being recorded at November 1983. The smoothing procedure we have used is a three-term moving average—it is also called a moving average of order 3. This series is tabulated in Table 6.1 in the third column. The curve is plotted in Figure 6.3.

So the third step in constructing the forecast of our original series is:

3. To use a moving average to smooth out the seasonally adjusted series, if necessary

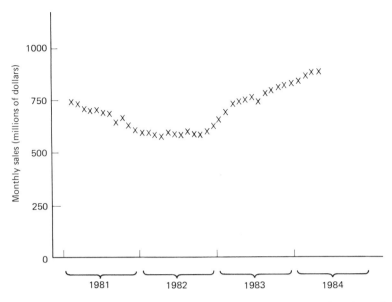

Figure 6.3 Result of applying moving average of order 3 to series shown in Figure 6.2.

The next step is:

4. To project the smoothed version of the seasonally adjusted series into the future

This is a bit tricky if we insist on an objective method, but only because it uses methods that are beyond the scope of this text. Perhaps we can be satisfied with a line projection, based on the 11 values of the smoothed series for 1983 (the moving average could not be calculated for December 1983). A regression line for these points is

$$Y = 647.2 + 15.44 \times X$$

from which one produces forecasts of the smoothed series as follows:

January 1984	848
February	863
March	879
April	894

(For this equation, $X = 1$ for January 1983, $X = 2$ for February 1983, and so on, so the forecasts are obtained by putting $X = 13, 14, 15,$ and 16.)

The last step in constructing the forecast is:

5. To apply the seasonality factors to the projections of the smoothed series

Sec. 6.1 Extraction of Trends and Seasonal Effects

That is

$848 \times 0.66 = 560$ is the projection for January 1984
$863 \times 0.81 = 699$ for February 1984
$879 \times 1.15 = 1010$ for March 1984
$894 \times 1.26 = 1126$ for April 1984

How does this forecast compare with the actual figures?

	Forecast	Actual
January 1984	560	650
February 1984	699	807
March 1984	1010	1085
April 1984	1126	1069

These forecasts may be evaluated by comparing them with the actuals in Figure 6.1. Certainly, a forecast that uses the seasonality adjustment is far superior to one that ignores it.

Note that the analysis just completed has produced two useful sets of numbers. One is the seasonal percentages themselves and the other is the forecasts. The process used to obtain the forecasts could be applied to the entire time series up to the end of 1983. The result is called the "fit" to the time series. It is calculated in Table 6.1, in the second-to-last column. This fit is graphed in Figure 6.4. It represents the best description we can get of the time series that is based on a forecast recipe. Note how similar this series is to the actual series of Figure 6.1. The net result of what we have done is to represent the data as the sum of two components:

$$\text{data} = \text{fit} + \text{residual}$$

Our aim has been to put the predictable part of the time series into the "fit" and hope that the residual is small enough that we can ignore it in our forecasts without being misled. We will discuss the evaluation of the residuals, and hence the fit, in Section 6.2.

Let us review the terminology introduced in this section. We began our analysis by calculating **seasonal effects:** a summary of the departure from the annual average at each point in the annual calendar. In the example, the departure was measured by a multiplicative factor. **Seasonal adjustment** of a time series is accomplished by removing the seasonal effects. The seasonally adjusted time series was subjected to a process called **smoothing:** the replacement of a time series by a new time series which is less changeable between successive times. One way to accomplish smoothing is to use a **moving-average** procedure: the replacement of a value in a time series by the average of a number of terms of the time series, selected from the same number of terms on either side of the time at the value replaced. The number of terms averaged

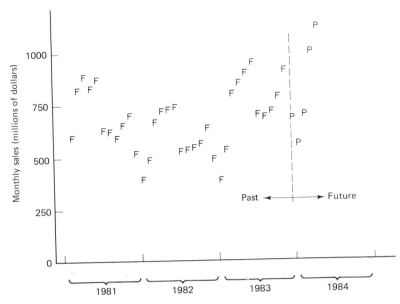

Figure 6.4 Fit to time series based on trend and seasonal effects. F, fit; P, projection (forecast).

for this moving average is called the **order** of the moving average. The calculation of a guess of future values of a time series is called **forecasting.** The smoothed long-run pattern of a time series is called the **trend.** The trend is usually considered to exclude any seasonal effects and any nonsmooth changes in the time series. Thus the smoothed, seasonally adjusted time series shown in Figure 6.3 is a bowl-shaped trend. The **fit** of the time series is the time series obtained by the application of the seasonal pattern to the trend.

Comments on the descriptive methods used

1. The replacement of the bowl-shaped trend by a straight line for 1983 was not a good idea, except that it provided an objective way to describe the forecast that was made. The straight line will be expected to forecast the trend too low, if the real trend we should be projecting is bowl-shaped. The forecast shows this: the January to March forecasts are too low. April may be anomalous—it may herald a change in the underlying process. It seems clear that the bowl-shaped trend cannot continue indefinitely; unfortunately, economic expansions have a habit of ending abruptly.

2. The suggestion that we replace all the complications of time-series analysis by regression procedures, that is, just to regress sales against year, is a bad suggestion. The phenomenon of serial correlation (which is assumed to be absent in the regression model) can lead to very poor fits using a regression approach. Exercise 5 at the end of this section should indicate why this is so. Serial correlation is discussed in detail in the next section.

3. We have shown how to fit the seasonal component in a time series. A seasonal effect is just one example of a cyclical effect. The problem of unraveling cyclical effects is much more difficult since we do not know what length of cycle to look for. Seasonal effects always have a 12-month cycle.

4. There is one trick for fitting linear or curved trends that is fairly simple when it applies. Let us consider two time series:

$$A: \quad 10, 13, 16, 19, 22, 25$$

$$B: \quad 100, 169, 256, 361, 484, 625$$

Let us "difference" each of these series: that is, subtract the first value from the second, then the second from the third, and so on. The results are:

$$\text{Diff } A: \quad 3, 3, 3, 3, 3$$

$$\text{Diff } B: \quad 69, 87, 105, 123, 141$$

At this stage it is obvious how to forecast series A, just add 3 each time to Diff A. The next few observations are 28, 31, 34, ... If we repeat the process on B, we get

$$\text{Diff Diff } B: \quad 18, 18, 18, 18$$

So the next few values of Diff Diff B are clearly 18, 18, 18, ... Therefore, Diff B can be forecast. The next few terms are 159, 177, 195, ... Thus the next few values of B itself are

$$625 + 159 = 784$$

$$784 + 177 = 961$$

$$961 + 195 = 1156$$

$$\text{etc.}$$

Now the only question is: When does a series have this nice property that when you difference it once or twice, the result is a string of constant values? The answer is that a linear trend will produce constant differences and a curved trend of a certain kind, called *quadratic,* will yield constant differences after the second differencing. We need not get into the details of quadratic curves: if the differencing seems to work, use it. In other words, if your trend seems, from "eyeballing" it, to be a smooth curve like a bowl or a cap, try differencing the series twice to see if the result looks fairly constant. If so, you can use this constant to rebuild the series in a systematic way, and this can be done outside the range of the data set. A forecast would be produced using this method.

In this section we have discussed methods for determining the fit and seasonality of a time series, in order to produce forecasts. In the next section we examine ways to assess the quality of time-series forecasts.

EXERCISES

1. Liquor industry sales include products from distilleries, breweries, and wineries. The following data report trends of each component of sales.[3]

	Value of output (millions of dollars)		
Year	Distilleries	Breweries	Wineries
1974	488	424	36
1975	500	476	44
1976	504	494	40
1977	567	587	49
1978	592	655	64
1979	617	747	72
1980	679	864	88

   (a) Forecast the 1985 figures for each component.
   (b) Forecast the combined total sales.
   (c) Forecast the percentage of the sales derived from wineries.
   (d) Which of the following two methods should be used to forecast the sales of wineries in 1985? Comment on your choice.

   1. Combine the forecasts from parts (b) and (c).
   2. Use the forecast from part (a) only.

2. The rate of therapeutic abortions, that is, the number reported per 1000 population, seems to be increasing.[4] Forecast the rates for the next five years. Comment on the reliability of your estimate.

Year	Rate
1972	1.8
1973	2.0
1974	2.1
1975	2.2
1976	2.4
1977	2.5
1978	2.7
1979	2.7
1980	2.7

3. Redo the forecast of the motor vehicle sales with a freehand smoothing of the seasonally adjusted data. Does your method produce better estimates? Do you think it usually would?

4. How good would a 12-month moving average be in forecasting the motor vehicle sales series? Discuss.

5. (a) Use ordinary linear regression to fit the total passenger vehicle sales time series of Table 6.1. Use only the 1981–1983 data for this.

(b) Compute SD_{pred}.
(c) Use the regression equation to forecast the sales values for January through April of 1984. Record the errors in these forecasts. Does SD_{pred} give a true indication of the size of the errors of forecast? (Compare the error with the forecast error achieved in the text.) Explain.

6.2 PATTERNS OF RESIDUALS

There is one other piece of information available from the analysis of the car sales data that we did not use. The difference between the "fit" series and the original series tells us about the anomalous monthly sales values. These differences are called **residuals.** Residuals from a time series are the differences between the fit and the actuals. Let us examine these now. See the last column of Table 6.1.

How can we tell if our forecast is a good one? The best way is to wait until the future happens, but the luxury of this option is seldom available in practice. We would like to check that we have done our best in fitting the data that we have at the time of the forecast, in our case based on the data up to December 1983.

The assessment of the fit can be done by examining the **pattern of residuals:** the predictability of each residual from previous residuals. Let us consider the residual series as a new time series. If we are able to detect some patterns in these residuals, our fit is capable of being improved. **Predictability of residuals** means that the fit can be improved. The reason is that the residuals represent the "errors" in our fit for the data we have as actuals. If these are predictable, we can adjust our fit and reduce these errors. The time series of residuals is shown in Figure 6.5.

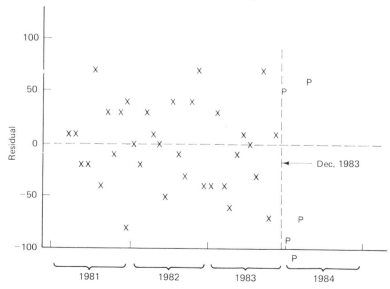

Figure 6.5 Residual time series.

The time series of residuals reveals an interesting phenomenon. The residuals seem to alternate between positive and negative values. This is a pattern, and we should be able to use it to improve our forecasts. First let us see how we can examine the pattern more carefully. The tendency for successive values of a time series to be on the same side of the overall average value is called **serial correlation.** It can be measured by the ordinary correlation coefficient. This is done by correlating the time series with itself, but one of the series is lagged by one time unit. If a time series is 1, 2, 3, 4, 5, the serial correlation is obtained by finding the correlation of the points $(1,2)$, $(2,3)$, $(3,4)$, and $(4,5)$. That is the pairs of values of X and Y that we would use to calculate the correlation of X and Y would be

Pair	X	Y
1	1	2
2	2	3
3	3	4
4	4	5

Let us now apply this idea to the data of Table 6.1. A scatter diagram of the residuals (of the residuals versus the lagged residuals) is shown in Figure 6.6. The correlation corresponding to this scatter is -0.40. In other words, when a value of the residual series is above its average, the following one tends to be below the average, and vice versa. It appears from the scatter diagram that to predict a residual based on the previous residual we should use a regression line with negative slope. The Y in the scattergram is indeed the subsequent residual corresponding to the X with which it is

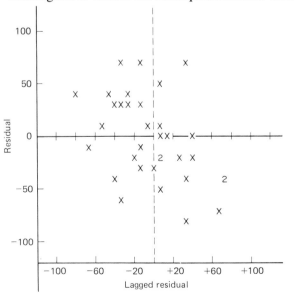

Figure 6.6 Scattergram revealing a negative serial correlation. The variable on the vertical axis is the residual series itself. The same series is shown, on the horizontal axis, but lagged by one time unit.

plotted. That is, we do want to predict the vertical-axis variable from the horizontal-axis variable.

The point of averages for the scatter in Figure 6.6 is not exactly zero, but it is nearly so; it is (−1.53, −2.82). Let us suppose that the prediction line goes through the origin (0, 0). Its slope is determined as usual by the correlation and the ratio of SDs. The slope in this case is

$$-0.40 \times \frac{38.4}{39.4} = -0.39$$

(Since the residual SD is almost the same as the lagged residual SD, the SD ratio will usually be very close to 1.0, so that the slope of this prediction line will usually be very close to r itself.) If a residual is, say, 50, we would predict the next residual to be -0.39×50 or about -20.

It was claimed that if the residuals have a pattern, the pattern should be exploited to improve the forecast. How does this work in this instance? The first residual we forecast was the one for December 1983. (We could not get this from the moving-average smoother since we are supposing, for the purpose of making the forecast, that we do not know the January value of the original series, and a moving average of order 3 would need this.) Our forecast was 50. We just showed that the residual time series suggested that the next residual should be about −20. Let us take our fit, which is 560 for January 1984, and assume that it would lead to a residual of −20. That means the actual would have had to be 580. So we use 580 as our improved forecast of the January sales. Next, we had a residual of −90 in our original forecast for January. This means that we now predict the February residual to be -0.39×-90 or about 35. So our improved forecast for February is $699 - 35$ or 664. Similarly, we get revised forecasts for March and April of 1052 and 1097. Are these really any better?

Month	Actual	Original forecast	Revised forecast	Improvement
Jan.	650	560	580	20
Feb.	807	699	664	−35
Mar.	1085	1010	1052	42
Apr.	1069	1126	1097	29

The improvement is not spectacular, averaging about 1 or 2 percent. However, the principle has potential. (Remember, also, that a curved projection of the smoothed, seasonally adjusted time series would certainly have improved things, but we chose not to use it for simplicity of exposition.)

For the record, the method we have used for taking advantage of the serial correlation in the residual is called *autoregressive modeling,* and the model we used was a first-order autoregressive scheme. However, these terms will not be used again in this text, nor will we be elaborating on these kinds of models.

If the examination of residuals does not turn up any serial correlation or other pattern, the reliability of the forecast is best represented by the SD of the residuals themselves. Since the average residual is usually zero or close to zero, the SD measures the typical size of deviations from zero, which is just the typical size of the residual itself. The reliability of the forecast is measured by the SD of the residuals once all attempts to improve the fit have ended.

The SD of the residuals from the original fit (let's assume that it was good enough to proceed) was $38.4 million. So our forecasts can be expected to be off target, typically, by about $38 million. This is actually a minimum typical error since when we move further into the future, we have less and less confidence in our forecast.

The negative serial correlation experience with the example is much less common than a positive serial correlation. A positive serial correlation makes the residual series move like a slow wave across the scatter diagram. An above-average value is often followed by another above-average value, and a below-average value is often followed by another below-average value. An example of a time series that has this property is the birthrate series alluded to in Chapter 1. We will take another look at these data now; see Figure 1.4.

The whole time series of birthrates is quite smooth: the fit determined by a moving average of low order would be almost indistinguishable from the original series. However, some investigators would be interested in inferring some long-term trends from this series. If it were hypothesized that birthrates follow a regular long-term cycle, some cyclical curve (such as a sine wave, or as some would say, a "sin" wave) might be fit to these data. The residuals to such a fit would show a marked positive serial correlation.

The examination of birthrates brings up another sticky problem in studying time series—the standardization of data so that the time trend will represent a time change that is meaningful. A time series of total numbers of births might be interesting for some purposes, such as planning school facilities, but in studying the reasons for the changes in the numbers of new births, it is better to express these numbers as rates. The problem is: What denominator is best?

The denominator for birthrates is the whole population, of both sexes and of all ages. Perhaps we should be looking at the rates for women of a specified age. The birthrate might be deflated by a bulge in the nonreproductive segment of the population. Table 6.2 gives an example of the birthrate data expressed as "fertility rates." This series is slightly different from the raw birthrate data, although the same general trends are there. It might be postulated that women are delaying the bearing of children to later ages, yet the fertility rate for older women has declined even more than for younger women aged 20 to 24.[5] The echo boom that has been predicted for some time as a result of the large number of parent-aged people in the population during the early 1980s, seems to have been modulated by the declining fertility rates.

The main points to remember from this section are:

1. Examination of residuals will tell us about the adequacy of the fit, and therefore of the forecast, of a time series.
2. When residuals have a pattern, this pattern can be exploited to improve the fit and the forecast.

TABLE 6.2 FERTILITY RATES FOR WOMEN AGED 20–24, EXPRESSED AS THE NUMBER OF LIVE BIRTHS PER 1000 WOMEN OF THAT AGE, 1921–1983[5]

Year	Rate	Year	Rate	Year	Rate
1921	165	1942	145	1963	226
1922	155	1943	147	1964	213
1923	145	1944	143	1965	189
1924	145	1945	143	1966	169
1925	139	1946	170	1967	161
1926	140	1947	189	1968	153
1927	140	1948	181	1969	148
1928	140	1949	182	1970	143
1929	140	1950	181	1971	134
1930	143	1951	189	1972	120
1931	137	1952	201	1973	118
1932	130	1953	208	1974	113
1933	118	1954	217	1975	113
1934	113	1955	218	1976	110
1935	113	1956	222	1977	108
1936	112	1957	227	1978	103
1937	114	1958	227	1979	102
1938	121	1959	234	1980	100
1939	120	1960	234	1981	96.7
1940	130	1961	234	1982	95.4
1941	138	1962	232	1983	92.4

3. Careful thought should be given to the way a variable is to be standardized so that the information from the time-series analysis will be relevant to the question of interest.

Finally, a word of warning. Much of this book is of an introductory nature, but this chapter on time series is especially introductory! If you plan to analyze a time series (other than the ones in this text), and the outcome is important to you, discuss it with someone experienced in the field. An "expert" in time series should have taken several courses devoted exclusively to this subject, or have had in-depth experience with time-series analysis over a number of years. Nevertheless, the principles discussed should help you to understand the basis of the time-series analyses of others, whether or not they are expert.

EXERCISES

1. The monthly consumption of soft drinks is seasonal, and the production of soft drinks is also. The following data give volumes of soft drinks produced during 1978–1980, in millions of gallons.[6]

Month	Volume	Month	Volume	Month	Volume
Jan./78	23.6	Jan./79	26.1	Jan./80	25.7
Feb.	24.1	Feb.	25.7	Feb.	25.7
Mar.	24.4	Mar.	24.7	Mar.	27.6
Apr.	27.7	Apr.	29.6	Apr.	31.9
May	29.1	May	36.0	May	29.7
June	37.5	June	36.9	June	37.6
July	34.1	July	38.8	July	35.8
Aug.	38.4	Aug.	40.3	Aug.	36.9
Sept.	29.6	Sept.	36.3	Sept.	38.5
Oct.	29.3	Oct.	34.0	Oct.	29.1
Nov.	28.4	Nov.	33.1	Nov.	21.8
Dec.	26.0	Dec.	29.9	Dec.	31.5

   (a) Forecast the production volumes for each month in 1981.
   (b) Determine the serial correlation of your residuals, and improve your forecast if possible.
   (c) How accurate is your forecast likely to be?
2. Exchange rates for the Canadian dollar into U.S. dollars declined markedly over the period 1964–1984:[7]

Year	U.S. dollars for one Canadian dollar
1964	107.98
1965	107.99
1966	107.60
1967	107.90
1968	107.75
1969	107.68
1970	104.40
1971	100.98
1972	99.05
1973	100.01
1974	97.80
1975	101.73
1976	98.61
1977	94.03
1978	87.70
1979	85.36
1980	85.54
1981	83.40
1982	81.03
1983	81.14
1984	76.50[a]

[a] Approximate.

(a) Smooth this time series using a moving average of order 3, and forecast the exchange rate over the next five years.
(b) How closely does the moving average approximate the series itself?
(c) How does the decade 1964–1973 differ from 1974–1983 in this exchange-rate series?

6.3 ANSWERS TO EXERCISES

Section 6.1

1. (a) The data seem to follow a fairly linear trend, especially since 1976. An eyeball fit, with the help of a ruler, produces forecasts of $860 million, $1254 million, and $152 million, respectively, for 1985.
   (b) A similar method but using only the summed sales figures gives $2.37 billion.
   (c) 7.3 percent, using the same method, but on the series of percentages of the market for wineries.
   (d) Using parts (b) and (c), we compute 7.3 percent of $2.37 billion or $173 million. This is a bit higher than the figure of $152 million from part (a). Which one is better?

   The real issue is whether the beer or spirits trends are predictive of the wine trend—using the market share and total market is a way of relating wine sales to the other sector's sales. In this case it is expected that wine and other alcoholic beverages are to some extent substitutes, so method 1 should be better.

   The forecast for method 1 is higher than for method 2, 172 versus 152. This is because linear increases in both market share and the total market are equivalent to quadratic increases in the wine-sector market. The linear projection of the wine-sector sales would probably underestimate future growth of this market, since the market share and the total market are apparently increasing linearly. But the quadratic nature of the wine-sector sales is not easy to identify from the wine-sector sales alone.

2. Any forecast must be based on an assumption of a continuing trend. In this series it is not clear what the trend is—a leveling of the abortion rate or a continued increase. One could project the next five years' rates to be 2.7, but there is only weak evidence to support this. The forecast would be unreliable.

3. Yes, it would be easy to improve on the regression fit that was used in the discussion of the example. Straight lines are quite restrictive for fitting trends; moreover, there is no reason to think that the relationship of sales with time is linear. The series of seasonally adjusted data appears curved, as noted previously. So the freehand fit will probably produce better estimates.

   There is a problem with freehand fits, however. We have a tendency to assume that patterns in data will persist into the future, even when these "patterns" have been caused by transient and inexplicable influences. If we look too closely at the data, we will always find a "pattern" of some sort. The statistical strategy to avoid being misled is to insist that any patterns used will be simple ones—such as linear or quadratic trends, or the smoothed fit produced by moving averages.

   In the car sales data, the trend appears to be quadratic, so a freehand curve would be a better fit than the regression line. The curve seems quite well established by the data.

216 Chap. 6 Descriptive Methods for Time Series

4. This 12-month moving average would be a good way to estimate the long-term trend, and it is automatically seasonally adjusted. However there are two problems involved in using it for forecasting the car sales. One is that for the last six months of 1983 (assuming that we are only using the 1981–1983 data to fit the trend), no moving-average value is possible, since we assume ignorance of January 1984 and beyond. So the trend is unknown at just the times that are most important for forecasting. The other problem is that for short-term forecasting, the seasonality is an important part of the forecast; we do not want to eliminate it from the forecast.

5. SD_{pred} would be too small for indicating forecasting errors if the regression line itself is used to do the forecasting. It is too large to indicate the errors of forecasting that would result by using the methods of this section. So the answer to the question in part (c) is "no."

Section 6.2

1. (a) Use the procedure outlined in Section 6.1.
   (b) The serial correlation of the residuals is 0.65.
   (c) Use the SD of the residuals.
2. (a) Estimates should increase slightly to about $0.80, in view of the curved fit.
   (b) Use the SD of the residuals.
   (c) The fit is declining at an increasing rate during 1964–1973, and at a decreasing rate during 1974–1983.

6.4 NOTATION AND FORMULAS

A time series for which the time parameter takes values corresponding to regular time intervals is denoted $\{X(t): t = 1, 2, 3, \ldots\}$, where $X(1)$ is the value of the time series at the first position (the earliest time).

The serial correlation of lag k is the ordinary correlation between $X(t)$ and $X(t + k)$. This is sometimes called the *autocorrelation for lag k*.

The simplest model for serial correlation of residuals $X(t)$ is $X(t + 1) = c \times X(t) + e(t)$, where c is a constant and $e(t)$ are independent errors. This model is called a *first-order autoregressive scheme*.

The moving-average process of order $2k + 1$ can be represented symbolically as

$$\hat{X}(t) = \frac{1}{2k + 1}[X(t - k) + X(t - k + 1) + X(t - k + 2) + \cdots + X(t + k)]$$

$\hat{X}(t)$ is the symbol for the fit. So

$$X(t) = \hat{X}(t) + e(t)$$

where $\{e(t): t = 1, 2, 3, \ldots\}$ is the residual time series of, hopefully, independent errors.

6.5 SUMMARY AND GLOSSARY

1. A time series may contain a seasonal or other cyclical component, a trend, and also residual variation that is not accounted for by cyclical patterns or trends. Each component must be estimated if useful forecasts are to be made.
2. After a first attempt to identify a trend and a cyclical component, if the residual series exhibits serial correlation, this can be used to improve the fit, and usually the forecast as well.
3. When a time series is replaced by a moving-average series, the result is smoother than the original. If we believe that the process we are describing is "smooth," the moving average may be a better description of it than the original time series.
4. Time series with a positive serial correlation move in slow waves. Those with a negative serial correlation move in a sawtooth-like pattern.
5. The residuals to a fit of a time series indicate how reliable the forecast is likely to be, assuming that the processes underlying the time series do not change. The SD of these residuals is a summary of the residual size.
6. There is often more than one way to standardize a time series: care must be taken to choose a standardization procedure suited to the purpose of the description.

Glossary

Seasonal effects are a summary of the departure from the annual average experienced at each point in the calendar. (6.1)

Seasonal adjustment of a time series is accomplished by removing the seasonal effects. (6.1)

The seasonally adjusted time series was subjected to a process called **smoothing:** the replacement of a time series by a new time series which is less changeable between successive times. (6.1)

Moving-average procedure: the replacement of a value of a time series by the average of a number of terms of the time series, selected from the same number of terms on either side of the time at the value replaced. The number of terms averaged for this moving average is called the **order** of the moving average. (6.1)

The calculation of a guess of future values of a time series is called **forecasting.** (6.1)

The smoothed long-run pattern of a time series is called the **trend.** (6.1)

The **fit** of the time series is the time series obtained by the application of the seasonal pattern to the trend. (6.1)

Residuals from a time series are the differences between the fit and the actuals. (6.2)

The **pattern of residuals** is the predictability of each residual from previous residuals. (6.2)

The tendency for successive values of a time series to be on the same side of the overall average value is called **serial correlation.** It is measured by the ordinary correlation of the time series with a lagged version of itself. (6.2)

PROBLEMS AND PROJECTS

1. Winning times for the 100-meter run at the Olympic games are noted below:[8]

Year	Winner	Time (sec)
1896	Thomas Burke, United States	12
1900	Francis Jarvis, United States	10.8
1904	Archie Hahn, United States	11
1908	Reginald Walker, South Africa	10.8
1912	Ralph Craig, United States	10.8
1920	Charles Paddock, United States	10.8
1924	Harold Abrahams,[a] Great Britain	10.6
1928	Percy Williams, Canada	10.8
1932	Eddie Tolan, United States	10.3
1936	Jesse Owens, United States	10.3
1948	Harrison Dillard, United States	10.3
1952	Lindi Remigino, United States	10.4
1956	Bobby Morrow, United States	10.5
1960	Armin Hary, Germany	10.2
1964	Bob Hayes, United States	10.0
1968	Jim Hines, United States	9.9
1972	Valeri Borzov, USSR	10.14
1976	Hasely Crawford, Trinidad	10.06
1980	Allan Wells, Great Britain	10.25
1984	Carl Lewis, United States	9.99

[a] This is Harold "Chariots of Fire" Abrahams.

Predict the time of the winner in the year 2000. (*Hint:* Take logs first. Also, consider why the winners since 1972 have done no better than the winners of 1964 and 1968.)

2. Health expenditures per person increased in Canada in the 1970s, but health expenditure as a proportion of personal income decreased:[9]

	Health expenditure	
Year	Per capita expense	Percent of personal income
1970	$295	9.4
1971	325	9.6
1972	350	9.3
1973	400	8.9
1974	450	9.0
1975	550	9.1
1976	600	9.1
1977	675	9.0
1978	725	8.9
1979	800	8.9

(a) Forecast both series and comment on the reliability of your forecast.
(b) Do you think the similar data for 1960–1969 would help you with this forecast? Discuss.

3. Alcohol consumption has been declining from a relatively high level in France, but has increased from a relatively low level in the United States:[10]

	Consumption (liters per person)	
Year	France	United States
1950	18.3	5.7
1955	20.0	5.3
1960	19.1	5.5
1965	18.3	5.9
1970	17.3	7.0
1975	17.2	8.8
1980	15.6	9.7

When will the United States and France have the same consumption of alcohol? Justify your prediction.

4. The table of random digits in Appendix 3 can be used to generate an artificial time series as follows:
   (a) List 50 random digits.
   (b) Form a new series of length 50 having only $+1$ or -1 at each position, by replacing 0, 1, 2, 3, or 4 in the original series by -1, and 5, 6, 7, 8, or 9 by $+1$. Now form partial sums as shown in the following example:

   $$3, 5, 9, \quad 2, \quad 3, \quad 3, \quad 1, \quad 4, \quad 9, \quad 7, \quad 8, 7, \ldots \quad \text{random digits}$$
   $$-1, 1, 1, \quad -1, \quad -1, \quad -1, \quad -1, \quad -1, \quad 1, \quad 1, \quad 1, 1, \ldots \quad \text{new series}$$
   $$-1, 0, 1, \quad 0, \quad -1, \quad -2, \quad -3, \quad -4, \quad -3, \quad -2, \quad -1, 0, \ldots \quad \text{partial sums}$$

   (c) Compute the serial correlation of this partial sum series.
   (d) Forecast the next 10 values of the partial sum series.
   (e) Compare your forecast with a continuation of the partial sum series, computed just as were the first 50 values. (But the continuation need only be done for 10 values.)
   (f) Is your partial sum series any different in character than a real series of share prices or interest rates would be? Might you be misled by the apparent regularity in a real time series to assume that its future movements (up or down) could be predicted accurately? Express in words the moral of this story. [*Note:* These partial sum series are called "random walks." The term refers to the partial sum as the net forward movement of a person taking steps forward ($+1$) or backward (-1) "randomly."] (This problem is continued as a computer project in Problem 7.)

5. The following series is the number of days per year from 1960 to 1982 in which one or more tornados were reported in Saskatchewan.[11] Forecast the 1985 figure.

   3, 4, 5, 9, 10, 3, 4, 2, 8, 9, 7, 10, 4, 15, 19, 28, 10, 18, 12, 16, 11, 4, 10

6. Discuss the implications of the trends apparent in Figures 1.4, 1.5, and 6.7 for population projections for 1984 and beyond.

Computer Project

7. (Continuation of Problem 4.) Longer random walks can be done quite easily using a computer package, especially if the "partial sum" feature is available. Generate 10 random walks of length 200. Describe the patterns that occur—if possible, categorize them into groups.

Sampling and Probability

Certainty generally is illusion, and repose is not the destiny of man.

Oliver Wendell Holmes (1841–1935)

I know of scarcely anything so apt to impress the imagination as the wonderful form of cosmic order expressed by the "Law of Frequency of Error." The Law would have been personified by the Greeks and deified, if they had known of it. It reigns with serenity and in complete self-effacement amidst the wildest confusion. The huger the mob and the greater the apparent anarchy, the more perfect is its sway. It is the supreme law of Unreason.

Francis Galton (1822–1911)

In Chapters 1 to 3 some mention was made of the usefulness of sampling; one can study a selected part of something and infer properties of the whole. Chapters 7 to 9 discuss the implications of random sampling; Chapter 7 discusses the tools we need to talk about randomness; Chapter 8, the method of generalization from a sample to a population; and Chapter 9, the methods for using samples to decide about the truth of claims concerning the population.

The method of sampling determines the method of making allowance for sampling variation and ultimately determines the method of generalization. Random sampling methods require probability methods to describe the link between the part and the whole. In this chapter we deal with the details we need about random sampling and probability.

Why are we abandoning further consideration of descriptive methods, which was the subject of Chapters 4 to 6? Although descriptive methods are very useful for the summary and exploration of data, there are many important questions that cannot be answered by these methods. Often when we are describing data, our real purpose is to learn something about the source of the data, not merely to describe the data itself. The logical connection between a data set and its source involves the notions of probability and sampling.

Key words are inferential statistics, descriptive statistics, statistical model, concrete population, hypothetical population, probability, probability distribution, experiment, trial, outcome, way, event, random, addition rule, multiplication rule, complementarity rule, independent, incompatible, binomial experiment, binomial distribution, estimation, parameter, statistic, expected value, standard error, sampling distribution, and correction factor.

7.1 INTRODUCTION TO INFERENTIAL STATISTICS

As a first example, consider the survey of higher education intentions of twelfth-grade students introduced in Chapter 1. The students were asked: "Do you plan to attend a postsecondary institution sometime in the future?" Fifty of the 2000 students are selected to respond to the question, and of the 40 who do respond, 65 percent answer "yes." Our descriptive summary of this information should certainly include the figure 65 percent, and the fact that this is based on the responses of 40 of the 50 registrants questioned will have some relevance as well (assessment of response bias). The issue to consider now is: Are we really interested in describing the 50 selected students, or are we interested in describing the population of twelfth-grade students? Clearly, the latter is the more relevant goal. The population of students is the focus of our interest, and this is the "source" of the data comprised of the 40 responses. The techniques used to infer information about the source of a data set, based on the data set itself, are the techniques of **inferential statistics,** in contrast to the techniques of **descriptive statistics** covered in Chapters 4 to 6. The category of statistical methods called descriptive statistics is made up of methods for describing the data set itself; but these data are only a subset of the population.

For inferential statistics, we must be concerned with how a sample is obtained from its source, and what this implies about the link between the sample and its source. The "population" is the technical word used for the "source" of data. Inferential statistics concerns methods for learning about populations from samples.

7.2 PROBABILITY MODELING

7.2.1 Statistical Populations

There are two kinds of populations used in statistical contexts: concrete populations and hypothetical populations. This section clarifies this distinction and describes how both kinds of population may be treated similarly in a statistical model. As usual, we begin with some examples.

In a political opinion poll, such as the ones that made George Gallup famous, the population of interest is usually the population of voters. We can conjure up the image of a large crowd of people, the population, from which a smaller crowd, the sample, is to be selected. We can do the same thing with the registrants from Aeio University. Populations that exist in the real world are called **concrete populations**—they are tangible collections of objects as opposed to figments of the imagination.

Voters and registrants constitute concrete populations. It may be surprising that any other kind of population can have practical value. But consider the following example.

One of the older arcade games consists of a driver's seat, a brake pedal, and a video display unit portraying a changing road scene, with periodic hazards, which require the driver to apply the brake. An electronic circuit establishes, for each hazard, the driver's reaction time: the time from the appearance of the hazard to the time of application of the brake. This is measured in thousandths of a second, so the reaction time for a driver will usually be different for each hazard. Now suppose that we want to estimate *the* reaction time of a certain driver. A reasonable procedure would be to average the driver's reaction times over, say, 10 hazards. If this whole test were to be repeated, the new average we generate would very likely be different from the old one. So we do not have *the* reaction time yet, but we may feel that we have an estimate of it.

The notion that there is a constant to estimate, which the sample never quite produces, leads us to devise a model in which this notion can be made precise. We can describe the constant as a property of a population from which the observed reaction times are a sample. However, this population is not concrete: it is one we imagine for our own convenience. This kind of population is called a **hypothetical population:** a population that is hypothesized for the convenience of describing a source of data. We do this so that we can describe without ambiguity what it is we are assuming about the sample, and what it is we are estimating. This hypothetical population will be a useful device if we can describe the variation in reaction times as variation that would arise in samples from the hypothetical population. The idea is to treat "samples" of measurements in the same way that we treat samples selected from concrete populations.

Under measurement conditions, it is reasonable to assume that the method of selection from a hypothetical population is simple random sampling; this sampling technique was defined in Chapter 2. The unknown constant that we are trying to estimate might reasonably be described as the population average. If we took a very large simple random sample, the sample average should be close to this constant. The description of the observation process in terms of a population and a sampling method is called a **statistical model.** It "models" the way a sample is produced.

The technical advantage of a statistical model is that we can make unambiguous statements, based on mathematical analyses, about the accuracy of the sample average as an estimate of a real-world characteristic. This advantage will be also a practical advantage when the model produces variability in the sample which mimics the real-world mechanism generating the data.

The notion of a constant reaction time that is the average of a hypothetical population requires further explanation. The hypothetical population for which the average is of interest is somehow defined by the measurement process itself. To describe how the measurement process determines a population, statisticians have introduced the idea of *long-run relative frequency*. If we imagine a subject producing thousands of reaction times, without any lasting trend in performance (in particular, the subject is not allowed to tire, nor the equipment to wear down), the histogram of these reaction times would portray the long-run relative frequencies of each interval of values. The statistical population for the experiment is the long list of reaction times, and for analytical purposes we can describe the population simply and completely by the histogram itself. The constant of interest is then the average of this list; we will often refer to this average as the "average of the histogram."

Statistical populations are usually populations of numbers, rather than populations of people, animals, or items. The set of numbers comprising a statistical population is usually described by stating its distribution (think of a smoothed histogram as a distribution). A characteristic of a statistical population, such as an average, is called a **parameter** of the population. Because a distribution is so often used to describe a statistical population, it is usual to refer to "parameters of a certain distribution" instead of "parameters of a population having a certain distribution."

We have described two very different types of statistical populations: concrete populations (voters) and hypothetical populations (reaction times). The analysis of samples for both situations is identical, but the process of modeling is quite different. In the next section we describe the requirements of the sampling process that make the sample amenable to standard statistical procedures. Figure 7.1 should help you to visualize the sampling process.

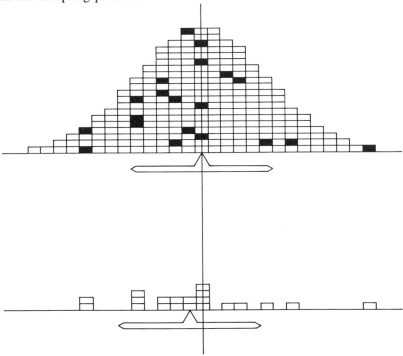

Figure 7.1 Schematic diagram showing relationship of a population and a particular random sample. Each element in the population has an equal chance of being selected into the sample. The fulcrum stand represents the average and standard deviation.

7.2.2 Variation in Samples

A woodlot owner who wishes to sell his trees may wish to assess the board feet of lumber that could be produced now if all the trees were cut down. Rather than do a comprehensive tree-by-tree assessment of the whole woodlot, he selects a sample of trees, assesses these, and then scales up these estimates to apply to the whole woodlot. The totality of trees in the woodlot is the population, and it is certainly a concrete population. There will be some variation in the board feet of lumber available from the various trees in the sample just as there is in the population. In fact, if there were very little variation among the sample trees, we would suspect that either the sample was not selected at random or that all the trees in the woodlot were quite similar. If we insist that the sample is a simple random sample, we expect that the variability in the sample should tell us whether the population is homogeneous or heterogeneous with respect to the available lumber per tree. This in turn will tell us whether or not our estimate will be precise.

Suppose that we are to use a sample of 10 trees for our estimate. If the trees in the woodlot are all close to the same size, all 10 sample values will be about the same. We would have quite a good estimate of the total board feet of the woodlot in this instance, since the sample trees will almost certainly be representative of this homogeneous population. On the other hand, if the size of the trees varies a great deal in the woodlot, our sample of 10 could yield an estimate of the total board feet that is much smaller than is typical of the woodlot, or much larger. The important principle here is that *the variability in the sample gives us a hint of how representative the sample is likely to be*. The sample not only provides estimates concerning the population, but also tells us how good its estimates are. This useful property is a consequence of random sampling procedures, which include in particular simple random sampling. See Figure 7.2.

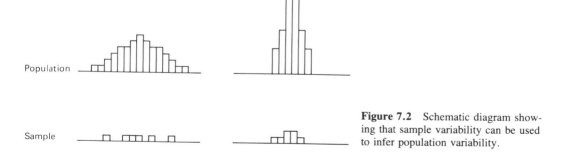

Figure 7.2 Schematic diagram showing that sample variability can be used to infer population variability.

Simple random sampling has two properties to recommend it:

1. Simple random samples are unbiased in the sense that they have correct averages in the long run (see Chapter 2).

2. Samples produced by simple random sampling provide information about how precise the sample averages are likely to be.

In the case of a hypothetical population the same properties of simple random samples hold. The only new difficulty is in guaranteeing that the sample is random. With reference to the example, how can we be sure that the 10 reaction times are values typical of a simple random sample of 10 from the population of all possible reaction times? Of course, the answer is that we cannot, but under certain conditions the assumption is reasonable. If the sequence of reaction times followed some trend, such as might be caused by improvement with practice, the sample could not be treated as if it were a simple random sample. If the hazards that the "driver" had to detect became more difficult to spot, this would again invalidate the simple random sample assumption. These complications require more sophisticated models for the observed variation. But when these complications are absent, we can use the simple random sampling model in this new setting of measurement. The properties of variation in the measurements can then be inferred from the sample, just as the variation in the concrete population can be inferred from its sample. In the measurement situation, the population is always a hypothetical one.

To decide whether the simple random sampling model for measurement error is reasonable in a particular application, one should consider the following characteristics of simple random sampling:

1. The population does not change during the time the sample is selected.
2. Each case in the population has an equal chance of being selected into the sample.
3. Each set of cases in the population has an equal chance of being selected into the sample.

The difference between properties 2 and 3 can be demonstrated as follows. Consider a population consisting of four elements A, B, C, D. We can select a simple random sample of two to get AB or AC or AD or BC or BD or CD, each being equally likely to be chosen. Now suppose that we select one item at random from among A and B, and one at random from among C and D. This latter sample still has property 2, but not property 3. The samples AB and CD cannot be chosen by the second method, so all subsets of A, B, C, D are not equally likely.

What sorts of measurement procedures have these characteristics? The first property holds when the measurement is repeated under the same conditions. The second property would not be satisfied unless the more typical measurements were specified to appear more often in the population than the less typical measurements. We must keep this in mind when we are considering reasonable histograms to describe our population. The third property implies that a measurement must not depend on previous or future measurements. If a measurement error tends to persist in subsequent measurements, the latter property would be contradicted. For example, time series are usually not well modeled as simple random samples—they require more complex modeling.

7.2.3 Sampling Models

The most basic random sampling models are those which have the property that all possible samples, of a given size, have the same chance of being selected. Sometimes such samples are selected **with replacement** and sometimes **without replacement.** This most basic random sampling model can always be visualized as a barrel of lottery tickets, and the sample consists of tickets drawn one at a time from the well-mixed barrel containing the population of tickets (see Figure 7.3). Usually, when a ticket is drawn, it is not replaced in the barrel before the next ticket is drawn—this is sampling without replacement. But if the sampling is done with replacement, the same ticket could, possibly, be selected more than once.

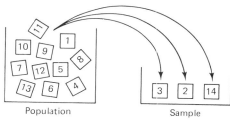

Figure 7.3 Schematic diagram of a statistical model for simple random sampling.

Simple random sampling is taken to mean that the sampling is without replacement. However, when the population is very large, the chance of selecting the same item more than once is negligible, so in this case random sampling with and without replacement produce the same kinds of samples. In such a case we call the sampling "random sampling." In the less common situation where random samples are selected from a small population (where "small" in this context means a population that is less than 20 times as large as the sample), and where the sampling is without replacement, some special adjustments to the simple theory are needed: these are discussed in Section 7.4.3.

The tickets in the sampling models discussed above may have a measurement written on them, or a code indicating a quality, such as sex or color. If this seems a little artificial, remember, it is supposed to be a model, not the real thing. Its utility rests on the fact that it is simpler than the reality it represents.

Once the sampling model is specified, we should be able to predict the kinds of samples we will get. This is the probability part of this chapter. If the sampling model is specified except for the population itself, we may still hope to infer what the population is like since we will have a sample from it generated by a known mechanism. This process is called estimation and is the subject of Chapter 8.

EXERCISES

1. Speedy George is employed by a messenger service to pick up and deliver documents from the service's customers. Before asking for a raise, Speedy decides to record the delivery times from source to destination for all of his business on a particular day. Describe the statistical population of which the recorded times are a sample.

2. The length of grass in a residential lawn is measured each day during a 30-day period in the summer. Should these measurements be treated as a simple random sample from a population if the aim is to describe typical grass lengths during the period?
3. At a peach canning factory, incoming truckloads of peaches are sampled by an inspector who selects the first five peaches to fall onto the conveyor belt. The diameter of the peaches is measured. Might it be reasonable to treat these five peaches as if they were a random sample from the truckload?
4. A sprinter is timed over a 100-m course three times on a particular day. Discuss the applicability of the three times as representative of the time that might be experienced by this sprinter one week hence in an actual race.
5. Ten sprinters are each tested three times as in Exercise 4. Of what population might the 30 measurements be considered a sample?
6. High school graduates in a particular school district are to be assessed for logical skills by a standard test. Describe how this study could be executed so that a simple random sampling model would be an appropriate model for the outcomes of the test.
7. (a) A grocer weighs an apple 10 times on scale A. The 10 weights are all different. What is the population from which the 10 measurements may be considered a random sample?
   (b) A grocer weighs 10 apples once each on scale B. Assuming that this weigh scale (scale B) is perfectly accurate, what is the population from which the 10 measurements may be considered a random sample?
   (c) A grocer weighs 10 apples once each on scale A. What is the population from which the 10 measurements may be considered a random sample?
   In each case indicate whether the population is hypothetical or concrete.

7.3 DETERMINATION OF PROBABILITIES

7.3.1 The Notion of Probability

To demonstrate that a coin is "fair," a gambler demonstrates that the coin turns up "heads" 50 times and "tails" 50 times in 100 tosses of the coin. The gambler wants to reassure us that the coin has no tendency to come up one side more than the other. Some people might be comforted by this evidence and join the gambling game using this coin. But in 100 flips of a fair coin, the gambler would need a great deal of luck, or skill, to produce *exactly* 50 heads and 50 tails. It is usual to have between 45 and 55 heads, but exactly 50 is not usual. In fact, if we rule out the possibility of controlling the flip, this happens about once every 12 times the 100-flip experiment is tried. Our intuition would have led us astray here. We need more careful language to describe events such as coin flips, and more generally for events that exhibit unexplained variation.

There is no "explanation" of the particular outcome of the flip of a coin, no reason to expect one or the other outcome. Thus the sequence of outcomes exhibits *unexplained variation*. What is the sampling model for this variation? The first thing to decide is the definition of the population. If we think of the population as consisting of two items, a "head" and a "tail," and we have in mind the usual kind of sampling, that is, simple random sampling, we run into a problem: there are no more items to sample by the third flip. Simple random sampling is sampling without replacement. With such a population, it would be impossible to get a sequence such as (head, head) for the first two flips. However, this suggests that we should fix the model, because the outcome (head, head) certainly will occur on occasion. There are two ways to fix the model, and we shall have occasion to use both of these devices.

One way is to define the population to be very large but comprised of heads and tails according to a certain relative frequency: 0.50 heads and 0.50 tails in this case would correspond to the assumption that the coin is fair. A random sample of 100 from this population would have the characteristics of real sequences of flips of a fair coin.

Another way is to sample with replacement; by this we mean that a lottery type of selection is used but the ticket is replaced after each draw. Thus we can model the sequence of outcomes of flips of a coin as a sequence of draws, using random sampling with replacement, from a population consisting of just two items, {H, T}. Such sequences would be indistinguishable from those generated by the large population using random sampling with or without replacement.

This second variant of the usual sampling model has been introduced because it is a little more convenient for describing some elementary notions of probability. To take advantage of this variant, it is important to realize that *simple random sampling from an infinite population can sometimes be exactly modeled as random sampling with replacement from a small population*. For example, in selecting a simple random sample of size 100 from a very large population, a population in which 60 percent of the items were of type *A*, the samples one would get could also be generated by a population of 10 items, six of which were type *A*, as long as sampling in the latter model was with replacement. You should pause at this point to review this idea if it is not clear already.

We will now proceed to use our new sampling model for further discussion of coin flipping. Let us consider some sequences of various lengths. It can be shown (see Section 7.5) that a sequence of 10 flips is likely to have somewhere between 3 and 7 heads, 100 flips between 40 and 60 heads, 1000 flips between 470 and 530 heads, and 10,000 flips between 490 and 510 heads. Note what this implies about the **proportion** of heads.

Number of flips	Proportion of heads
10	0.30–0.70
100	0.40–0.60
1000	0.47–0.53
10000	0.49–0.51

From these results one can speculate that the **long-run relative frequency** of heads would be 0.50. Conversely, if we had observed this phenomenon for a real experiment (rather than deduced it from a model), we would infer that the coin was fair. The long-run relative frequency seems to be telling us a property of the coin, a characteristic of each and every flip. The application of a long-run relative frequency to a single experiment is a logical leap: we invent a new word, **probability,** to describe this property. We say that the probability that a single flip results in a head is 0.50. This statement is motivated by, but different from, the statement that the proportion of heads in a large number of flips will be close to 0.50.

The most widely used definition of probability requires us to imagine an experiment (like tossing a coin) that can be repeated indefinitely. We observe the outcome of the experiment (a head or a tail), and consider a particular outcome (such as "head"). The probability of a particular outcome in a repeatable experiment is the long-run relative frequency of the outcome.

Now let us review the sampling model with this new perspective. We suppose we have a population, possibly a very small one, from which we take a random sample. The description of this sampling process may be characterized by stating the probabilities of each outcome. The samples from such a process should have variation characteristics that depend only on the list of population items and their associated probabilities: this two-entry list is called a **probability distribution.** So the sampling model can be described by a probability distribution.

What has been described above is the replacement of our sampling model {population, method of sampling} by a probability distribution. See Figure 7.4. The remainder of Section 7.3 examines the properties of certain probability distributions.

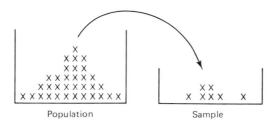

Figure 7.4 Schematic diagram showing the equivalence of a sampling model and a probability distribution.

7.3.2 Counting the Ways

The calculation of probabilities can sometimes be reduced to an exercise in counting. The idea is that since a probability is a relative frequency, if we count up the number of ways a particular event can happen, and we also count the total number of ways that the experiment could result (with or without the event happening), the ratio of these two counts gives the probability that the event would happen in a single trial. But for this approach to work, we must have the "ways" listed so that they are each equally likely.

For example, if a subcommittee of two were to be selected at random (simple random sampling) from a committee of three men and two women, what is the probability that the subcommittee consists of the two women? Every subcommittee has an equal chance of being selected and only one consists of the two women, so in this case, to calculate the long-run relative frequency of this special subcommittee, all we have to know is how many subcommittees are possible. Let us list them assuming the committee members are W_1, W_2, M_1, M_2, M_3.

$$W_1, W_2$$
$$W_1, M_1$$
$$W_1, M_2$$
$$W_1, M_3$$
$$W_2, M_1$$
$$W_2, M_2$$
$$W_2, M_3$$
$$M_1, M_2$$
$$M_1, M_3$$
$$M_2, M_3$$

That is, there are 10 such subcommittees, and only one of them is W_1, W_2. If we were to select the subcommittee over and over again, one-tenth of the subcommittees would be W_1, W_2. The probability of this event is therefore one-tenth, or 0.1. In words, we might describe such a selection as "unusual." (If the mandate of the subcommittee is to recommend a policy concerning maternity leave, we might describe the outcome as "unbelievable" and doubt that the selection was in fact done by simple random sampling—but this is encroaching on the subject of hypothesis testing and Chapter 9.)

Next let us calculate the probability that the subcommittee consists of two men. Three of the 10 subcommittees are of this type, so the required probability is 0.30. Such a result would be expected to happen 30 percent of the times that a committee was so chosen. The use of counts to compute probabilities requires that the items counted be of equal probability; in our case this was guaranteed by the specification that simple random sampling was to be used (see the definition in Section 2.3).

Next consider the calculation of the probability that three flips of a coin result in one head and two tails, in any order. The possible outcomes are:

Three heads
Two heads and one tail
One head and two tails
Three tails

So the probability of one head and two tails is 1/4, or 0.25, right? No! This is incorrect because the events listed above are not equally likely. If we want to count equally likely events, we should use the following list:

HHH

HHT

HTH

THH

TTH

THT

HTT

TTT

from which we can see that the required probability is 3/8, or 0.375. The events listed here are outcomes of three coin flips, with the order included as a feature of the outcome. The fact that these events are equally likely should be intuitively plausible, but we will cover this in detail shortly.

Next consider the following counting problem. A furniture company produces five kinds of dining-room tables and three kinds of dining-room chairs. How many combinations of tables and chairs are possible in making up a single table–chair combination? Now for each table type there is a choice of three chair types, and there are five table types, so the answer has to be $3 \times 5 = 15$ possible combinations. In fact, *if you have m varieties of one thing and n varieties of another thing, there are always $m \times n$ ways to form pairs of the two things.*

A certain password for a computer account consists of two alphabetic letters and two digits, 0 to 9. Thus passwords look like AA11, CW80, ZV02, and so on. If I pick a password at random from all possible passwords, what is the chance that I would get one of the 1000 valid passwords (and gain access to the account)? To answer this question, we first try to decide how we list the possible selections of a password such that each selection is equally likely to be selected. (I am selecting the password at random.) The list of all such passwords is such a list. Next, we count how many of these are valid passwords: that is, 1000. Also we need the total number of passwords in the list. This is where our pairing rule comes in. There are 26 ways to choose the first letter and 26 ways to choose the second letter, so there are $26 \times 26 = 676$ ways to choose the two-letter pair. Similarly, there are $10 \times 10 = 100$ ways to choose the two-digit pair. Finally, we can see that there must be $676 \times 100 = 67,600$ ways to choose the letters–digits combination making up the whole password. As 1000 of these are valid, if we pick one at random from the 67,600, the probability that it will be valid is 1000/67,600, which is about 0.015. In other words, our chance of hitting a correct password by a single random selection is about 1.5 percent.

A word on terminology. The term "chance" is used in many contexts, but when it is used to refer to the relative frequency with which something happens, it is usually multiplied by 100 and called a percent. We hear that the chance of heads is 50 percent, or of winning a door prize is 1 percent. We will follow this usage in this text, reserving the word "probability" to refer to the relative frequency directly. That is, the probability of a head is 0.50 or of a door prize is 0.01. Watch out for this source of confusion. A small percentage such as 0.0123 percent can look a lot like a probability, whereas if it were to be expressed as a probability, it would be written 0.000123.

Often it is not possible to write down a list of equally likely ways in which an experiment can result. In such circumstances it is usually necessary to manipulate probabilities according to a few simple rules—and that is the next topic.

7.3.3 Combination of Probabilities

Let us first reiterate the words we will use to describe our probability experiments.

Experiment: the framework for a chance outcome
Trial: one of many similar elementary experiments
Outcome: the result of an experiment or trial
Way: one of several equally likely chance outcomes
Selected at random: selected in a manner that makes each choice equally likely

Now let us consider an experiment consisting of the following. We observe the weather at a particular location for a 24-hr period. We record the presence of snow, or of rain, if either of these outcomes occurs. Now let us suppose that based on a meteorologist's evaluation, the chance of rain is 10 percent, the chance of snow is 5 percent, and the chance of both is 1 percent. If we do not want to get either wet or cold, we hope that it neither rains nor snows. We might want to know the chance of rain *or* snow: that is, rain, snow, or both. How can we figure this out from what we are already given?

It helps to think of the given chances as long-run relative frequencies even if we are interested in a single day's weather. Of the 10 percent of the time that it rains, only 1 percent refers to times when it also snows. Thus the percent of time that it rains but does not snow is 9 percent. Similarly, it snows but does not rain 5 percent − 1 percent = 4 percent of the time. So we have rain only, 9 percent, snow only, 4 percent, and both rain and snow, 1 percent, as we were initially given. So the total percent of the days that rain or snow would occur is 9 percent + 4 percent + 1 percent, or 14 percent (i.e., the chance of rain or snow is 14 percent).

You may have been tempted to add the 10, 5, and 1 percent to get this chance, or perhaps even just 10 percent + 5 percent, but you can see that this would have been incorrect. What is the principle that can keep us from making these errors?

One rule is that we can add the chance of one event to the chance of another event to find the chance of one or the other event happening *if* the events are *incompatible*, that is, if the occurrence of one event means that the other one cannot happen. Now {snow} and {rain} are not incompatible events: it is possible to have both in one day. That is why 10 percent + 5 percent did not give the right answer (14 percent). On the other hand, the events {snow but not rain}, {rain but not snow}, and {rain and snow} are incompatible, so we could add up the chances of each of these to get the correct chance, 14 percent, based on the chance for each component event.

Now let us consider another situation in which we might want to combine probabilities. From past experience we know that the chance that a neighbor will water his lawn on Saturday is 75 percent, and we also know that the chance that it will rain on Saturday is 10 percent. What is the chance that it rains *and* the neighbor waters his lawn on a particular Saturday? One way to look at this is that of the 10 percent of the Saturdays that it rains, 75 percent of them will also see the neighbor watering his lawn, and 75 percent of 10 percent is 7.5 percent. We might tentatively conclude that 7.5 percent is the chance that both events happen. But it may occur to us that the neighbor is surely less likely to water when it rains than if it does not rain, and in fact it is possible that the correct chance is much smaller than 7.5 percent. What is the guiding principle in this example?

The probability for two events to both occur is the product of the probabilities of the individual events provided that the events are *independent*. Events are independent if the occurrence of one event does not change the chance of occurrence of the other.

In the lawn-watering example, the events could only reasonably be considered independent if the neighbor felt that the rain should not affect his watering plans at all. If we believe that, the calculation of 7.5 percent is correct. If not, we would need more information about the neighbor's behavior to calculate the required probability.

Note that the multiplication rule was stated in terms of probabilities: If we multiply chances, 10 percent $\times$ 75 percent, do we get 7.5 percent or 750 percent? The latter is nonsense in this context. But multiplication of probabilities does work: $0.10 \times 0.75 = 0.075$. It is a little easier to keep things straight if the arithmetic is done with probabilities rather than with chances. However, answers are more generally understood if they are expressed in percentages. This is why we use both.

There is one other rule that is the simplest of all and quite useful. If the chance of rain is 10 percent, what is the chance of there not being rain? {Not rain} must occur all the time that {rain} does not occur, so the answer is 90 percent. The event {not rain} is called the complement of the event {rain}. Complementary events have chances that add to 100 percent. Of course, the probabilities add to 1.0. Given the probability of one event, you can easily calculate the probability of the complementary event.

To see how useful this last rule is, let us calculate the chance that a sequence of 10 coin tosses turns up at least one head. That is, 1 head, or 2 heads, . . . , or 10 heads. The simple way to approach this problem is to ask what the complementary chance is: that is, what is the chance that a sequence of 10 flips does not turn up at least one head. This means that it must have turned up no heads. The probability of no heads on two flips is $1/2 \times 1/2$ since the flips are independent, and for 10 flips is

$$(1/2)^{10}$$

which is about 0.001. This represents a chance of about 0.1 percent. Thus the chance of one or more heads is 100 percent $-$ 0.1 percent or 99.9 percent. It would have been much harder to calculate this chance directly using the chance for 1 head, 2 heads, . . . , up to 10 heads, and combining these.

In summary, three useful rules for calculating certain probabilities in terms of other known probabilities are:

1. The *addition rule:* If A and B are two events, the probability of A or B occurring is the sum of the probability that A occurs and the probability that B occurs *if* the events A and B are incompatible.
2. The *multiplication rule:* If A and B are two events, the probability of A and B occurring is the product of the probability that A occurs and the probability that B occurs, *if* the events A and B are independent.
3. The *complementarity rule:* If A is an event, the probability that A does *not* occur is 1 − the probability of A.

These rules are illustrated schematically in Figure 7.5.

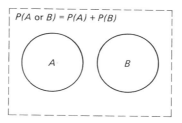

The addition rule
(Valid if A and B are incompatible)

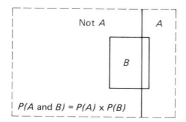

The multiplication rule
(Valid if B 's chance is the same when A occurs as when A does not occur)

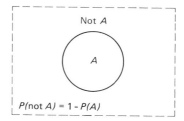

The complementarity rule
(Always valid. Either A or not-A occurs 100% of the time)

Figure 7.5 Schematic diagrams illustrating three rules for calculating probabilities. These diagrams are called Venn diagrams. The complementarity rule always holds, but the other two each have the condition indicated in the diagram. See Section 7.8 for details.

The following example requires the use of most of the techniques that have been presented for calculating probabilities. Let us assume that the sex of a child is determined by a mechanism that is adequately modeled by the flip of a coin; however, this coin is biased in favor of heads, the chance of a head being 55 percent, and this is to correspond to the preponderance of male births. Furthermore, the independence of one flip of a coin from the next flip does seem to suit the independence we observe among the sex of successive children in families. (In fact, this supposed independence has been disputed on the grounds that families may keep having children until a boy is born.) In other words, we could experiment with the sex ratio in families of various sizes with this model, without inconveniencing the large number of families it would take to do the experiment in real life. It will be seen that we do not even have to flip the coin! We will be able to find out whatever we need by thinking about our model. Let us focus on the following question: What proportion of families of three children will have at least one boy?

We will avoid the shortcut of the complementarity rule just to illustrate the details of the use of the other rules. Now the event we are to examine is something that occurs in families of three children. We have three spots that we have to fill with either a *g* or a *b*. Let us reserve the first spot for the oldest child, the second for the middle child, and the last spot for the youngest child. Thus {*ggb*} describes one possible sequence of children for a family with three children, and it is modeled by three successive coin flips resulting in tail, tail, head. Our model for generating families of size 3 is now completely described.

The next step is to list all the possible outcomes. If you look back at the coin-flipping example discussed earlier, you will see there are again eight possible outcomes: *ggg*, *ggb*, *gbg*, *gbb*, *bgg*, *bgb*, *bbg*, and *bbb*. If these were equally likely events, our evaluation of the chance of at least one boy would reduce to a simple count of these outcomes. [The answer would be $(7/8) \times 100$ percent, or 88 percent.] However, let us assume that the chance of a boy at any particular birth is 55 percent, so certainly *bbb* is more likely than *ggg*, and *bbg* is more likely than *bgg*. In fact, the probability of an outcome *ggg* is $0.45 \times 0.45 \times 0.45$ since the births are assumed to be independent of each other. Similarly, the outcome *gbb* may be computed to have a probability $0.45 \times 0.55 \times 0.55$. (We are using the multiplication rule here.) The following table shows the results of these calculations.

Outcome	Probability
ggg	0.091
ggb	0.111
gbg	0.111
bgg	0.111
gbb	0.136
bgb	0.136
bbg	0.136
bbb	0.166
All outcomes	1.000

The direct approach to obtain the probability that a family with three children have at least one boy is to consider the evaluation of all of the outcomes that have this property: namely, the last seven outcomes listed. The outcomes are all incompatible:

no two of them can happen in one family of three children. So their probability is obtained by their sum: 0.111 + 0.111 + ... + 0.166 = 0.909. (We are using the addition rule here.) Thus the chance of a family with three children having at least one boy is 90.9 percent.

It is easy to see that the complementarity rule would have obtained this more quickly. The event that the family with three children have at least one boy has a complementary expression that the family not have zero boys. So the probability is 1 minus the probability of zero boys: 1.00 − 0.091 = 0.909.

Note that the events *ggb*, *gbg*, and *bgg* were equally likely: we could have evaluated the probability of an event {exactly one boy} by counting up the number of ways this could happen and multiplying this by the probability of each one (i.e., 3 × 0.111 = 0.333). This approach would have yielded a table like the following:

Outcome	Probability
0 boys	0.09
1 boy	0.33
2 boys	0.41
3 boys	0.17
All outcomes	1.00

Since the outcomes listed are again incompatible, one could add 0.33, 0.41, and 0.17 to get the required probability of at least one boy: 0.91. The table above is a special case of a probability model which we shall discuss in detail in Section 7.3.5: the binomial probability distribution.

Note that the addition and multiplication rules each have a condition that must be satisfied if they are to be valid. The addition rule does not work when the events are not incompatibile as was shown in the "rain and snow" example. Similarly, if two events are dependent, the multiplication rule does not work. This was illustrated by the "lawn-watering" example. More complicated rules, and additional information, are required for the calculations in these cases.

You will need some practice to become familiar with these methods for combining probabilities. The exercises at the end of this section should help.

7.3.4 Probability Models

The subject of probability has both a charm and a utility in its own right. However, in this text we limit the discussion of probability to those ideas that are necessary for statistical applications. Our main use of probability is to describe the relationship between samples and the populations from which they are drawn. In statistics we are usually inferring information about populations from samples, but we are enabled in this by our knowledge of the reverse process: the properties of samples selected from known populations.

We have discussed how a basic sampling model, consisting of a population and a method of sampling, provides a way to explain and predict the amount of variation in certain data sets. We have also discussed, in Chapter 4, how a histogram or frequency distribution can provide a complete description of a sample, as long as the order of recording the data is unrelated to the data values themselves. This is the case with random samples. Moreover, the construction of histograms and frequency distributions applies to populations as well as to samples. Our statistical task then becomes the inference of a population frequency distribution from a sample frequency distribution.

In case we are only interested in a characteristic of the population distribution, rather than the whole population distribution, the task is to infer information about this characteristic from the sample distribution. We have defined the word "parameter" as a characteristic of a population distribution. We will be focusing in future sections of the text on *making inferences about (population) parameters based on sample distributions*. See Figure 7.6. But in the focus on frequency distributions in the following few sections, one should not lose sight of the basic sampling model, and always be aware of the role of the random sampling assumption. The inference to population from sample depends critically on the sampling method.

Figure 7.6 A sample reveals its population.

The probability model that arises most often in applied statistics is the normal distribution, and this will be discussed much more in future chapters. Some mention will also be made of other distributions, especially the binomial distribution (which we introduce in Section 7.3.5), the t distribution (Chapter 8), and the chi-square distribution (Chapter 9). In all these models, one should make an effort to identify the sampling model that produces the distribution, and keep in mind that all the results depend on the sampling model operating as postulated. It is easy to lose sight of this model, and it is also perilous to do so.

The normal distribution was discussed in detail in Chapter 4, but its purpose there was to describe samples, no matter how they were obtained. There was no attempt to relate the data at hand to a larger population. The normal distribution was not used as a sampling model, nor did we ever refer to it as a probability distribution. It was merely a convenient device to summarize a data set. In the remainder of the text, we will be focusing on the inference from sample to population, so that simple description of the sample will not be enough. This is why probability models play such an important role in statistics: they allow us to describe populations using information from samples.

Apart from these named distributions, we will discuss inferences from populations that contain just a few different numerical values, but where the sampling is with replacement. A number of statistical phenomena are explained by such models.

7.3.5 Probabilities for Binomial Experiments

Consider the following chance experiments:

1. A coin is tossed 10 times and the number of heads recorded.
2. A family has a sequence of seven children, and the number of girls is recorded.
3. Twenty students take a test and the number who pass is recorded.
4. A batch of 100 calculators is tested and the number with malfunctions is recorded.
5. Twenty light bulbs that have been in use for 1000 hours are observed to be still functioning or not: the number still functioning is recorded.

The similarity in all these examples can be made explicit. They are all examples of a **binomial experiment,** which may be described as follows:

1. A certain number of "trials" is specified for the experiment.
2. Each trial can result in one of two outcomes, A or B, say.
3. The chance of outcome A at each trial is some constant. This chance applies to each of the specified trials.
4. The trials are independent of each other.
5. The final outcome recorded is the total number of occurrences of event A.

You should check that these specifications really do apply to the examples above. If you doubted the constancy of the chance of passing in experiment 3, you are probably correct in this assessment. The model probably does not apply in this case. The other examples are appropriately modeled by a binomial experiment. Having the list of specifications should make checking this out a routine task.

The nice thing about binomial experiments is that we can use a standard method for calculating the probabilities relating to its outcomes. Although the context changes, the calculation is the same. In fact, since there are tables available for the probabilities for a binomial experiment, it is not even necessary to do any calculation. So the application of this technique just depends on checking that the assumptions of the binomial experiment really apply, and using the table. Table A4.2 presents binomial probabilities for experiments with 15 or fewer trails and various values of the constant probability for individual trail outcomes.

Let us determine the probabilities for 0, 1, 2, or 3 boys in a family with three children. We enter the table at $n = 3$ trials, and the probability (of a boy) at each trial of $p = 0.55$, and find the probabilities for 0, 1, 2, or 3 boys to be 0.09, 0.33, 0.41, and 0.17. The details of using this table are covered in the answers to exercises (Section 7.7). The explanation and formula for the binomial probabilities are given by a problem, so will require a bit of work. As a practical matter, the formula is unnecessary since the formula is tabulated, so the problem (see Problem 38) should be viewed as optional.

A binomial experiment can be viewed as the sampling model for the variation in sample percentages due to random sampling. Since we can determine the probabilities associated with the number of boys in a family of a given size, it is a very minor modification to write down the probabilities associated with the various possible percentages of boys in families of the given size. For example, in families with three children, we have:

Number of boys	Percentage of boys	Probability
0	0	0.09
1	33	0.33
2	67	0.41
3	100	0.17

This calculation shows that the distribution of sample percentages is known exactly once the population percentage (55 percent in the example) is known. However, we shall see in Section 7.5 that the normal approximation for binomial probabilities is much easier to use and is often a good enough approximation. The normal approximation works well when there are at least 10 trials and when the probability at each trial is not too close to 0 or 1 (say, between 0.1 and 0.9). So the number of boys in families with three children is one in which the exact binomial calculation, or table, is needed. A graphic display of the binomial distribution just calculated is shown in Figure 7.7.

The application of the binomial distribution is not limited to the artificial situations we have described so far. Consider the following example. A lumber company produces a kit for assembling a garden shed. The construction calls for the use of 10 nut-and-bolt sets. The quality of the hardware supplied with the kit is such that only 80 percent of the nut-and-bolt sets are functional. How many of these nut-and-bolt sets

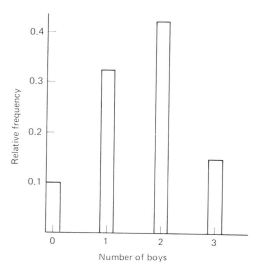

Figure 7.7 "Histogram" for the binomial distribution of the number of sons in families with three children, where the probability of a child being a son is 0.55. (See the text for assumptions.) This representation of a frequency distribution is different from the curve histograms. Here the bar height represents relative frequency (probability) rather than relative frequency per unit. This diagram is sometimes called a bar chart. It is used for the frequency distribution of a discrete variable.

should the lumber company include in its kit if it wants 95 percent of its kits to have enough functional sets?

The number of functional sets among a given number of sets has a binomial distribution. (Check the assumptions of the binomial experiment.) We just have to find a binomial table that gives 95 percent as the probability that 10 or more sets are functional. The binomial based on 15 sets gives a 93 percent chance of providing the 10, and for 16 sets the probability is about 97 percent. To be on the safe side the lumber company should put 16 sets in its kit (or else find a nut-and-bolt supplier with better quality control). The binomial distribution has many such applications in management science and in many other areas.

The main use of the binomial distribution in this text will be to model the variability in sample percentages. In a survey of a sample of 150 members of a political party delegation of 3000, the number in favor of decreased international trade tariffs might be 90, so the percentage in favor is 90/150 or 60 percent. Now the percentage of the 3000 might be 60 percent also, but it could well be more or less than 60 percent. How much more or less might it reasonably be? Since the binomial models the number of a certain type in a certain number of independent trials, it also models the *percentage* of a certain type in a certain *number* of independent trials. The binomial model should help us to answer our question.

We address such questions in detail in Chapter 8, but one important ingredient in our analysis will be the knowledge of the relationship between the SD of the sample percentage and the size of the sample percentage itself. For example, if we knew that the population percentage were 50 percent (i.e., if we knew that the whole delegation of 3000 would vote 50 percent in favor of the decreased tariffs), we can deduce (using Chapter 8) that the sample percentage in the 150 would be somewhere between 35 and 65 percent. To get a better sample estimate, one that is more precise, we would have to use a larger sample. So the 60 percent in the sample is not really a clear indication that a majority of the 3000 favor the tariff decrease.

The point of the tariff example is that the binomial model is applicable to sample percentages, and that a key element in its usefulness is that the variability of the sample percentage is predictable from the model. Both these ideas will be detailed further in Chapter 8.

EXERCISES

1. A die with the numbers 1 to 6 indicated on its six faces is rolled. What is the chance that it lands with a 1 or a 2 face up?
2. When two dice are rolled, the outcome is usually deemed to be the sum of the two upward faces. What is the chance of a total of 7 in one roll of two dice?
3. Mrs. H has just bought a new car. The car salesman cannot remember which of five models or seven colors Mrs. H ordered. The salesman asks his secretary to select some combination at random and order it for Mrs. H. What are his chances of ordering the correct combination?
4. What is the probability of getting exactly three heads in four flips of a fair coin?
5. Find the probability of exactly three heads in five tosses of a fair coin.
6. What proportion of families with 6 children have four or more boys, assuming that the chance of a boy at any particular birth is 52.5 percent?
7. Pilots in World War II offensives were estimated to have a 95 percent chance of returning safely from each mission. What is the chance of returning from 20 missions safely?
8. A group of 10 rats seem to play follow-the-leader. The probability that any one rat makes its way through the maze in a specified time is 0.10. But if the lead rat gets through, so do all its followers. So the probability that the 10 rats get through is 0.10. Is this a binomial experiment with 10 trials? Explain why or why not.
9. The probability that a homeowner forgets to put his garbage out on the correct day of the week is 0.25. The probability that during any given week, the garbage has an obnoxious odor that permeates the home is 0.10. Is the probability that the homeowner forgets to put out the garbage when it is smelly 0.10×0.25, or more than this, or less? Explain.
10. A dating service advertises the following information about its male registrants: 35 percent are tall, 75 percent are dark, and 60 percent are handsome.
    (a) What is the chance that a randomly selected male registrant is tall, dark, and handsome? Or is it impossible to calculate from the information given?
    (b) What is the chance that a randomly selected male registrant is tall, or dark, or handsome, or some combination of these traits? Or is it impossible to calculate from the information given?
11. You are about to toss a coin to decide who buys lunch, you or your friend. Your friend has a habit of calling "heads" 80 percent of the time. You know that the coin cannot be heads more than 50 percent of the time, so you reason that the coin-tossing procedure must be favorable to you if your friend is doing the calling. Is this reasoning correct? Explain.

12. An objective exam consists of 25 questions which each have a choice of one answer from five possible answers. If an examinee knows 10 correct, and selects the rest at random by using a sequence of random digits from 1 to 5 (i.e., 1 indicates choose answer 1, etc.), what is the chance of the examinee getting at least 60 percent of the questions correct?

13. You will need a thumbtack for this exercise. Toss it in the air 50 times, noting whether the point lands up or down. Record the outcomes in order of their occurrence. Your record should look something like

    Trial: 1, 2, 3, 4, 5, 6, 7, ···
    Outcome: D, U, D, D, U, U, D, ···

    For each group of five trials, note the percentage of D's; these percentages have to be 0, 20, 40, 60, 80, or 100 percent. Indicate how often, out of the 10 groups, each possible outcome percentage occurs. Compare this with the prediction of the binomial distribution. (Use the binomial table for an experiment of length 5, assuming that the proportion "down" is the proportion observed in the 50 tosses.)

14. The probabilities of surviving the years between age 95 to 100 are small, even for those who reach age 95. The probabilities below refer to the survival through one year, starting at each birthday listed under "age."

Males		Females	
Age	Probability	Age	Probability
95	0.78	95	0.83
96	0.76	96	0.80
97	0.70	97	0.75
98	0.59	98	0.63
99	0.43	99	0.46

    (a) What is the probability that a male alive at his 95th birthday is alive at his 100th birthday? State what assumption you need to do this calculation, and whether you think it is reasonable.
    (b) What is the probability that a couple, both having the same day and year of birth, and alive at their 95th birthdays, survives to age 100? State what assumption you need to perform this calculation, and whether you think it is a reasonable assumption.

7.4 ESTIMATION OF PARAMETERS

7.4.1 Parameters and Statistics

Estimation is the process of using frequency distributions in samples to tell us about frequency distributions in populations. The histogram of the sample data is itself an estimate of the population histogram, provided that the density scale is used for both (see Section 4.2). However, for many purposes, the entire frequency distribution of the population provides more information than we really need. Instead, we focus on the estimation of certain characteristics of the frequency distribution called **parameters** of the distribution. Note that parameters are characteristics of the population, not of the sample. The most frequently used parameters are the **population average** and the **population standard deviation.**

Consider a very small population consisting of just three possible numbers: 2, 3, and 7. This population has an average of 4.0 and a SD of approximately 2.2. Thus 4.0 and 2.2 are examples of parameters of this population.

How are these two parameters, the population average and the population SD, to be estimated from the sample data? It is natural to suggest the average of the sample and the SD of the sample as estimates. (This choice is widely accepted by theoretical statisticians in the case of the average, but is somewhat controversial in the case of the standard deviation. Problems 28 and 39 help you explore the most popular option.) For example, if our random sample of five observations, sampled with replacement, is 7, 2, 2, 3, 7, the population average would be estimated to be 21/5 or 4.2, and the population SD would be estimated to be 2.3. Now, if a person knows that the population is {2, 3, 7}, he or she would not normally be interested in these estimates based on the sample. But if all one has is the sample, and one wants to know about the population, these estimates 4.2 and 2.3 can be useful.

There is still one question that should be in your mind at this point: How close to the population values can the estimates 4.2 and 2.3 be expected to be? If the sampling process is really as we have modeled it, we should be able to say something about how much variation it produces. The key to this is that the amount of variation in the sample gives us a hint about how much variation there is in the population. The sample SD is an estimate of the population SD. But even if we knew the population SD, would we know how much the sample average would deviate from the population average? The answer to this question is simple but by no means obvious. We need a bit more terminology to provide this answer.

When we are discussing estimation, a procedure for calculating a summary number that is based on the sample is called a **statistic.** The sample average is a statistic. The sample standard deviation is another statistic. We use statistics to estimate parameters. It is useful to distinguish the calculation procedure, which is the statistic, and the number that the procedure produces, which is called the *value of the statistic*. The difference between the value of a statistic and the parameter it estimates is a quantity that will vary from sample to sample, but we would like to know, roughly, how big this difference is likely to be. Is 4.2 likely to be within 0.1 of the population average, or can we only count on it to be within 1.0 of the population average?

What is needed to proceed with this question is some mathematical properties of the statistics we are interested in using for estimation. In this chapter we use only one such property, called the central limit theorem, concerning one statistic, the sample average. To describe this result we need to define two terms that describe properties of a statistic: the expected value and the standard error.

7.4.2 The Expected Value and the Standard Error

A statistic is a calculation procedure, and the value that it produces will generally vary from sample to sample. In this section we discuss two important properties of a statistic, and then examine these properties with respect to a particular statistic: the sample average.

The simplest statistic of all produces from a sample of size 1 a single sample value. It is only defined for samples of size 1. What can be said about this statistic? Remember that the statistic is a method of producing a number from a sample, not the sample value itself. Can we say something about the average value of the proposed simple statistic, or about how much it might vary on repeated sampling?

The question almost answers itself—the key is to consider what would happen on repeated random sampling, with replacement. If we imagine a large number of repetitions of this sample of size 1, the average value of these sample values should be equal to the population average. This long-run average of the values of the statistic obtained by repeated random sampling, with replacement, is called the **expected value** of the statistic.

So far we seem to have a trivial idea with a new name. The usefulness of the new name is that we can equally apply it to more complicated statistics, usually for sample sizes larger than 1, by considering the one value of the statistic as a random sample of size 1 from the hypothetical population which describes the variability of the statistic. We will elaborate on this idea (the expected value of a statistic with a certain sampling distribution) shortly. First let us examine a direct application of the expected value idea to the evaluation of a lottery.

A new lottery was launched in the summer of 1984, called Lotto West. The information shown in Figure 7.8 was provided in the initial advertising. The com-

LOTTO WEST is easier to WIN

	LOTTO 6/49	LOTTO WEST
Numbers drawn	6 + bonus	8 + bonus
Player's choices	6	6
Total numbers to choose from	49	56
Chances of winning the jackpot with 6 correct numbers in any order	1 in 13,983,816	1 in 1,159,587
Chances of winning with 5 correct numbers plus bonus number in any order	1 in 2,330,636	1 in 579,794
Chances of winning with 5 correct numbers in any order	1 in 55,491	1 in 12,336
Chances of winning with 4 correct numbers in any order	1 in 1032	1 in 411
Chances of winning with 3 correct numbers in any order	1 in 57	1 in 34

Figure 7.8 This table appears in the Lotto West Media Fact Book.[1]

parison table refers to Lotto 6/49, which was discussed briefly in Chapter 1. The point is that the brochure provided a lot of information about the two lotteries, but one question is not answered: What is the long-run average return on the purchase of a single ticket? In other words, if I were to buy one ticket each week, at the cost of $1, for several years, what would be my average weekly prize? If you examine the brochure, you cannot calculate this quantity because the prize amounts are not given; the reason they are not given is because they depend on the number of tickets sold. The more sales, the bigger the prize pool and the bigger the prizes. However, Lotto West freely admits that it puts 50 percent of its retail dollar receipts into the prize pool; the rest goes to administration costs and various public causes.

Lotto West number selection mechanism. (Photograph courtesy of Western Canada Lottery Foundation.)

This percentage is the key to answering our initial question. It may be checked that the "expected value" of a ticket is 50 cents for this lottery. If we play the lottery

Sec. 7.4   Estimation of Parameters   247

often enough, we will end up with 50 cents for every dollar we spend on tickets.*

If more people knew about expected values, maybe lotteries would advertise this quantity. Then again, maybe they would not. It is argued by many that the purchase of one hope for one week has to figure into the value of a lottery ticket.

As an aside, what do you think would be the chances of winning a major prize in Lotto West if you bought one ticket every week for 40 years? Let us define a major prize as either the one that has 1 chance in 1,159,587 or the one with 1 chance in 579,794. These prizes would often be $100,000 or more. The chance works out to be about 1 in 300 (roughly): 1/3 of 1 percent chance. Faced with these facts, it is a wonderful thing that so many people contribute their money to public causes via the public lotteries.

Let us return once more to our very modest population $\{2, 3, 7\}$. Imagine the generation of many samples, each of size 5, say, from this population. For now we will consider sampling with replacement. One hundred samples of size 5 would produce 100 averages. Can we predict the average of these 100 averages? The answer is that it would be very close to 4.0, the population average. This is just the unbiasedness property of the sample average, which is intuitively plausible and is mathematically verifiable. (Verification in this instance is demonstrated in Problem 28.) The long-run average of sample averages from a population is called the **expected value** of the sample average from the population. This constant (4.0 in the example) is called the expected value of the average not because we expect the average to be that value for a particular sample (of 5) but because we expect the average of the averages, in the long run, to be equal to that value.

Note that the expected value may be calculated without actually producing the long-run sequence by finding the average of the population: 4.0 is calculated as

---

*Suppose that the prize pool is a percentage ps of sales. Let us denote the various prize amounts as $P_1, P_2, \ldots$, and the number of tickets winning each of these prizes as $n_1, n_2, \ldots$, respectively. The total prize pool, T, say, must equal

$$(n_1 \times P_1) + (n_2 \times P_2) + \cdots = T$$

Now if N tickets are sold at $1 each, the income from sales is N, and the right side of the equation above is $ps \times N$ dollars. So

$$(n_1 \times P_1) + (n_2 \times P_2) + \cdots = ps \times N \qquad (1)$$

But $N - (n_1 + n_2 + \cdots)$ won nothing. With this in mind we can construct a model of the lottery that is like the elementary models we have been discussing. View the purchase of a ticket as a random selection from a population of size N, and the population consists of n_1 numbers of size P_1, n_2 numbers of size $P_2, \ldots$, and $[N - (n_1 + n_2 + \cdots)]$ zeros. What is the expected value of our ticket (i.e., the long-run average value on repeated sampling)? It is just the average of this population, which is

$$\frac{(n_1 \times P_1) + (n_2 \times P_2) + \cdots + [N - (n_1 + n_2 + \cdots)] \times 0}{N} \qquad (2)$$

since N is the number of tickets sold, and the numerator is the sum of the population values. This calculation gives the population average, which is also the expected value of a ticket. But if we substitute (1) in (2), we find that the expected value of a ticket is ps dollars.

(2 + 3 + 7)/3. For this reason the expected value of the sample average generated by sampling with replacement may be calculated from a parameter of the population: namely, the population average. We highlight this result: *The expected value of the average of a random sample from a population is the population average.*

Another characteristic of the sample average is the difference there is, on the average, between the sample average and the population average. The average of five draws from {2, 3, 7} will vary from one experiment to the next, but the amount of variation in the averages should be predictable in a rough way from our knowledge of the population. If the population were {4, 4, 4}, the average of 10 draws would certainly be 4.0. But if the population is {2, 3, 7}, the average might be 3.0, or 4.5, or possibly, 7.0. However, an average like 7.0 would be unlikely, since it would require the random draw to be 7 five times in a row. What are usual averages like? How far from 4.0 might they reasonably be?

The variability of averages is measured by the **standard error of the sample average for averages of samples of size** n, which is defined as the standard deviation of the long-run distribution of averages of samples of size n. "Long run" here is meant to imply a very large number of averages, but not that the sample size is large for any particular average. In fact, we calculate the standard error for averages of given sample size, such as averages of samples of size 5.

The standard error of a sample average tells us how far from the population average the sample average is likely to be. To see this, recall that the standard deviation of a list of numbers tells us how far these numbers are from their average. Now, when the list of numbers is a list of averages, the SD of this list tells how close the averages are to *their* average. But the average of the averages is close to the population average. This justifies the interpretation proposed for the standard error.

We have indicated that the variability of an average will be large or small depending on the variability in the population. But the variability in the sample is very closely linked to the difference between the sample average in one sample and the expected value of that average. There is a very important and useful relationship between the variability in a population, measured by the population SD, and the size of typical deviations between the sample average and the expected value of the sample average: *The standard error of the sample average of a sample of size n is the standard deviation of the population divided by the square root of the sample size.*

Now the standard deviation of the population is calculated in the same way as the standard deviation of any data set. For the population {2, 3, 7}, it is about 2.2, so the standard error of averages of samples of size 5 is 2.2 divided by the square root of 5, which works out to about 1.0.

The interpretation of the standard error of the average is that it represents a typical deviation that the average of the sample might be from the population average. In other words, the claim is that the averages of samples of size 5 will tend to be about 1.0 from the population value of 4.0. Of course, some such averages will have a zero deviation, and some perhaps 1.0 or even 2.0, but usually the deviations will be close to 1.0.

Now we are set to answer completely the questions posed in the preceding section. When we use a sample average to estimate a population average, can we say how close the estimate is likely to be? We have claimed that the standard error of the average measures the amount that a sample average would typically deviate from the population average. So if we know the standard error, we can say something about the accuracy of our estimate of the population average; it is typically "off" about 1 standard error. If the standard error is small enough for practical purposes, we are content with our estimate.

Why was the standard deviation of the sampling distribution of the average equal to

$$\frac{\text{population SD}}{\sqrt{n}} ?$$

It is useful to know the basic mathematical result that implies this; we will need it again in Chapter 9. First we need to define a population parameter called **variance.** It is just the square of the SD:

$$\text{population variance} = (\text{population SD})^2$$

Like the SD, the variance measures variation but it does so in rather unnatural units—squared units. So variation in measurements of height would be measured by the variance in square feet, and variation in weights would be measured in square pounds. This is obviously a mathematical invention. Its use in statistics depends on the fact that *the variance of the sum of two independent variables equals the sum of the variances*. We will accept this mathematical result without explanation. What it means is that if we were to sample repeatedly (with replacement, so the sample values are selected independently) two weights from a population of weights, the variance of the sum of the weights would be equal to the sum of the variance of the weights. Since the two weights are drawn from the same population, the two variances are equal, so the sum of the two variances is just twice the population variance. So the variance of the sum of the two weights is two times the variance of one of the weights. By repeating this argument, we arrive at the following: The variance of the sum of n independent variables, each with the same variance, is n times the common variance.

The application of this strange result to applied statistics is that *the SD of the sum of n independent variables, each with the same SD, is just the square root of n times the common SD*. The SD of an average (a sum divided by n) is just the SD of the individual variables times the square root of the sample size, divided by the sample size. This simplifies to: The SD of a sample average from a population is the SD of the population divided by the square root of the sample size, n. This is exactly the result quoted earlier.

The purpose of this last demonstration is twofold:

1. To remind you that the standard error formula applies to averages of *independent* samples from a population
2. To record a result (variance of sum = sum of variances, for independent variables) that we will need in Chapter 9

There is one little problem to resolve. The calculation of the standard error depends on the population SD, which we usually would not know. However, we have an estimate of this population SD available: namely, the sample SD. So we just plug in the sample SD instead of the population SD in the description of the standard error of the sample average.

The standard error of a sample average of a sample of size n may be estimated by dividing the sample SD by $\sqrt{n}$. Suppose that we obtain the sample $\{7, 2, 2, 3, 7\}$ from an **unknown** population. What can we say about the population average? The sample average is 4.2 and the sample SD is 2.3. The standard error of the sample average is estimated to be

$$\frac{2.3}{\sqrt{5}}$$

which works out to about 1.0. Thus the population average is estimated to be 4.2, although we realize that this estimate will typically be off by about 1.0 or so. We write as a shorthand for this that the estimate is

$$4.2 \pm 1.0$$

We have defined the standard error of the sample average in terms of the population standard deviation and the sample size. In so doing, we managed to avoid saying explicitly what "standard error" really means. To describe this tricky concept more clearly, we will use a very simple and very artificial example.

We suppose that the population from which samples will be taken is the population $\{1, 2, 3, 4\}$. We will select a sample of size 2 from this population using the usual lottery model; first we will draw our samples with replacement. Our objective is to describe how much the sample average errs from the population average. The population average is $(1 + 2 + 3 + 4)/4 = 2.5$. The samples that are possible are:

1, 1	1, 2	1, 3	1, 4
2, 1	2, 2	2, 3	2, 4
3, 1	3, 2	3, 3	3, 4
4, 1	4, 2	4, 3	4, 4

Note that we have considered 1, 2 to be a different sample than 2, 1. This was done so that the 16 samples listed would be equally likely. Because of the way the samples are selected (think of drawing tickets and replacing them) the sample 1, 2, in that order, must have the same chance as 1, 1. But so does 2, 1, in that order. So the selection of the numbers 1 and 2 into the sample of 2 must have twice the chance of the selection of 1, 1. But the 16 ordered samples listed above are equally likely.

Now, we consider the sample averages in each of these possible samples. Listing them in the same format as the samples themselves, they are:

1.0	1.5	2.0	2.5
1.5	2.0	2.5	3.0
2.0	2.5	3.0	3.5
2.5	3.0	3.5	4.0

Collecting these together, we have:

1.0	with frequency	1
1.5		2
2.0		3
2.5		4
3.0		3
3.5		2
4.0		1

Our sample average is more likely to be 2.5 than to be 1.0 or 4.0. The distribution displayed above is called the *sampling distribution* of the sample average. A statistic with a particular kind of sampling method always generates a sampling distribution; here we have one example for a special population of the sampling distribution of the average for random sampling with replacement. The **sampling distribution** of a statistic is the frequency distribution of the statistic over a great many samples.

What are the average and SD of this sampling distribution? They may be calculated directly from the distribution itself to be 2.5 and 0.79. These are quite useful numbers: they tell us that the average sample average is 2.5 and the SD of the sample average is 0.79. The fact that the average of the sample average is 2.5 is comforting because if we use the sample average to estimate the population average (the latter was calculated earlier to be 2.5), we will be correct on the average. This is the "unbiasedness" property of a sample average. Moreover, the amount of error in this estimate may be described by saying that its SD is 0.79. The only problem is that calculating these numbers can be quite tedious, especially for realistic populations which may be very large. (Of course, realistic populations are usually not known, which is why we are sampling them, but this problem is discussed in Chapter 8.)

Is there an easy way to get the 2.5 and the 0.79? Yes. The average of the sampling distribution of the sample average is always equal to the population average. The SD of the sampling distribution of the sample average is the population SD divided by the square root of the sample size—it is exactly the quantity we have defined as the standard error of the sample average. Let's check it in the example. The SD of $\{1, 2, 3, 4\}$ is 1.12. The square root of 2 is 1.41. So the standard error, or SE, is $1.12/1.41 = 0.79$. This is the same as the result that we obtained the long way. The SE of the sample average is the SD of the distribution of the sample average.

In the example, the SE of the average of a sample of size 2 was 0.79. How big would the sample have to be to cut this SE down to one-half this value, 0.395? The answer is that the sample would have to be of size 8, four times as big as the previous sample. The SD of the population is the same in both cases, and the square root of 8 is 2.83, so the SE is $1.12/2.83 = 0.395$ approximately. The point is that one cannot expect twice the precision from an average based on twice the sample size; one needs four times the sample size for this.

In summary, *the average of the sampling distribution of the sample average is the population average, and the SD of the sampling distribution of the sample average is the standard error of the sample average.*

If we combine these statements with the fact that the SE of a sample average, in sampling with replacement, is

$$\frac{\text{SD}}{\sqrt{n}}$$

we obtain a very useful result: The sample average tends to be closer and closer to the population average as the sample size gets larger and larger. The situation is described graphically by Figure 7.9. In estimating a population average using a sample average, we can make the estimate as accurate as we wish by ensuring that the sample size is large enough. This idea is developed further in Chapter 8.

It is instructive to apply the results about the average and standard error to the binomial distribution. To do this we assume that we are sampling a population of 0's and 1's, with replacement, such that the proportion of 1's corresponds to the probability of the specified event at each trial. For example, the binomial distribution of the number of successes in 25 trials, when the probability of success at each trial is 0.40, would be modeled as the selection of a random sample of size 25, with replacement, from the population $\{1, 1, 0, 0, 0\}$.

Now we can compute the population average, which is 0.40, and the population SD, which is 0.49. However, when the population is made up of just 0's and 1's, the average and SD depend only on the *proportion* of 1's. (This is why it did not matter how big we made the population, as long as the proportion of 1's was correct.) Let us call the proportion of 1's p, and the number of trials, n. So $n = 5$ and $p = 0.40$ in the example. It can be shown in general for populations of 0's and 1's that the population average is p and the population SD is

$$\sqrt{p \times (1 - p)}$$

So we could have calculated the 0.49 this way:

$$\sqrt{0.4 \times 0.60} = \sqrt{0.24} = 0.49$$

Now the average of a sample from a population of 0's and 1's is exactly the same as the proportion of 1's in the sample. The standard error of this proportion is

$$\frac{\text{SD of population}}{\sqrt{n}}$$

which in the example is $0.49/5 = 0.10$ approximately. So the sample proportion from

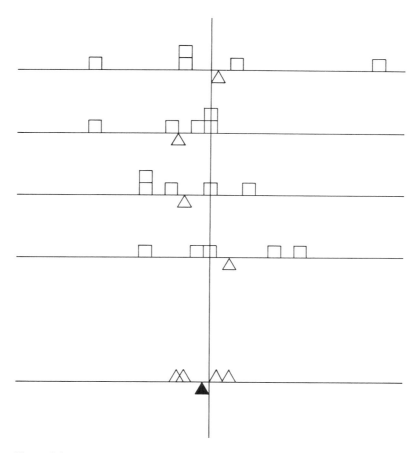

Figure 7.9 Four random samples from the population shown in Figure 7.1. The last distribution shows the distribution of the four sample averages. Note the smaller SD of the distribution of averages, and also the closeness of average of all 20 observations, indicated by the dark triangle, to the population average, indicated by the vertical line.

the population shown has a sampling distribution that has average 0.40 and SD 0.10. Put another way, the sample proportion has expected value 0.40 and standard error 0.10.

The sample proportion from a population consisting of a proportion p of 1's and the rest 0's has expected value p and standard error

$$\sqrt{\frac{p \times (1-p)}{n}}$$

We will use this result often in the remainder of the text. The key step in the modeling was to construct a population of 0's and 1's which had the correct population proportion of "successes."

All the above assumed that sampling was done with replacement. Section 7.4.3 explores what happens to the sampling distribution of the sample average when sampling is done without replacement.

We have introduced the terms *parameter, statistic, expected value, standard error,* and *sampling distribution* and have given an example of how these concepts are used to provide estimates of characteristics of populations based on samples. In Sections 7.5 and 7.6, we describe some additional probability models, involving populations that are commonly met in applications. The problems of estimation for all these models are discussed in Chapter 8.

7.4.3 The Standard Error in Sampling Without Replacement

As was mentioned in the discussion of sampling models, we usually assume that the population is much larger than the sample, so that sampling with or without replacement are the same thing. This simplifies matters, as the following example will point out.

If we have a population of 100 items, and we sample all 100 of them, the sample average and the population average are the same number, and this is true for every such sample. The difference between the sample average and the population average is always zero in this case. So the standard error that measures this variation should be zero. The formula we have so far is just the SD divided by the square root of the sample size, SD/10 here. This is not necessarily zero, but it should be. Where did we go wrong?

The formula for the standard error just quoted is correct only for sampling with replacement. For infinite or very large samples, it will also be appropriate since the difference between sampling with and without replacement is negligible in this case. In general, the standard error for sampling without replacement will be less than for sampling with replacement by a **finite population correction factor**

$$\sqrt{\frac{N-n}{N-1}} = \sqrt{1 - \frac{n-1}{N-1}}$$

where n is the size of the sample and N is the size of the population. The table below shows the relationship. Note that $(n-1)/(N-1)$ is approximately the proportion that the sample size is of the population size.

$\frac{n-1}{N-1}$	Correction factor
0.50	0.71
0.40	0.77
0.30	0.84
0.20	0.89
0.10	0.95
0.05	0.97
0.01	0.99

Sec. 7.4   Estimation of Parameters   255

This factor is about 0.97 when $(n - 1)/(N - 1)$ is $1/20$ and closer to 1 when $(n - 1)/(N - 1)$ is smaller. So the correction factor is of little consequence when the sample is very small relative to the population. A rule of thumb is to ignore the correction when n/N is less than $1/20$.

There are situations where the correction factor must be used. Suppose that the 100 houses comprising a neighborhood are to be appraised by a developer. He might sample 50 of them and go to the trouble of appraising these. The estimate of the average house value among the 100 would be based on the average value of the 50, and the error in this estimate would depend on the standard error. The correction factor in this case is about 0.7. This means that the ordinary standard error calculation must be reduced to 0.7 of its original value. This decrease in the error of the estimate of 30 percent may be very helpful to the developer. Note that the estimate itself is not changed; the average of the sample is still the estimate. But of course the use of this estimate may well depend on how accurate it is deemed to be.

Let us continue now with the artificial example mentioned at the end of the preceding section, in order to explore the effect of sampling without replacement from a small population. The samples of size 2 that are possible from the population $\{1, 2, 3, 4\}$ are:

Sample	Average
1, 2	1.5
1, 3	2.0
1, 4	2.5
2, 3	2.5
2, 4	3.0
3, 4	3.5

These samples are equally likely. Recall that sampling without replacement is the same as simple random sampling, and simple random sampling was defined as the method that generates all samples with equal probability. The average and SD of the distribution shown—of the sampling distribution of the sample average for samples of size 2—are 2.5 and 0.65, respectively. Again the average sample average is the same as the population average; however, the SD is no longer 0.79 (which was calculated as $SD/\sqrt{n}$), but 0.79 times a factor of less than 1. That factor is the correction factor:

$$\sqrt{\frac{\text{population size} - \text{sample size}}{\text{population size} - 1}} = \sqrt{\frac{N - n}{N - 1}}$$

In the example it is

$$\sqrt{\frac{4 - 2}{4 - 1}} = 0.82$$

Since $0.82 \times 0.79 = 0.65$, we have verified that this correction factor works exactly.

To calculate the SE for sampling without replacement, we just multiply the SE for sampling with replacement times the correction factor. This shortcut avoids the listing of possible samples and the calculating of the standard error the long way.

We will not dwell on this factor further in this book except to note one thing. Since the factor is approximately 1.0 when the *ratio* of the sample size to the population size is small, this implies that the size of the population is not an important issue. In such situations it is the size of the sample that is the key. The ordinary formula for the standard error does not even use the population size. The average of a population of 200 million can be estimated just as accurately as the average of a population of 1000, by a sample of size 50, for example, if the two populations have the same SD and if both populations can be sampled randomly. This is why the Gallup Poll can be so successful even though relatively small samples are taken.

7.4.4 Stratified Sampling and Cluster Sampling

In this section we shall digress to touch on a very important practical issue: improving sample estimates by using knowledge about the population. Gallup's polling organization seldom uses simple random samples. Although we will not be able to describe exactly the design of the Gallup Poll, we will at least indicate two strategies that such polls frequently use: cluster sampling and stratified sampling. The ideas of cluster sampling and stratified sampling were introduced briefly in Chapter 2. We are now in a position to clarify some of the points made there.

An example will illustrate the need for sampling designs more complex than simple random sampling. Suppose that we plan to survey undergraduate students at a college about the advisability of setting up a "women's studies" degree program. If the purpose of the survey is to see whether a majority of students are in favor of such a program (rather than to see whether women are in favor of it), we will have to be careful to use a sample with the correct mix of men and women. But if our sample size for the survey is to be 25, say, we are not automatically guaranteed that a simple random sample would produce the correct proportions. If the student body is 40 percent men and 60 percent women, our sample may or may not have 10 men and 15 women. Ideally, the sample should have these proportions (10/25 = 40 percent, 15/25 = 60 percent).

Of course, if we do not know the proportion of men and women among undergraduate students, we would have to accept the proportion that occurred in a simple random sample (SRS) as adequate. But if we do have this information about the population, we should use it to improve the representativeness of the sample. The method used to take advantage of this kind of knowledge about the population is called **stratified sampling.** The distinction between simple random sampling and stratified random sampling is that the stratified sample is formed by aggregating several simple random samples, one from each stratum. For example, the student survey would include a SRS of size 10 from the population of male undergraduates and a SRS of 15 women from the population of female undergraduates. The 25 students so selected would be the stratified random sample.

Sec. 7.4   Estimation of Parameters   257

We shall use a very simple example to illustrate the improvement obtained from stratifying a sample. We begin with a population {1, 2, 3, 4} from which a sample of size 2 is to be selected. Our aim is to see how well the sample average estimates the the population average.

For an SRS of size 2, we have already shown in Section 7.4.3 that the standard error of the sample average is 0.65.

Now suppose that we have a way of dividing the population into subpopulations, or **strata**; this could result in strata such as {1, 2} and {3, 4}, for example. An SRS of size 1 from each stratum would give us our stratified random sample of size 2. The possible samples we could get (they are all equally likely) are as follows:

Sample	Average
(1, 3)	2.0
(1, 4)	2.5
(2, 3)	2.5
(2, 4)	3.0

The SD of these four averages is 0.35. The standard error of the sample average based on the stratified sample is 0.35. This is less than the SE of the sample average based on the SRS; the SE was 0.65. For both samples the expected value of the sample average was 2.5. (For the stratified sample, just average the four sample averages to verify this.) Clearly, the stratified sample provides the more precise estimate of the population average.

To what extent does this improvement depend on the way we choose the strata? To explore this, suppose that the strata were {1, 4} and {2, 3}. We have:

Sample	Average
(1, 2)	1.5
(1, 3)	2.0
(4, 2)	3.0
(4, 3)	3.5

The SD of these four averages is 0.79. Thus the SE of the sample average for this stratification is 0.79. This is larger than 0.65. Stratification provided a worse estimate in this case. The method of choosing strata is clearly important. (If the population is large enough that the correction factor can be ignored, stratification will always reduce the standard error. It is still necessary to choose the strata wisely so that the maximum reduction in the standard error is attained.)

What principle should guide us in choosing strata for stratified sampling? *The strata should be as homogeneous as possible,* certainly more homogeneous than the population itself. (Homogeneity is gauged by the SD: the smaller the SD, the more homogeneous.) The strata {1, 2} and {3, 4} were more homogeneous than {1, 2, 3, 4}.

(The strata SDs were 0.5, whereas the population SD was 1.12). The strata $\{1, 4\}$ and $\{2, 3\}$ had a less clear-cut relationship to the population: certainly, $\{1, 4\}$ was less homogeneous than $\{1, 2, 3, 4\}$; it turned out that this was enough to make this stratification a poor one for estimation.

What would be the wrong way to stratify the sample of the 25 students? First we must assume, for this argument, that male undergraduates have a different attitude toward women's studies than do female undergraduates: for example, that females are more in favor of the proposed program than are males. If we form strata by, say, date of registration, we will have a mix of males and females in each stratum. Such strata would be expected to contain widely varying proportions of students in favor of the proposed program because they would have widely varying proportions of males and females. Thus in this case the samples could contain widely varying proportions in favor of the program—and would produce a less reliable estimate of the population average than one based on homogeneous strata.

Let us now consider another alternative to simple random sampling: cluster sampling. Again we will consider the population broken into subpopulations: first we will use $\{1, 4\}$ and $\{2, 3\}$. A **cluster sample** is obtained by taking a simple random sample of clusters. Thus our sample of size 2 can be one of two choices: $\{1, 4\}$ with an average of 2.5, or $\{2, 3\}$ with an average of 2.5. In this case we have an SE of the sample average of 0.0. The comparable SE for an SRS is still 0.65. The cluster sample does very well indeed.

On the other hand, if we had chosen the clusters as $\{1, 2\}$ and $\{3, 4\}$, the two possible sample averages would be 1.5 and 3.5, so the SE of the sample average in this case is 1.0, which is greater than 0.65. What principle will help us to choose clusters properly? That is, what principle will help us to choose cluster that provide precise estimates? *Clusters should be chosen to be as heterogeneous as possible.*

Cluster sampling is most commonly used in situations where SRS is not feasible. Surveys based on household interviews are usually done by city block. A random sample of city blocks is selected from a city and all residents of the blocks that are selected are interviewed. It is hoped that the blocks encompass a diversity of life-styles so that the cluster sample will provide precise information about the entire city. However, it is likely that the city blocks are more homogeneous than the city itself, since neighborhoods would be expected to have fairly constant levels of socioeconomic status. In other words, this seems to be a situation where cluster sampling should not be used. But our comparisons were based on comparing samples of the same size, not necessarily of the same cost. City block sampling is far less expensive per interview than would be simple random sampling. This effect offsets the lesser precision of the cluster sample in this case.

An example where clusters would be heterogeneous would be a survey of population age using households as the clusters. Each household would be expected to contain persons of various ages (children, parents, grandparents), so a random sample of households might well be better for estimating the average age of the population than an SRS of the same number of persons from the population.

It is easy to confuse the sampling procedures for cluster samples and stratified

Sec. 7.5   Probabilities from a Normal Approximation   259

samples: both start with a partition of the population into subpopulations. But the stratified sample consists of parts of each and every stratum, whereas a cluster sample consists of entire clusters, but only some of them.

EXERCISES

1. A simple version of the game of roulette is as follows. A ball is swirled round and round a wheel with 37 positions, labeled 0, 1, 2, . . . , 36, and a player places his bet, say $1, on one of the 37 spaces on a mat indicating one of the positions on the wheel. If the ball stops at the same position as that indicated by the bet, the player wins $36 for each dollar bet; otherwise, the player loses his bet. What percentage of the money bet does the "house" keep, in the long run?

2. In the game of roulette described in Exercise 1, how much can a person expect to lose in 10 separate spins of the wheel, assuming that he bets $1 each spin? Give an idea of how much this loss might vary from its expected value.

3. A population consists of the integers $\{1, 2, 3, \ldots, 10\}$. A sample of size 20 is selected with replacement from this population. What is the expected value and the standard error of this sample? What are typical values of this average?

4. Repeat Exercise 3 but using sampling without replacement.

5. List all the equally probable samples of size 2, selected with replacement, from the population $\{1, 2, 3\}$. Compute the average of each sample. Compare the average average with the population average and the SD of the averages with the standard error of averages of size 2 from this population. Comment on your findings.

6. Repeat Exercise 5 using sampling without replacement.

7. A list of 10,000 incomes has an average of $50,000 and an SD of $10,000. A sample of 100 of these incomes has an average of $49,000. Might the sample be a random sample, or is this unlikely?

8. A population of 200 children is sampled using a simple random sample of size 25. The IQ of the children sampled is determined, and the average is 110 and the SD of the sample values is 10. How close do you think the average, 110, is to the average IQ of the 200 children?

9. A contraceptive device is advertised to be 95 percent effective, meaning that in one year's use, the chance of pregnancy is 5 percent. If 10 couples use this method of contraception, what is the chance that at least one pregnancy will occur by the end of a year's use?

7.5 PROBABILITIES FROM A NORMAL APPROXIMATION

7.5.1 Normal Probabilities for Averages

In statistics we often wish to make some statement about the reliability of a sample average as an estimate of a population average. When samples are random samples, such statements are possible using the language of probability. For nonrandom samples, we are usually forced to rely on guesswork and faith in our own intuition.

Even when we do have random samples, and when probability statements are appropriate, there is the risk that our probability model may be in error—the probability model seems to require some knowledge of the population distribution in order to predict the variability of a sample average. This would be a serious drawback to the application of statistical models were it not for the following result: *The probability distribution of an average of independent measurements is approximately normal, and the approximation can be made as close as desired by increasing the sample size on which the average is based.* Rigorous versions of this result are called **central limit theorems** and they are among the most important contributions of mathematics to applied statistics. An important point to appreciate about this result is that the population probability distribution may be anything at all. The result is still true.

The central limit theorem is the result which so impressed Galton 100 years ago. It inspired the tribute quoted in this chapter opening. Galton called it the "law of frequency of error."

We will refer to the italicized statement above as the central limit theorem, to distinguish it from other theoretical results of this section. The role of the independence condition will be explained shortly. First let us see what the statement means.

Consider a population that is clearly not a normal distribution. Our example of such a population will consist of 40 percent 0's and 60 percent 1's, and we will model random selection of a sample from this population by sampling with replacement from the population {0, 0, 1, 1, 1}. Initially we will examine what happens to the distribution of averages when we take samples of size 3 from this population. For example, the first few samples produced by a lottery-like process are:

$$
\begin{array}{ll}
1, 0, 1 \quad \text{with average} & 0.67 \\
1, 1, 1 & 1.00 \\
0, 1, 1 & 0.67 \\
0, 0, 1 & 0.33 \\
1, 1, 1 & 1.00 \\
1, 0, 0 & 0.33 \\
1, 1, 0 & 0.67 \\
\end{array}
$$

So we are beginning to discover the relative frequencies for the various possible values of the average:

Value	Frequency	Relative frequency
0.00	0	0
0.33	2	2/7
0.67	3	3/7
1.00	2	2/7

Sec. 7.5 Probabilities from a Normal Approximation

There is a suggestion of 0.67 being a popular outcome. Let's see what happens when we generate 100 such samples of size 3. The distribution turns out, in the 100 draws, to be as follows:

Value	Frequency	Relative frequency
0.00	5	0.05
0.33	30	0.30
0.67	41	0.41
1.00	24	0.24

Is this a normal distribution? Certainly not, but it is more like a normal distribution than the population we started with, which was:

Value	Relative frequency
0.00	0.40
1.00	0.60

At least we have a hump near the middle. But we do not have the symmetry of the normal, and furthermore there are still big gaps in the list of possible values; these gaps do not appear in a real normal population.

The result about the distribution of averages tells us that the distribution of these averages becomes more like a normal distribution as we increase the size of the sample on which the average is calculated. Let's see what happens if we use samples of size 10 instead of size 3. The results of an experiment in which 100 samples of size 10 were generated, and the averages calculated, are as follows:

Value	Relative frequency
0.0	0.00
0.1	0.00
0.2	0.02
0.3	0.05
0.4	0.07
0.5	0.21
0.6	0.22
0.7	0.25
0.8	0.13
0.9	0.04
1.0	0.01

The distribution has a slight skew to the left. This disappears if we increase the sample size to 30, as is demonstrated by the following result of a similar experiment based on the same population of 0's and 1's as before.

Value	Relative frequency
0.37	0.01
0.40	0.02
0.43	0.02
0.47	0.02
0.50	0.07
0.53	0.09
0.57	0.12
0.60	0.23
0.63	0.15
0.67	0.12
0.70	0.07
0.73	0.05
0.77	0.02
0.80	0.01

Figure 7.10 shows these frequency distributions on a comparable scale, the density scale. The approximation result is clearly illustrated by these experiments. The histogram for a sample size of 1 is exactly the distribution of the population itself. No matter what this distribution is, the distribution of averages of samples from this population will approximate a normal curve, the approximation getting better as the number of values that are averaged gets larger.

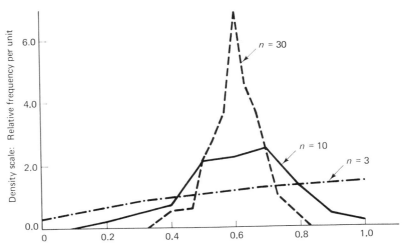

Figure 7.10 Empirical sampling distributions for averages of samples of size 3, 10, and 30. The population sampled is {0, 0, 1, 1, 1}. Each curve is based on 100 random samples. The approach to normality is apparent.

There is another feature of this sequence of experiments shown in Figure 7.10 which may have caught your attention. Not only is the distribution of the average becoming more like a normal distribution, but it is also getting less and less disperse;

Sec. 7.5 Probabilities from a Normal Approximation

the SD of this distribution is decreasing. The average seems to stay at about the same level. The average and SD of the distribution of the sample average can be shown, with a bit of mathematics, to be simply related to the average and SD of the population. In fact, we have the useful result that *the probability distribution of the sample average has an average equal to the expected value of the sample average and an SD equal to the standard error of the sample average.* But the expected value of a sample average is just the population average, and the standard error of the sample average is the SD of the population divided by the square root of the sample size. In other words, the average and SD of the sampling distributions that the experiments above were producing are known completely once the population average and SD, and the sample size, are specified. The experimental results could have been predicted by the theoretical results that we have already stated.

Now we will verify numerically these last claims for our artificial population $\{0, 0, 1, 1, 1\}$. This population had an average of 0.60 and an SD of 0.49. The expected value of the average is 0.60, and $0.49/\sqrt{\text{sample size}}$ is the standard error of the average. These calculations are recorded as follows:

Sample size	Expected value	Standard error
3	0.60	0.28
10	0.60	0.15
30	0.60	0.09

The histograms in Figure 7.10 have an average and SD equal to the expected value and standard error shown in the table above. The average of the probability distribution of the sample average stays the same, but the SD of the probability distribution of the sample average gets smaller. These theoretical results have important implications for estimation of population averages, to be discussed in Chapter 8.

Let us check that the results are believable in view of the experimental results we have produced. Do the frequency distributions from our experiments have averages and SDs close to the expected value and standard error that we have just calculated.

First we look at the 100 averages from samples of size 3. The average of these 100 values is 0.61 and their SD is 0.28. These and the results for the other samples are displayed in the following table:

Experiment sample size	Experiment		Theory	
	Average	SD of averages	Expected value of average	Standard error of average
3	0.61	0.28	0.60	0.28
10	0.61	0.16	0.60	0.15
30	0.60	0.08	0.60	0.09

The agreement here between the experimental and theoretical values is very close. Had we used 1000 instead of 100 samples in each experiment, the agreement would have been even better.

In Chapter 4 we alluded to the fact that the MAD (the mean absolute deviation) was an inappropriate measure of spread to incorporate into the theory of the distribution of averages. The narrowing of the distribution of averages depends on the standard deviation of the population, not on the MAD. To see this, consider the two populations A and B defined as follows:

	MAD	SD
A: 0, 2	1.0	1.0
B: 0, 2, 2, 4	1.0	1.4

Now the distribution of the sample average from the two populations does not have the same dispersion. For averages of samples of size 100, the SD of the sampling distribution of the average is 0.10 for A and 0.14 for B. The MAD of this same sampling distribution would also be 40 percent larger for B than for A. The population MAD cannot be used for describing the relationship between a population and the sampling distribution of an average from that population. This is why the SD is more often used in statistics than the MAD. The MAD still has a claim to being a simpler descriptive measure of populations, but it is less useful for studying statistical phenomena of averages.

Now we will return to the independence condition that was included in the central limit theorem statement. One form of sampling that we have encountered that violated the independence condition is random sampling without replacement. If we took samples of size 4 from our population $\{0, 0, 1, 1, 1\}$, without replacement, our sample average would be 0.5 or 0.75, with these two values occurring equally frequently. This is not anything like a normal distribution. Furthermore, larger samples than size 5 are impossible, so the central limit theorem clearly does not apply. In sampling without replacement, the sample values are not independent, so the condition of the theorem is not satisfied in this case.

The central limit theorem does not apply to the sampling of finite populations when the sampling is done without replacement. Nevertheless, as long as the sample size does not become close to the population size, the sampling distribution of the average will become more normal as the sample size increases. The applicability of the normal approximation to this sampling distribution would still depend on the sample size, however, just as in sampling with replacement (or from infinite populations).

In assessing the applicability of the central limit theorem to averages of moderate sample sizes, the following result is useful: *the sample average from a normal population has* (*exactly*) *a normal distribution*. The central limit theorem tells us that the effect of increasing the sample size is to reduce the effect of any nonnormality in the population distribution. If the population is almost normal, then even an average

Sec. 7.5 Probabilities from a Normal Approximation

of a small sample of five observations will be close to normally distributed. But when the population is far from normal (e.g., the population of 0's and 1's), the distribution of the average will not be close to normality until the sample size is 30 or more. So the sample size needed for the normal approximation to work depends on the shape of the frequency distribution of the population.

The sampling distribution of the sample average was shown to have an SD that is less than the population SD: in fact, it is the population $SD/\sqrt{n}$. This result assumed that sampling was with replacement. What is really needed for this result is independence of the sample values, just as the central limit theorem needed independence. Here is an example to show what confusion can arise if we forget about this independence condition.

Imagine a college course in which 10 tests are given. The course grade is to be determined by the average score on the 10 tests. The marking of the tests is such that on each test the average mark is 60 percent and the SD is 15 percent. The division between an A grade and a B grade is decreed to be 75 percent. Suppose that you get 70 percent on each and every test, so that your overall term mark is 70 percent. Presumably, since you have a B on each test, you should get a B for the course. However, the central limit theorem apparently predicts that the *average* of the 10 grades has a normal distribution (approximately) and that its average for the whole class would be 60 percent and have an SD of

$$\frac{15 \text{ percent}}{\sqrt{10}} = 4.7 \text{ percent}$$

So a mark of 70 percent, expressed in standard units, would be $(70 - 60)/4.7$ or 2.1 standard units. This puts you in the 98th percentile, and your rather B performance on the tests looks like it is qualifying you for an A+ on the course.

This "mark accumulation" paradox is easily explained. If your performance on one test was independent of your mark on any other test, the seemingly paradoxical result would be valid. Under these conditions it would be exceptional for a student to be above average on each and every test. However, most courses are of the type where performance on part of the course is a good guide to performance on another part of the course. This means that the test results are not independent. This is why the central limit theorem cannot be counted on to boost your cumulative grade.

Of course, there will likely be a bit of a boost—insofar as each test does examine some different aspect of your relative ability in the class, the narrowing of the distribution of the average score, relative to the distribution of the individual test scores, will occur to some extent. The degree of dependence between two tests can be summarized by the correlation coefficient: the greater the correlation coefficient between tests, the more dependent the tests and the less the narrowing of the distribution of the average relative to the distribution of the individual tests.

Figure 7.11 displays the effect that independent tests would have on a final grade, for the simpler case of three tests. The test distributions are all assumed normal in the figure, so that the distribution of the average is normal even though only three tests are averaged.

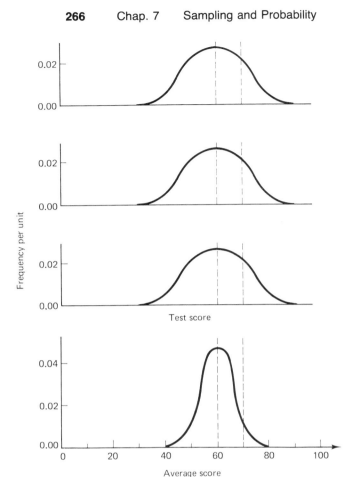

Figure 7.11 This figure shows the effect that independence of performance on term tests would have on the distribution of the final average grade. A student obtaining 70 percent on each of three tests would be in the 75th percentile on each test but in the 88th percentile for his or her average score of 70 percent. This phenomenon is usually less marked in practice because of the dependence of test grades.

The experiments of this section have so far shown several statistical phenomena. We started with a population that was known but not of any particular kind. We studied what would happen if we repeatedly selected samples with replacement, of a certain size, from this population. We discovered experimentally that:

1. The distribution of the sample average had the shape of a normal distribution, with this phenomenon becoming more marked as we used larger samples.
2. The sample averages varied somewhat but clustered around a central value that was the same as the population average.
3. The dispersion of the sample averages was less than the dispersion of the population sampled, with this phenomenon becoming more marked as the sample size increased. In fact, the SD of the distribution of the sample averages turned out to be equal to the standard error.

These experimental results are predicted by the earlier theoretical results. You should reread the theoretical results and ensure that you understand their meaning.

The practical use of these results is that the distribution of the sample average can be estimated from mere knowledge of the sample average and sample SD, something that can be calculated even when the population is completely unknown. The details of this process are outlined in Chapter 8.

7.5.2 Other Normal Approximations

Normal approximations are not limited to distributions of averages. Here is an example of a different use of the normal distribution.

A student who has a lengthy drive to his college wishes to determine how late he can leave in the morning and still reach his 8:30 A.M. class on time. The unpredictability of the traffic and the difficulty in finding a parking place make his time quite uncertain. He measures the time from his home until his arrival in class on three days to be 49 minutes, 57 minutes, and 37 minutes. When is the last minute he should leave home in order to have a 95 percent chance of being on time for class.

This is clearly not an easy problem. In the first place, 57 minutes before 8:30 is probably not enough time, even though it would have been good enough on the three days tested. Just looking at the wide variation among the three times should suggest that times over 57 minutes are not too uncommon. Another problem is that we do not know if the traffic is getting better or worse or staying the same, on average. We might make a tentative assumption that it will be about the same over the near future. Perhaps we may make the assumption that the three measurements we have are a random sample from the measurements that we would experience over the next few weeks. But we still have the problem that we do not know whether the 37 and 57 are nearly extremes, or whether the extremes are much broader than this.

In this situation, statisticians often make a brave assumption: that the shape of the population distribution is normal. Unless we imagine that the times are some sort of average, we have no reason to think that this assumption is valid. Let us suppose that the times are sums of times each of which are affected by independent "accidents" or vagaries of the rush-hour traffic. Since a sum is just an average times a constant, a sum and an average will have the same shape of sampling distribution. In this case it would be reasonable to make the normality assumption on the grounds that the measured time actually is a kind of average. Then we can use a simple calculation to get a rough answer to our original question about the last-minute leave-time.

From our data we estimate the average and the SD; the average is about 48 and the SD is about 8. A normal distribution with this average and SD has 1.65 as the value, in standard units, that is exceeded only 5 percent of the time. But 1.65 in standard units would be estimated to be $(1.65 \times 8) + 48$ or about 61 minutes. In other words, if the student leaves home at 7:29 A.M., we expect he would be late for class only about 5 percent of the time.

What we have done in all this is to demonstrate a possible use of the normal distribution as an approximation to distributions whose shape is unknown. This is a risky business but sometimes it is better than doing nothing. Of course, it might well be argued that it would be better for the student to leave at 7:00 A.M. for a while until he can collect enough data to see whether the distribution looks normal or not, and also to get better estimates of the average and SD. Decisions such as the one involved here plague the applied statistician. However, awareness of the issues involved will help in the decision-making process.

One other situation where the normal approximation is quite useful is in approximating the probabilities associated with other *known* probability distributions. Sometimes the normal approximation is easier to do, and accurate enough. The only example of this we will discuss is the normal approximation to the binomial distribution. First let's look at an example.

A fair coin is to be tossed 100 times. We wish to know the probability that exactly 58 heads will turn up in the 100 tosses. Now it happens that tables do exist for such probabilities, but they are lengthy tables and not widely distributed. One reason for this is that the normal approximation is adequate for most purposes. To find this approximate probability, we first have to find the normal distribution that is closest to the binomial distribution we are approximating. The way to do this is to match average and SD. The average number of heads in 100 tosses of a fair coin is 50 heads. The SD of the number of heads is 100 times the SD of the proportion of heads. In Section 7.4 the SD of the distribution of the proportion was shown to be

$$\sqrt{\frac{p \times (1 - p)}{\text{sample size}}}$$

So the SD of the number of heads in 100 tosses is

$$100 \times \sqrt{\frac{0.5 \times 0.5}{100}} = 5.0$$

The normal distribution with average 50 and SD 5 should be a good approximation to the binomial distribution for the number of heads in 100 tosses of a fair coin.

But there is one little problem with this. Our binomial distribution can only take values 0, 1, 2, . . . , 100, whereas the normal distribution can take any value, such as 58.3 and 57.6. The normal distribution will almost never be exactly 58. The approximation of the normal to the binomial is illustrated in Figure 7.12.

The approximation looks close, but how do we use it to find something like the probability that the number of heads is exactly 58? The key here is that it is the area under the normal that indicates probability: the area *near* 58 is what we want. More precisely, the area between 57.5 and 58.5 will approximate the probability we want. Now 57.5 is 7.5/5.0 = 1.5 standard units above the average and 58.5 is 8.5/5.0 = 1.7 standard units above the average. The area between these, from Table A4.1, is 0.022, so the chance of 58 heads is approximately 2.2 percent. This is actually the same that one would get from exact binomial tables. The approximation is very good for binomial experiments with at least 20 trials.

A similar device enables one to calculate the chance of getting 58 heads or more using a normal approximation. One simply computes the standard units for 57.5 and

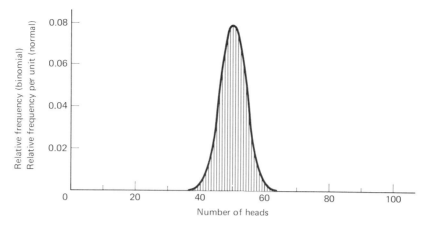

Figure 7.12 Normal approximation to the binomial. The number of heads in 100 tosses of a fair coin.

uses the area to the right of this under the normal curve. The standard units are $(57.5 - 50)/5.0 = 1.5$, and the chance turns out to be, from Table A4.1, about 7 percent. This procedure is called the **normal approximation to the binomial with continuity correction.** (It might have been called ". . . with discreteness correction.")

The normal approximation to the distribution of a sample proportion works well only when the sample is large enough that the central limit theorem can be expected to work. For population proportions between about 0.1 and 0.9, and sample sizes of at least 10, this approximation is fairly accurate.

EXERCISES

1. A class of 200 students of statistics participates in a classroom demonstration, not a mass confrontation but of a rather more intellectual nature. Each student selects a dollar bill from his or her wallet and is asked by the instructor to average the digits of the serial number to come up with a number between 0 and 9 as the average digit. (Students may indicate by a show of hands what their number is, to the nearest integer.) If the dollar bills have seven digits in the serial number, describe the distribution that would result from this experiment.

2. A population has a normal distribution with average 1.0 and SD of 0.0. What proportion of its values are between -1 and $+1$?

3. The freshman class at Sun U drinks a lot of orange juice: the girls drink about 20 oz and the boys drink about 40 oz each day. If there are equal proportions of boys and girls, what proportion of students drinks more than 50 oz? Assume that the SD of the amounts drunk for both groups is about 10 oz.

4. One hundred mice of the same genetic strain are weighed at age three weeks. They have been fed on identical diets and kept in similar environments. Would you think that these weights might be normally distributed? Explain.

5. When the sample size is doubled, the difference between the sample average and the population mean tends to
   (a) Become twice as large
   (b) Become 40 percent larger
   (c) Become 30 percent smaller
   (d) Become 50 percent smaller
   (e) Stay about the same size

6. One hundred subjects take an ESP (extrasensory perception) test. Their scores appear normally distributed (at least the distribution is reasonably symmetrical and unimodal), with average 55, and an SD of 15. The highest score attained is 90. Do you think the person who has attained 90 has demonstrated ESP powers? Or does the "normal" distribution suggest that no subjects have this power?

7. The following total scores are typical of the scores obtained by honors candidates at Cambridge University in the 1860s, according to Sir Francis Galton's book *Hereditary Genius*[2]: 500, 750, 750, 750, 1250, 1750, 2500, 3500. (A small sample of Galton's data is used here; the data values have been rounded since the source of data was a grouped histogram.) Estimate the percentage of candidates who would get in excess of 3500 marks. (The empirical evidence suggests that it is about 6.5 percent.) (*Hint:* Use a log transformation and then assume normality.)

7.6 NONPARAMETRIC PROBABILITIES

The point of this section is that the concept of probability is not limited to situations in which the model satisfies the assumptions of one of the named probability distributions, such as the "normal" or the "binomial" distribution. As usual, we begin with an example.

A light bulb manufacturer tests one bulb selected at random from the batch of bulbs manufactured each day. The test consists of subjecting the bulb to a sustained high-powered electrical source, and the bulb is deemed to be defective if it burns out within 1 hour. The length of time the bulb lasts before burning out is recorded. The manufacturer has carried out these tests for several months. Any data that reveal a problem in the manufacturing process are eliminated from the data set. Thus the data retained describes the variation in the test performance under normal manufacturing conditions.

The technical manager would like to set an automatic cutoff level that would signal to the process engineer that the equipment should be inspected. This is a costly process and the manager does not want to do it more than about 1 day in 10 if the process is operating normally. How should he determine the cutoff level?

Suppose that the most recent 100 good test results have a mean of 1 hour and a SD of 0.2 hr. Should the cutoff be set at a certain number of SDs below the average according to the normal probability table? That is, since $1.0 - (1.28 \times 0.2) = 0.74$ hr, this cutoff might be expected to be greater than just 10 percent of the test scores when the operation is running normally. But this calculation is based on the normality assumption, which in this case is probably false. With 100 past data values one can examine the histogram of values and pick off the 10th percentile, which would be a better method than one using an assumption of normality (i.e., use the 10th from the smallest test value). Assuming that the test scores have a certain shape is called a **parametric** assumption.

Using the observed data as the best indication of the distribution is a **nonparametric** procedure. Parametric methods are used when there is some knowledge that the population will have a distribution of a certain shape. Such knowledge is seldom certain enough to overrule the data when there are enough data to determine the population distribution reasonably well.

In situations where averages are being studied, the central limit theorem will usually ensure that normal distributions are applicable. But for percentiles or ratios of quantitative data, the resulting distributions will not be normal and may be of unknown shape, especially for small samples. Moreover, some data are ordinal: only the relative size of the observations has meaning. For example, the position in a preference list of 10 automobiles would be ordinal data. Averages and SDs of such data are usually not very informative. Techniques have been developed to handle ordinal data but are not detailed here. Some further information on nonparametric procedures is given in Chapter 9.

7.7 ANSWERS TO EXERCISES

Section 7.2

1. One population that would do to answer this question in a technically correct way is the population that consists of the same numbers as the sample. However, this is not very useful to Speedy. He wants to portray his measurements as "typical" in some sense. He might argue that the one day is like other days, or at least like other weekdays, and that his times might be considered a simple random sample from all times he is experiencing "these days." Of course, it is unlikely that the manager of the messenger service would believe this; he may well suspect Speedy of unusual hustle for this particular day. This discussion illustrates why study design is so much a part of statistical work.
2. No. The measurements include a component due to the growth, and sudden cutting back, of the grass. This could be detected in the measurements, and this source of variation separated from the unexplained variation.
3. Probably not, since peaches sort themselves by size to some extent in the truck, and the first ones on the conveyor may not be typical of the truckload.
4. If the sprinter has been in training for some weeks, it is unlikely that his times are still changing in any systematic way; rather the variation might be considered random in this case. However, there may be some systematic difference between practice times, without competition, and the actual race time. Also, his third sprint may be slower than his first as a result of fatigue. The latter point might invalidate the use of the sample in predicting the race outcome.
5. It is a complicated sample, each three times coming from the hypothetical population pertaining to a particular sprinter. The population sampled would be a composite of the 10

hypothetical populations. (But the 30 measurements are certainly not a simple random sample from it.)

6. A list of all high school graduates would have to be compiled. If the names were numbered, a sample of random numbers in the same range would select a simple random sample of the students. Any process equivalent to the lottery selection method would be acceptable. (It is usual to use a random number generator of a computer.) The students selected would write the exam and the papers would be marked to produce the sample scores. The scores of the population of graduates could then be based on the scores of the selected students, using the sampling model to provide the link.

7. (a) Since the same apple is being weighed each time, and its actual weight is constant, the population must be the collection of values that would be generated by the scale with this particular apple over many weighings. The variation in these values would be determined by the imprecision of the scale. Hypothetical population.
   (b) The population of actual apple weights for the source of these apples. Concrete population.
   (c) The population in this case would be the hypothetical population of measurements produced by all the apples at the source and many weighings of each apple. The population is a mixture of a concrete and a hypothetical population, which is therefore hypothetical.

Section 7.3

1. Each face is equally likely, so each face has probability 1/6. The occurrence of the 1 face up is not compatible with the occurrence of the 2 face up, so the probability of a 1 or a 2 is 1/6 + 1/6 or 1/3. The chance is 33 percent.

   An alternative solution is: The chance of a 1 or a 2 is the same as the chance of a 3 or 4, and also the same as the chance of a 5 or a 6. These three events are incompatible, so the chance for each of them is 33 percent.

2. The equally likely outcomes for this experiment are the 36 possible combinations on the two dice. Note that 1, 2 is not the same outcome as 2, 1 in this list, where the first number refers to the one particular die, the second number to the other. (Think of a red die and a green die, if this helps.) Now the outcomes that would give a sum of seven are as follows:

Die 1	Die 2
1	6
2	5
3	4
4	3
5	2
6	1

   In other words, there are 6 of the 36 equally likely outcomes that sum to 7, so the probability of rolling a total of 7 is 6/36, and the chance is about 17 percent. (The 36 equally likely outcomes must each have probability 1/36, so that the sum of the probabilities of all the incompatible outcomes is 1.0. Adding up the probabilities for the six outcomes listed above gives $1/36 + 1/36 + \cdots + 1/36$ or $6/36$. This is a cumbersome way of verifying that the procedure of counting the number of ways a particular event comes out, and dividing it by the number of ways any outcome can happen, gives the probability desired, as long as the "ways" in the list are equally likely.)

3. There are 5×7 or 35 possible combinations and they are all equally likely if the secretary really chooses one at random. Only one of these is correct, so the probability of a correct choice is $1/35$. The chance is about 3 percent.

4. The outcomes may be represented as HHHT, HTHT, and so on. These outcomes are equally likely because the coin is fair. There are $2 \times 2 \times 2 \times 2 = 16$ possible outcomes, and four of these outcomes have exactly three heads: HHHT, HHTH, HTHH, THHH. Thus the required probability is $4/16$, or in other words, the chance is 25 percent.

5. Here we have a binomial experiment, $n = 5$ and $p = 0.7$. We enter Table A4.2 at $x = 3$ to find the required probability of 0.3125. In other words, 31.25 percent of experiments that consist of five tosses of a fair coin would turn up exactly three heads.

6. Again we have a binomial experiment in which $n = 6$ and $p = 0.525$. However, Table A4.2 does not have an entry for $p = 0.525$; the nearest entries are 0.50 and 0.55. So we use these and split the difference. That is, we have, from Table A4.2:

x	$p = 0.50$	$p = 0.55$
4	0.23	0.28
5	0.09	0.14
6	0.02	0.03
4 or 5 or 6	0.34	0.45

So the required probability is about 0.395. (A computation using the exact formula yields 0.392, so this interpolation is adequate for practical purposes.) That is, about 40 percent of families of 6 children would have four or more boys. (Note that the probability of having three boys, and three girls, is only 0.30 when the probability of having a boy is 0.50; it is a popular myth that such equal splits are quite likely.)

7. Here $n = 20$, and $p = 0.95$, and the number of A outcomes is to be 20. But Table A4.2 does not include $n = 20$. One way to find the answer is to calculate the probability directly: $(0.95)^{20}$, which is (if you have the right kind of calculator) easily computed to be 0.36. So the pilot's chance of surviving twenty missions is 36 percent.

8. No. The successes of the 10 rats are not independent of each other. If they were, the probability that all 10 rats get through would be $(0.10)^{10}$, which is a very tiny probability (0.0000000001).

9. The probability is greater than 0.10×0.25 if the odor helps to remind the homeowner. The 0.10×0.25 would be correct only if the obnoxious odor had no influence on the homeowner's chances of remembering, that is, if the events were independent. The multiplication rule should not be used here. We need more information to calculate the probability exactly. (In fact, the information we need is the probability that the homeowner remembers to take out the garbage in the event that the garbage is smelly. If this probability were, say, 0.90, the required probability would be 0.10×0.90.)

10. (a) Impossible to calculate without further information since the traits are not necessarily independent.

(b) Impossible to calculate without further information since the traits are not incompatible.

11. The reasoning is incorrect. Even if the friend called "heads" 100 percent of the time, he would still be right 50 percent of the time.

12. Of the 15 answers selected at random, the chance that any one is correct is 20 percent. To get 60 percent in total, the examinee needs 5 or more of the 15 correct. The binomial tables give 0.836 as the sum of the probabilities of 0, 1, 2, 3, and 4, so the probability of 5 or more is 1 − 0.836 or 0.164. So the chance of the examinee getting 60 percent or more is 16.4 percent (even though the examinee only knew the answer to 40 percent of the questions).

13. Your distribution may be quite different from what the binomial model predicts as long-run relative frequencies since 10 experiments is not a very long run. If the percentage down in the 50 experiments were 70 percent, and we use this in our binomial model, the long-run relative frequencies (i.e., probabilities) are as follows:

Outcome %	Long-run relative frequency	Equivalent frequency
0	0.00	0
20	0.03	1.5
40	0.13	6.5
60	0.31	15.5
80	0.36	18.0
100	0.17	8.5
All outcomes	1.00	50.0

14. **(a)** If one multiplies the five probabilities together, the result is 0.11, which is the required probability. This is a correct computation, but not because of the independence rule, which clearly is not valid here. The events "survive through 96th year" and "survive through 97th year" are dependent. This is an example of the combination of probabilities of dependent events (see Section 7.8).

(b) The figure for females comparable to 0.11 for males is 0.14. One would have to assume independence of survival of the husband and wife to calculate the probability of their joint survival to be 0.11 × 0.14 = 0.015. If life-style affects survival at this age, the independence assumption may not be justified. (The implication of dependence is that joint survival could be as high as 0.11.)

Section 7.4

1. The relevant population for this problem is one with 36 0's and one 36. A random sample of 1 is selected from this population: its expected value is 36/37, representing the amount that the player will win per $1 bet. Thus the house keeps 1/37 dollars or about 2.7 cents for each dollar bet, on average. Thus the house keeps about 2.7 percent of the money bet.

2. The expected loss in 10 spins is just 10 times the expected loss in one spin (i.e., 54 cents). The standard error of this average loss is (population SD)/$\sqrt{10}$. The population SD is 5.83, so the standard error is 1.84. The estimated loss after 10 games is 54 cents plus or minus

about $1.84. [In other words, there is a fair chance of a net gain (a negative loss) but the tendency is to realize a loss.]

3. The average and the SD of the population are 5.5 and 2.9, respectively. Typical values would be expected to be in the range 5.5 plus or minus $(2.9/\sqrt{20})$, that is, 5.5 ± 0.65.

4. It is impossible to take a sample of 20 from a population of 10 when sampling is without replacement.

5.

Possible samples	Average
1, 2	1.5
1, 3	2.0
2, 1	1.5
2, 3	2.5
3, 1	2.0
3, 2	2.5

Average average = 2.0, which is equal to the population average

$$\frac{1 + 2 + 3}{3} = 2.0$$

The SD of the averages is

$$\sqrt{\frac{(1.0 - 2.0)^2 + (1.5 - 2.0)^2 + \cdots + (3.0 - 2.0)^2}{9}} = 0.58$$

The standard error of the sample averages of size 2 is the SD of the population divided by the square root of 2. The SD of population is the

$$\sqrt{\frac{(1 - 2)^2 + (2 - 2)^2 + (3 - 2)^2}{3}} = 0.82$$

So the SE is $0.82/1.41 = 0.58$. Again the compared quantities are equal. These equalities are guaranteed by the properties of the sampling distribution of the sample average. See Section 5.4.

6.

Possible samples	Average
1, 1	1.0
1, 2	1.5
1, 3	2.0
2, 1	1.5
2, 2	2.0
2, 3	2.5
3, 1	2.0
3, 2	2.5
3, 3	3.0

The average average is 2.0, as is the population average. The SD of the averages is 0.41.

The SD of the population is 0.82, as before (Ex. 5). The SE of the average is (0.82/1.41) times the correction factor:

$$0.71 = \sqrt{\frac{3-2}{3-1}}$$

so the SE is $(0.82/1.41) \times 0.71 = 0.41$. See the comment on Exercise 5.

7. The SE is $10{,}000/10 = 1000$, so 49,000 is typical and not unlikely if the sample were a random sample.
8. The closeness should be gauged by

$$\text{SE} = \left(\frac{10}{\sqrt{25}}\right) \times \sqrt{\frac{200-25}{200-1}} = 1.9$$

So the average of 110 is probably within about 1.9 of the average IQ for the 200 children.

9. This event is the complement of the event that no pregnancies occur. The probability is $1 - (0.95)^{10}$, which is 0.40. So the chance is 40 percent that at least one pregnancy occurs.

Section 7.5

1. The serial numbers would contain digits that were like a random sample of size 7, selected with replacement, from the population $\{0, 1, 2, \ldots, 9\}$. The average digit would be expected to have a distribution that is approximately normal, with average approximately equal to the expected value of the average, 4.5, and with SD equal to the standard error of the average, which would be the SD of the population, 2.9, divided by the square root of the sample size (sample size = 7) of 2.65, which is 1.1, that is, a normal distribution with average 4.5 and SD 1.1. The hand count would be a discretized version.
2. 100 percent!
3. 50 is 1 SD above average for the boys, and it is 3 SD above the average for the girls. Using the table for the normal distribution, since this seems a reasonable assumption in the absence of other information, about 16 percent of the boys and almost 0 percent of the girls exceed the 50-ounce limit. So the answer is one-half of 16 percent, or 8 percent.
4. The sources of variation in the weights must be multiple and small, and the measured weights may well be a sum of "errors" of independent sources. The weights might well be normally distributed. Of course, a little empirical evidence to the contrary should be enough to induce one to give up the assumption of normality, but in situations like this it would be reasonable to proceed as if the distribution were normal in the absence of evidence to the contrary.
5. The typical size of the difference mentioned is measured by the standard error, which becomes smaller by a factor $1/\sqrt{2}$ of 0.7. Thus it is smaller by 30 percent. Option (c) is the answer.
6. Neither suggestion is justifiable. The score of 90 is fairly typical of the largest score in 100 scores with average 55 and an SD of 15; since 90 is 2.33 SDs above the average, the normal table predicts that about 2 out of 100 would be this high. So the 90 is not that exceptional. On the other hand, if all subjects had ESP to some degree, it is quite possible that the distribution would still be normal. "Normality" of a distribution does not mean that the subjects are "normal."
7. The logarithms of the scores are

2.7, 2.9. 2.9, 2.9, 3.1, 3.2, 3.4, 3.5

The average and SD of these are 3.08 and 0.26. $\log_{10} 3500 = 3.54$, so 3500 is $(3.54 - 3.08)/0.26$ or 1.77 standard units above the average in the transformed data. Assuming normality, this value is exceeded by about 8 percent of the candidates.

7.8 NOTATION AND FORMULAS

The addition and multiplication rules are special cases of general formulas. The general formulas are

$$P\{A \cup B\} = P\{A\} + P\{B\} - P\{A \cap B\} \tag{1}$$

$$P\{A \cap B\} = P\{A\} \times P\{A \mid B\} \tag{2}$$

$$P\{\bar{A}\} = 1 - P\{A\} \tag{3}$$

In these formulas, A and B represent events. $P\{\cdot\}$ *means "the probability of."* $A \cup B$ defines an event in terms of A and B: it is the event that A, or B, or both A and B, occur. $A \cap B$ defines the event: A and B both occur. The symbol $P\{A \mid B\}$ means "the probability that A occurs given that B is known to occur." The symbol $\bar{A}$ means the complement of A, that is, "not A."

With this symbolism the addition rule is just (1) with $P\{A \cap B\} = 0$. The latter is guaranteed by the condition that A and B be incompatible. The multiplication rule is just (2) but with $P\{A \mid B\} = P\{A\}$. This condition is guaranteed by the independence of A and B. Formula (3) is the complementarity rule.

Venn diagrams are helpful in understanding these formulas.

The number of ways of ordering n things is $n \times (n-1) \times (n-2) \times \cdots \times 2 \times 1$ and is called "factorial n" and written $n!$. For example, when $n = 4$, $n! = 24$. The objects A, B, C, and D can be ordered in 24 sequences:

$$ABCD, ABDC, ACBD, ACDB, ADBC, ADCB$$

$$BACD, BADC, BCAD, BCDA, BDAC, BDCA$$

$$CBAD, CBDA, CABD, CADB, CDBA, CDAB$$

$$DBCA, DBAC, DCBA, DCAB, DABC, DACB$$

The number of ways of selecting k things from n things, without regard to order, is

$$\frac{n!}{(n-k)! \times k!}$$

For example, the number of ways of selecting two things from four things is

$$\frac{4!}{2! \times 2!} = \frac{24}{4} = 6$$

That is, from the four things A, B, C, and D one can select AB, AC, AD, BC, BD, or CD.

A sequence of n independent trials in which each trial is recorded as a success (S) or a failure (F), and the probability of success at each trial is p, is a binomial experiment. Any particular outcome of the sequence of n trials that results in k successes will have a probability equal to

$$p^k \times (1 - p)^{n-k}$$

since the probabilities of independent events are multiplied to determine the probability of all of them happening, and since the probability of failure is $(1 - p)$. The trials may be numbered 1 to n. The selection of the k trial numbers that result in successes can be done in

$$\frac{n!}{(n - k)! \times k!}$$

ways, and these ways are incompatible. To find the probability that one or other of these events occur, we can add the probabilities. Since the probabilities are all the same, namely,

$$p^k(1 - p)^{n-k}$$

the total probability of these events is

$$\frac{n!}{(n - k)! \times k!} \times p^k(1 - p)^{n-k}$$

This is the general formula for the probabilities in a binomial experiment. That is, the probability that a sequence of n independent trials with probability of success $= p$ has k successes is equal to the expression shown. These probabilities are tabulated in Appendix A4.2

Probabilities from distributions may be symbolized in general as indicated by the following examples: $P\{X < 2\}$, $P\{X = 2\}$, and so on, to represent the probability that the variable takes a value less than 2, or the probability that it takes the value 2, and so on. If X has a standard normal distribution, we can state that $P\{-2 < X < 2\} = 0.95$.

The traditional symbols used for certain population parameters and their sample-based statistics (or estimators as they are often called) are as follows:

Parameter name	Parameter symbol	Estimator name	Estimator symbol
Population average	μ	Sample average	$\overline{X}$
Population standard deviation	σ	Sample standard deviation	SD or s
Population standard error of average	$\sigma_{\overline{x}}$	Estimated standard error of average	SE or $\dfrac{\text{SD}}{\sqrt{n}}$ or $\dfrac{s}{\sqrt{n}}$

The formula relating the standard error to the SD is

$$SE = \frac{SD}{\sqrt{n}}$$

The SE in sampling without replacement, where n is the sample size and N the population size, is SE multiplied by the correction factor

$$\sqrt{\frac{N-n}{N-1}}$$

7.9 SUMMARY AND GLOSSARY

1. When we want to infer something about the source mechanism that has generated some data, we must carefully specify two things: what population has been sampled, and what is the method of selection of the sample from the population.
2. The sampling may be random or nonrandom. In the latter case we cannot make valid probability statements about the relationship of the sample to the population. Statements that do not use probability in describing this relationship are usually noninformative or unjustified.
3. Random sampling may be with or without replacement. For large populations, there is no practical difference in the samples produced by these methods. For small populations, sampling without replacement is more precise. (See the discussion of "finite population correction factor" in Section 7.4.3.)
4. The probability of an event is its long-run relative frequency in a sequence of identical independent trials where selection is random and with replacement. The histogram of all possible outcomes from such a sequence is the probability distribution of the outcome.
5. The combination of probabilities may often be simply achieved using the rules for incompatible, independent, or complementary events.
6. The binomial distribution is the exact probability distribution of the number of "successes" in a binomial experiment. It depends on the probability of success at each trial and on the number of trials in the experiment.
7. A statistic is a quantity computed from the sample data values. An example is the sample average or the sample SD. Statistics are used to estimate parameters; parameters are characteristics of the probability distribution, or, equivalently, of the population. Examples of parameters are the population average and the population SD.
8. A sampling distribution of a statistic is the distribution that the values of the statistic would have if the sampling procedure that generated the data from which the statistic was calculated were repeated many times.
9. The standard error of an average measures how far from the expected value of the average the sample average is likely to be. It is the SD of the probability

distribution of the sample average. It is estimated by dividing the SD by the square root of the sample size, and so decreases as the sample size increases.

10. The sampling distribution of an average approximates the normal distribution in large samples. Its average is the same as the population average, and its standard deviation is the standard error of the average.

11. When sampling is done without replacement, the standard error of the average is reduced. This factor is called the correction factor. It is only important when the sample size is more than 5 percent of the size of the population.

12. A sample proportion may be viewed as a sample average from a population of 0's and 1's. As such its average and standard error are expressible in terms of the population proportion of 1's. The normality of the sampling distribution of the sample proportion is realized in large samples just as it is of sample averages in general.

13. The normal approximation to a probability distribution is determined by first expressing the variable in standard units and then determining any probabilities needed as if the variable had a standard normal distribution. The probabilities calculated in this way can then be reexpressed in terms of the original units of the data.

14. It is not necessary to assume a particular probability distributional form (such as normal or binomial) when the data set is large enough to estimate the whole probability distribution well.

Glossary

Descriptive statistics is the area of statistical method concerned with the description of data sets. (7.1)

Inferential statistics is the area of statistical method concerned with inferences about a population based on sample data that have been selected from the population. (7.1)

The description of the observation process in terms of a population and a sampling method is called a **statistical model.** It "models" the way a sample is produced. (7.2)

Populations that exist in the real world are called **concrete populations;** they are tangible collections of objects as opposed to figments of the imagination. (7.2)

A **hypothetical population** is a population that is hypothesized for the convenience of describing a source of data. (7.2)

The **probability** of a particular outcome in a repeatable experiment is the long-run relative frequency of the outcome. (7.3)

A **probability distribution** is a list of population items and their associated probabilities. (7.3)

The terminology of probability experiments (7.3):

Experiment: the framework for a chance outcome

Trial: one of many similar elementary experiments

Sec. 7.9   Summary and Glossary

Outcome: the result of an experiment or trial
Way: one of several equally likely chance outcomes
Selected at random: selected in a manner that makes each choice equally likely

Rules for calculating probabilities:

The addition rule: If A and B are two events, the probability of A or B occurring is the sum of the probability that A occurs and the probability that B occurs, if the events A and B are incompatible. (7.3)

The multiplication rule: If A and B are two events, the probability of A and B occurring is the product of the probability that A occurs and the probability that B occurs provided the events A and B are independent. (7.3)

The complementarity rule: If A is an event, the probability that A does *not* occur is

$$1 - \text{the probability of } A \quad (7.3)$$

Independent events are events for which the occurrence of one event does not change the chance of occurrence of the other(s). (7.3)

Incompatible events have the property that the occurrence of one event means the other(s) cannot happen. (7.3)

A **binomial experiment** is be characterized by the following elements:

1. A certain number of trials is specified for the experiment.
2. Each trial can result in one of two outcomes, A or B, say.
3. The chance of outcome A at each trial is some constant. This chance applies to each of the specified trials.
4. The trials are independent of each other.
5. The final outcome recorded is the total number of occurrences of event A. (7.3)

A **binomial distribution** is the probability distribution of the number of occurrences of outcome A in a binomial experiment. (7.3)

Estimation is the process of using frequency distributions in samples to tell us about frequency distributions in populations. (7.4)

A **parameter** is a characteristic of a frequency distribution or its associated population. (7.4)

A procedure for calculating a summary number that is based on the sample is called a **statistic.** The calculated summary number is the value of the statistic. (7.4)

The **expected value** of a statistic is the long-run average value, on repeated random sampling with replacement. (7.4)

The **standard error** (or **SE**) **of the sample average for averages of samples of size** n is defined as the standard deviation of the long-run distribution of averages of samples of size n. (7.4)

The **sampling distribution** of a statistic is the frequency distribution of the statistic, based on a fixed sample size over a great many samples. (7.4)

The standard error for an average in sampling without replacement from a finite population must be scaled down by a **finite population correction factor**

$$\sqrt{1 - \frac{n-1}{N-1}}$$

where n is the size of the sample and N is the size of the population. This factor is multiplied by the standard error that would apply if the sampling were with replacement, or if the population had been infinite. (7.4)

The **continuity correction** is the adaptation necessary to approximate the probability distribution of a discrete variable by that of a continuous variable. (7.5)

PROBLEMS AND PROJECTS

1. Construct a sampling model for the rolling of two dice: the outcome recorded is the sum on the upward faces of the two dice. Compute the expected value of the average sum for 10 rolls of the two dice. What is the chance that the average sum from the 10 rolls is 10 or better?

2. Suppose that a population consists of an unknown number of 0's, which we will call n_0, and an unknown number of 1's, which we will call n_1. The entire population may be assumed to consist of these 0's and 1's. A sample of size 25 is selected at random, with replacement, from this population. It consists of 10 1's and 15 0's.
    (a) What is the sample average and SD?
    (b) What is the standard error of this sample average, based on the sample information?
    (c) How close is the sample average to the population average?
    (d) What are reasonable values for n_0 and n_1 if the population size is known to be 100?
    (e) Estimate a range of values for the proportion $n_0/(n_0 + n_1)$.

3. A proud high jumper clears 2 m on the first jump of the day with probability 0.70. If he fails on the first try, he clears 2 m on the second try with probability 0.90. If he fails on the second try, he does not jump again the same day. What is the probability that the high jumper clears 2 m on any particular day? (Assume that the probabilities never change from day to day.)

4. While you are attending a party, a charming fellow offers you the following "deal." He shows you three special cards which are colored red and black: one card is red on both sides, one is red on one side and black on the other side, and the third one is black on both sides. The cards are shuffled and placed in a hat, and you select one and place it on the table without looking at the bottom side. Neither you nor your acquaintance has seen the down face of the card on the table, nor have either of you seen the cards that remain in the hat. As it turns out, a red side is showing face up on the table. Your acquaintance says: "Which card do you think is on the table: red-red or red-black; it is obviously not the black-black one. The two cards that are possible each have an equal chance of selection, so I guess the bet should be an even one. I'll bet you $2 that the hidden side of the card on the table is red." Is this a fair bet? Explain.

5. There are sixty words in the paragraph that make up this problem. Select a random sample of size five that will allow you to estimate both the mean and the standard deviation of the word length distribution in this paragraph, and calculate these estimates. ("Word length" is the number of letters in a word.) Explain your sampling procedure in detail.
6. Calculate the actual error of the average used in Problem 5. Compare this with the standard error of the average. Comment on any discrepancy between the two.
7. Which is more likely: to get 6 heads in 10 tosses of a fair coin, or to get 60 heads in 100 tosses of a fair coin? (This problem is continued as a computer project in Problem 33.)
8. Three hundred students of a statistics class are asked to keep a diary of their activities for one week. Five diaries are selected, anonymously and randomly, for analysis. Nonclass times spent studying the statistics material for these five students turn out to be

    6.2, 8.5, 3.4, 4.8, 7.2 hr

    On the basis of this information alone, how many students do you think spend
    (a) More than 6.0 hr?
    (b) More than 10.0 hr?
9. Seven squash players of comparable skill at the game all want to play for the club team, which has five positions on it. The coach decides that the fair way to form the team is to select five at random for each day the team is to meet with another club: at the end of the season the success of each particular selection of five players will be compared as to performance. The coach plans to use this information for selecting the team to play a special European tour at the end of the regular season. He believes that individual player scores do not tell the whole story since there is considerable interaction and advice between players participating in any given meet. If the regular season consists of 21 meetings, how will the coach's plan work out? That is, does his method of analysis sound like a good one for selecting a winning team?
10. A coin is tossed repeatedly and the outcome recorded, head or tail. Verify that the expected value of the number of heads in a sequence of any given length is one-half the number of tosses, and also that the expected value of the percentage of heads in the sequence is 50 percent. If the sequence of tosses is very long, will the percentage of heads be close to 50 percent? Will the number of heads be close to one-half the number of tosses? Explain. (This problem is continued as a computer project in Problem 34.)
11. Sample averages tend to get close to their expected values as the sample size gets larger. What about sample totals? (That is, is the total income in a sample of 1000 employees close to the expected value of that total? The expected value is the population average income times the sample size.) Assume that the sample size never gets close to the population size—ignore the finite population correction factor.
12. An "objective" examination is one whose questions are answered by indicating which of several alternatives is the correct answer. A statistics exam of this type has 100 questions, and each question has five choices, only one of which is correct. Five students take the exam with the following results indicating the number of correct answers: 0, 23, 45, 66, and 77. Which student shows the best, and which the worst, understanding of statistics based on this test? Explain.
13. A statistics professor asks students to guess his weight. He records the distribution of guesses and notes that the average is 173 lb and the SD is 20 lb. The professor displays the results from the same course in the previous year, for which the results were an average of 176 lb and an SD of 22 lb. Most of the 300 students in each class have contributed their

guess. What point about precision and bias is the professor attempting to make when he reveals that his actual weight is 200 lb?

14. A population consisting of the 10 integers $\{1, 2, \ldots, 10\}$ is to be sampled randomly in two ways:

- A random sample of size 5 selected with replacement
- A random sample of size 5 selected without replacement

Perform this sampling process four times using each method. Calculate the SD of the averages of the four samples for each method. Do the two methods produce SDs for the averages in accordance with the theory? What would happen if samples of size 10 were selected instead of size 5? (This problem is continued as a computer project in Problem 35.)

15. (a) Suppose it is known that 52 percent of North Americans alive today are female, and that 20 percent of North Americans alive today are over 65 years of age. If possible, compute the percentage of North Americans alive today that are both female and over 65 years of age. If this is not possible, explain why.
    (b) Suppose it is known that 50 percent of North Americans alive today are aged 25 to 65 years, and 10 percent of North Americans alive today are over 60 years of age. If possible, compute the percentage of North Americans alive today that are aged 25 or more. If this is not possible, explain why.

16. While speeding down the highway, you are flagged down by the police radar unit. The officer asks you if you know how fast you were going and you say "about 105 km/hr" (the speed limit at this spot is 100 km/hr). Does the measurement error in your estimate "105 km/hr" contain bias, or chance error, or both, or neither? Explain your answer.

17. A student is given a test of 100 true–false questions with the understanding that the grading scheme is as follows:

Letter grade	Number of correct responses
A	85–100
B	65–84
C	45–64
F	0–44

    Suppose that the student answers each question by tossing a fair coin to decide between true and false. What is the chance that this student gets a C?

18. An opinion poll concerning student activities at ABC University is to be performed. The university undergraduates are sampled as follows. A random sample of 35 males (out of 3500) is combined with a random sample of 30 females (out of 3000), to make a total sample of 65 students. Is this total sample a random sample of 65 from the 6500? Explain why or why not.

19. Five brothers all take an IQ test. The test is scored in a way that is supposed to make the IQ result depend on mental potential but not on age. The test administrator is surprised that, in spite of the foregoing, the IQs increase with age in this family; that is, the eldest brother has the highest IQ, the next-to-eldest brother has the next to highest IQ, and so on. What is the chance that this particular ordering of IQs would occur just because of measurement

error (i.e., even if all the brothers have the same potential as measured by such tests, so that all orderings are equally likely)?

20. A charitable lottery is set up in which 1000 tickets are sold for $10 each. A simple random sample of 25 tickets is selected to receive prizes of $100 each. A company buys 100 tickets for its employees as a Christmas bonus.
    (a) *True or false:* The company has a $(25/1000)^{100}$ probability of holding all the winning tickets. Explain.
    (b) Estimate the amount of prize money won by the company's 100 tickets. Indicate the precision of your estimate.

21. Scores on a graduate-school entrance exam for a group of 64 ABC University students average 650 and have an SD of 100. A rumor has spread that a score of 700 is necessary for entrance to the top 10 schools in North America. If the ABC University students taking the exam can be assumed to be a random sample of the "population" of individuals who took the exam, estimate the percentage of the population whose exam scores would be over 700.

    If you are then told that 18 of the 64 ABC University scores (i.e., about 28 percent) were over 700, would you adjust your previous estimate? Justify your answer.

22. A survey of recent high school graduates who are first-semester students at ABC University is to be undertaken to determine the number of computing science courses they expect to take before they receive their undergraduate degree. It is suggested that the random sample of 64 students who are to be interviewed be selected as follows:

    - 25 students from those who are currently registered in a computing science course
    - 39 students from those who are not currently registered in a computing science course

    These proportions, $0.39 = 25/64$ and $0.61 = 39/64$, are known to be the proportions of these two groups of students in the target population. Is the suggested method likely to be better than simple random sampling? Explain.

23. Indicate whether the following statements are true or false. Explanations must be provided.
    (a) Large random samples provide estimates of population averages that are both precise and unbiased.
    (b) A sampling distribution of a sample percentage has an SD that depends on the population percentage. In typical surveys, this SD is usually not known.
    (c) Random sampling with replacement of a population consisting of two items can always be characterized as a binomial experiment.
    (d) A binomial experiment with a large number of draws (e.g., tossing a coin 1000 times) results in an outcome (e.g., number of heads) that is very close to its expected value.
    (e) If a population distribution is strongly skewed, the sampling distribution of the sample average may be far from normal for small samples.

24. You are offered the chance of playing the following game, for a price that has yet to be determined. A calculator is programmed to produce a sequence of digits indistinguishable from random digits (i.e., it produces digits as if drawn with replacement from a box containing the digits 0, 1, 2, . . . , 8, 9 with equal relative frequency). If the next three digits produced by the calculator are identical, you are paid an amount equal to this identical value, in dollars. Any other outcome pays you nothing.
    (a) How much should you pay to play this game if your aim is to break even, on average,

provided that you play enough times? (This amount is to be paid for each set of three numbers.) (*Hint:* Construct an appropriate sampling model for the game payoff.)

  (b) If you were to play this game 1000 times, how much money would you expect to win? Ignore your payment for playing the game in this calculation. Indicate a range of values in your answer. (This problem is continued as a computer project in Problem 36.)

25. The following questions relate to random sampling from a population $\{1, 2, 5, 6\}$. A random sample of size 2 is selected without replacement from this population.
  (a) List the six possible samples and compute the sample average for each one. Are these samples equally likely?
  (b) Compare the SD of the six sample averages with the SE of the sample average for a sample of size 2 selected without replacement from the population. Explain the outcome of this comparison.
  (c) If the sampling had been done with replacement, would the SE of the sample average be larger or smaller than the SE of the sample average for sampling without replacement? Explain.
  (d) What is the SE of the sample average for a random sample of size 4 selected without replacement? (This problem is continued as a computer project in Problem 37.)

26. In political opinion polls that use random sampling, SE estimates do not depend on whether sampling is done with or without replacement. Explain why you do, or do not, agree with this statement.

27. A squash team of five players is to be selected from a roster of seven players. If the five are a simple random sample from the seven, and if the seven are all of unequal ability, what is the chance that the five best players are selected? (This problem is continued as a computer project in Problem 38.)

28. This problem guides you through an exploration of sample estimates of a population average and standard deviation. Consider the small population used in the text: $\{2, 3, 7\}$. List all possible samples of size 2 that can be selected from this population, assuming that the sampling is done *with* replacement. The list must include nine possible samples, listing $(2, 3)$ and $(3, 2)$ separately, in order that each sample in the list has the same chance of occurring. $(2, 2)$ is listed once.
  (a) For each of the nine samples, calculate the average and SD. Note that SD of two numbers is one-half their range.
  (b) In a statistical experiment a large number of samples of size 2 are selected at random, with replacement, from the small population. What is the connection between the long list of experimental outcomes and the list of nine samples constructed in part (a)?
  (c) Calculate the average of the sample averages in part (a) and compare it with the population average of 4.00. Is the sample average an unbiased estimate of the population average?
  (d) Calculate the average of the sample SDs and compare it with with the population SD of 2.16. Is the sample SD an unbiased estimate of the population SD?
  (e) In a statistical procedure called analysis of variance (which we introduce in Chapter 9), the population parameter of interest is the population variance, which we will call V. The connection with the population SD is that

$$V = (\text{population SD})^2$$

In other words, for the small population defined above, $V = 4.67$. Verify numerically that SD^2 is not an unbiased estimate of V but that $S^2 = SD^2 \times [n/(n-1)]$ is an unbiased estimate of V. (n is the sample size here, so $n = 2$.)

(f) Is S an unbiased estimate of the population SD?

(g) The population SD is computed by calculating the differences of the population values from the population average. The sample SD is computed by calculating the differences of the sample values from the sample average. If we were to calculate a new statistic using the differences of the sample values from the population average, call it X, how would it compare, on average, with the sample SD? (Just think about this to get an answer.)

(h) Compare numerically the average value of X^2 and S^2 as estimates of the population SD^2. (This problem is continued as a computer project in Problem 39.)

29. One version of "solitaire parcheesi" is the following. A course of 100 spaces is to be covered in as few turns as possible. A "turn" consists of rolling a die as many times as one wishes and advancing your game piece by a number of spaces equal to the sum the rolls of the die. The only constraint is that if a 1 is rolled, you must score zero for the whole turn, and your game piece stays at the same position it was in at the start of the turn. Subsequent rolls begin a new turn.

What strategy would you use to play this game—that is, to reach 100 in as few turns as possible? Assume that you have time to play the game only once.

30. One hundred tickets are given out at a company picnic for the purpose of distributing a cash prize. Ticket holders write their name on the ticket and put it in a barrel. Each ticket carries a 10-digit number, and each digit in this number has been determined by random sampling of the digits 0, 1, 2, . . . , 8, 9. The amount of the cash prize is to be the sum of the digits (in dollars) on the one ticket that is selected from the barrel, and the prize is given to the person whose name appears on the ticket.

Before approving this prize scheme, the president asks her assistant to estimate how much the prize is likely to cost. If you were this assistant, what information would you provide the president? Be specific.

31. On a slow day in the city, you decide to play the following game. You flip a coin and move forward one sidewalk square if it is a head, and move backward one square if it is a tail.
  (a) After 10 flips, what are your chances of being at exactly the same place that you started?
  (b) After 25 flips, what are your chances of being at least five steps ahead of where you started?
  (c) After 250 flips, would you expect to be farther away from your start position, or closer, or about the same distance, compared with 25 flips?

32. A coin has the property that it turns up heads 2/3 of the time, in the long run. It is to be tossed 25 times.
  (a) Construct a probability model for the number of heads in 25 tosses of this coin (i.e., an idealized description of the mechanism which produces a certain number of heads in 25 tosses).
  (b) It is usually assumed that coin tosses are independent. Do you think that this assumption is valid? Explain.
  (c) Determine approximately the chance that fewer than five tails appears in the 25 tosses.

Computer Projects

33. (Continuation of Problem 7.) Generate the following series of binomial experiments, where in each one the probability of a head is 0.50.
  (a) 100 experiments of 10 trials each. Note the number of experiments in which there are 6 heads or more.
  (b) 100 experiments of 100 trials each. Note the number of experiments in which there are 60 heads or more.
  Use your results to answer Problem 7.

34. (Continuation of Problem 10.) Using a computer-generated sequence of 0's and 1's, where the chance of a 0 at any position is 50 percent, record at each position in the sequence the following two numbers:

  - The difference between the number of 0's and one-half of the position number
  - The difference between the proportion of 0's and 0.50

  (a) Summarize the behavior of these two sequences of numbers.
  (b) What is the main difference between the two sequences?
  (c) What general rule can you infer about the behavior of samples of data in which the data consist of records of the presence or absence of some quality?

35. (Continuation of Problem 14.) Do Problem 14 except that the sampling experiments should be repeated 25 times instead of 4 times.

36. (Continuation of Problem 24.) Using Monte Carlo simulation, play the game described in Problem 24 1000 times and record your winnings (ignoring the payment for playing). Repeat this entire procedure as many times as is practical, say 100 times, to attempt to determine the long-run average value of the winnings in 1000 games. Should this number be close to the answer to Problem 24(a)? Explain.

37. (Continuation of Problem 25.) Generate 100 samples of size 2, selected without replacement, from the population $\{1, 2, 3, 4\}$. Compute from formulas the expected value and the standard error of the average of samples of size 2 from this population. From the samples generated compute the sample averages and the SD of each of the 100 samples. Compute the average of the 100 sample averages and the SD of these 100 averages. Comment on the relationships expected among the numbers you have computed.

  Repeat all of this assuming that the samples are selected with replacement, using samples of size 2 as before.

  Answer parts (c) and (d) of Problem 25.

38. (Continuation of Problem 27.) (This problem does not require a computer, although the use of a computer may simplify the calculations. A calculator should be used, however.) The number of ways of ordering n distinguishable objects is the product of $1, 2, 3, \ldots, n$. For example, there are six ways to order the letters A, B, C: $ABC, ACB, BAC, BCA, CAB, CBA$, and $6 = 1 \times 2 \times 3$.
  (a) Compute these products for up to 10 distinguishable objects.
    If n objects are ordered, and they are all distinguishable except m of them which are identical, then for each ordering of the n, one can perform $1 \times 2 \times 3 \times \ldots \times m$ reorderings of the identical things that will produce an ordering of the n things that looks the same. For example, if the objects are A_1, A_2, B, where A_1 and A_2 are identical, A_1, A_2, B and A_2, A_1, B would look the same and $2 = 1 \times 2$.

(b) Explore these ideas and report your findings.
(c) If a sequence of n things consists of only two different kinds of things, say n_a of A and n_b of B, how many ways can this sequence be rearranged so that it looks the same? How many ways can it be rearranged so that it looks different (i.e., the order of the A's and B's is different)?
(d) A coin is tossed 10 times. The outcome is recorded as {H, H, T, H, ... , T}. How many different outcomes are there that include exactly three heads and seven tails? If the coin is fair, what is the probability of three heads? Does this relate to the binomial distribution? Explain. What if the coin is not fair?

39. (Continuation of Problem 28.) See the definitions of Problem 28. Using computer-generated samples of size 2, compute the averages of:

> The sample average
> The sample SD
> The square of the sample SD
> S
> V
> X

Comment on the relationships between these estimates and the population parameters already introduced: average, SD, and V.

Estimation

The golden rule is that there is no golden rule.

G. B. Shaw (1856–1950)

Man can learn nothing unless he proceeds from the known to the unknown.

Claude Bernard (1813–1878)

This is the second of three chapters that discuss inferential statistics. The theory was covered in Chapter 7, and Chapters 8 and 9 discuss applications of the theory. There will be more definitions and procedures, but most developments follow directly from theoretical results about the sampling distribution of an average.

The application of the sampling theory to be discussed in this chapter is called "estimation." Its aims are similar to the aims of descriptive statistics but with the big difference—we now attempt to describe a population, using a sample, rather than to describe the sample itself. The link provided by random sampling is all-important.

Key words are confidence interval, confidence level, t statistic, and t procedure.

8.1 INTRODUCTION

A carefully selected random sample does not produce a perfect description of the population sampled. Random sampling can produce unbiased estimates, whereas alternative sampling procedures do not; but we have to live with the sample-to-sample variability that accompanies random sampling procedures.

The opinions sampled in a Gallup poll are used to describe the opinions of the population sampled. Measurements of reaction times for a person are taken to describe a characteristic of the person. We do not expect these descriptions to be perfect: we must be prepared to tolerate the fact that the description provided by the sample will contain "error." Any two samples will yield different descriptions and they cannot both be perfect descriptions of the same population.

In this chapter we are concerned primarily with ways to detect and report the amount of error in a sample, based on information contained in that very same sample. This is a bit like pulling yourself up by your own bootstraps, and it is amazing that it can be done at all. The principal tool used to report the error in our estimates is the "confidence interval." It will be seen to be, simultaneously, the estimate and an indication of the error in the estimate.

One point to bear in mind: all the techniques of Chapters 7 to 9 apply only to random samples. If a pollster walked the streets of several towns and cities selecting subjects of all varieties and persuasions to respond to a poll, the results would simply be a conglomerate of opinions representing no particular population. Random sampling is not the same as sampling a wide variety. Nor is it the same as sampling haphazardly or without regard to the expected response. It is sampling done according to a particular prescription. For the elementary forms of random sampling, the prescription requires that the chance of selecting a sample of a certain size has the same chance of being selected as any other sample of the same size. The lottery procedure is one that produces samples with this property.

In Chapters 4 to 6, which dealt with "descriptive statistics," we discussed methods for describing whole distributions and for describing the center and spread of those distributions. When we are starting with random samples, the analogous tasks are to describe the population distribution, or the parameters of this distribution. In Sections 8.2 and 8.3 the parameter of interest is the population average, and in Section 8.6 the population distribution is to be estimated. Section 8.4 discusses the question of sample size: How big must the sample be so that the variation introduced by random sampling has an acceptably small influence on our estimates? Section 8.5 introduces a new sampling distribution, the t distribution, which can sometimes be used to describe sample averages when the sample is too small to provide a reliable estimate of the population standard deviation.

8.2 ESTIMATION OF POPULATION AVERAGES

In Chapter 4 the term "accurate" was reserved to describe measurements that were both unbiased and precise. Random samples, whether drawn with or without replacement, have the property that the sample average is an "unbiased" estimator of the population average: that is, the expected value of the sample average is equal to the population average. This means that a random-sample average is accurate if it is precise.

The precision of an estimator may be gauged by the standard deviation of the distribution of this estimator over many samples. This standard deviation can be estimated from a single sample when the estimator is the sample average. As discussed in Chapter 7, the SD of the distribution of the sample average is the standard error.

The standard error is estimated from a single sample by (sample SD/square root of the sample size).

Figure 8.1 emphasises an important point: Sample averages do not vary as much as the population from which the sample is selected. In fact, if the SD of a population is 1.0, as in Figure 8.1, the sample SD will also be about 1.0, but the SD of the distribution of averages of size 16 will be $1.0/\sqrt{16}$ or 0.25.

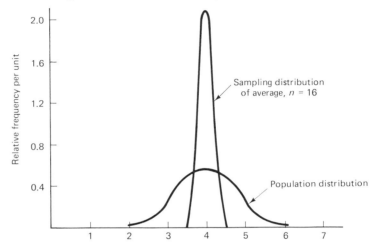

Figure 8.1 Population with SD = 1.0 and the sampling distribution of averages of samples from this population, based on samples of size 16, with a smaller dispersion than the population.

The name for the distribution of a statistic is the "sampling distribution." Be sure to distinguish these sampling distributions, which apply to statistics over many samples, from the sample distribution, which is the distribution observed in a single sample. The "sample SD" means the SD calculated from a sample—it estimates the population SD. But the SD of the sampling distribution of the average is smaller than the sample SD—the former measures the dispersion among averages of many samples. It is this SD of the sampling distribution that is given the name standard error, which we abbreviate as SE.

The only way to avoid confusion when we are discussing SD and SE is to keep asking "SD of what?" and "SE of what?". The SE describes the variability of a statistic, and different statistics have different SEs. In this text we will be concerned primarily with the SE of averages. The SD describes the variability of a frequency distribution, either a population distribution or a sample distribution.

To see if you have followed the the last two paragraphs, ask yourself whether the following statement seems obvious, or bewildering: The SD of the sampling distribution of the average is less than the SD of the population sampled, as long as the sample size on which the average is based is greater than 1. If this is not yet clear, you should reread the last four paragraphs.

The final element that completes the theory of the distribution of the sample average is the central limit theorem. It tells us that the distribution is approximately normal, no matter what the population distribution is, as long as the sample size is sufficiently large. When the sample is not sufficiently large, the theory is more complicated: one strategy, which uses the t distribution, is discussed in Section 8.5. For now we concentrate on the large-sample theory.

We can summarize our large-sample theory about the sample average as an estimator for the population average as follows: *The sampling distribution of the sample average is approximately normal with average equal to the population average and standard deviation equal to the standard error.* Of course, this result would not be very helpful if one really had to know the population average and standard deviation to use it; in estimation problems, the whole point is to learn about the population parameters from the sample statistics. However, the result above can be combined with our knowledge of the sample average and SD to yield useful statements about the population parameters.

To explore this process of estimation of a population average, let us consider a sample of pulse measurements from 10 long-distance runners:

$$\text{Pulse (beats/min): } 45, 54, 47, 58, 62, 48, 52, 45, 60, 49$$

The average of these is 52.0 and the SD is 5.9. We clearly will use the sample average 52.0 as our estimate of the unknown population average. However, since this number 52.0 is based on a random sample, we do not expect it to be exactly the same as the population average, and if we are interested in the population average, we must obtain some idea of how accurate the estimate is. The standard error tells us this: The standard error of the average, which we will denote SE, is the standard deviation of the sample average among the averages of many samples of the same size. We do not know it exactly, but an estimate of it is

$$\frac{\text{sample SD}}{\sqrt{n}}$$

where n is the sample size. For the example it is 5.9/3.16 or about 1.9. Our conclusion would be that the population average is about 52.0 ± 1.9 beats per minute.

What do we really mean by ±1.9? First let us ignore the fact that the 1.9 is an estimate of the SE—rather, we will suppose temporarily that it is exact. The statement that the SE is 1.9 means that *if* one were to repeat the estimation procedure many times, 68 percent of the time the sample average would be within 1.9 of the population average. Of course, it is also true that 68 percent of the time, the population average would be within 1.9 of the sample average, and this gives a slightly more useful statement when we know the sample average. After all, our aim is to describe the population, not the sample.

Now, the interval 52.0 ± 1.9 is just one such interval that might have been generated from the same population. If we imagine several samples of the same size, the procedure for calculating this interval stays the same but the interval itself will change from sample to sample. When the SE is known exactly, only the center of the interval, determined by the sample average, varies from sample to sample. The intervals all have equal width. In this situation, the statement that the population average is within 1.9 of the sample average is equivalent to the statement that the interval "sample average ±1.9" contains the population average. This interpretation, sometimes called the *coverage property*, makes the interval itself an estimate of the population average. This is an example of a "confidence interval." In fact, it is called a "68 percent confidence interval" for the population average.

Figure 8.2 illustrates typical intervals that could be generated from several samples. Sixty-eight percent of them should enclose the population average.

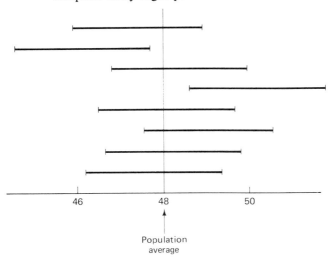

Figure 8.2 Varying confidence intervals for a population average, based on several samples from a single population. In 68 percent of samples, these 68 percent confidence intervals will include the population average. Note that two of the eight do not have this property. In a large number of confidence intervals generated in the same way, the percentage of such "misses" would be 32 percent.

The concept of a confidence interval does not depend on knowledge of the standard error (the 1.9 of the example), and this is why we needed the coverage interpretation of the confidence interval statement. We will describe confidence intervals of varying length in Section 8.5. The procedure described in this section does assume that the SE is known. When the SE is estimated by substituting the sample SD for the population SD in the formula for the SE, the procedure is not quite correct. For large samples this error will have negligible effect. As a rule of thumb, a sample size of 20 should be adequate to get an estimate of the SE that is close enough to the population SE that the difference can be ignored, at least for the practical application of the confidence interval procedure described here.

Let us now define confidence intervals more formally. We must distinguish between the intervals and the calculation procedure that produces the intervals. An **L percent confidence interval procedure** for a parameter is a method of calculation that produces intervals from random samples; these intervals include a certain population parameter value for L percent of the samples. This percentage is called the **confidence level** of the confidence interval procedure. An L percent **confidence interval** for a population parameter is an interval produced by an L percent confidence interval procedure for that parameter.

This definition of a confidence interval is a bit complicated. (So is any correct definition of a confidence interval.) We will now review the various aspects of the definition.

There is a right way and a wrong way to interpret confidence interval statements, and the difference has practical importance. A 68 percent confidence interval is not an interval which will contain 68 percent of the sample averages that would be obtained on repeating the sampling procedure. Such an interval would most naturally be centered at the population average, which we do not know, whereas a confidence interval is centered at the sample average. Nor is a particular 68 percent confidence interval the interval in which the population average will fall 68 percent of the time; the population average is a constant and does not vary, either it is in the interval or it is not. Rather the 68 percent confidence interval is an interval that is calculated according to a certain procedure that would produce a different interval for each sample, and 68 percent of these intervals would overlap the population average. It is the intervals that are varying, not the population average. Figure 8.1 depicts several confidence intervals to demonstrate where the 68 percent comes in. Of course, if we have only one sample from which to estimate the population average, which is the usual situation, we would be able to calculate only one of these intervals.

Confidence intervals are interval estimates. They produce a whole interval as an estimate for a population parameter, instead of just a single value. Single-value estimates are sometimes called point estimates: 52.0 is a point estimate in the example. Point estimates by themselves do not have much practical value. For example, if the purpose of the pulse estimation for the long-distance runners is to compare their average pulse with the "norm" for ordinary healthy people, it is important to know whether the population average for the runners could reasonably be 72.0, given that the sample average turned out to be 52.0. This kind of information depends on knowledge of the SE, not merely of the point estimate itself.

How do we specify the confidence level in practical applications? When an interval is calculated which is claimed to contain the value of the population average, how sure must we be that the claim is correct? The more certainty we want in this claim, the wider the interval will have to be. But a very wide interval estimate may not be very useful because it fails to identify the population parameter closely enough for practical purposes. We apparently have a dilemma: we will have to trade off the precision of the estimate (the narrowness of the confidence interval) for the degree of certainty that the interval contains the population parameter value. There is a way to avoid this awkward choice, and that is to increase the sample size; this is discussed in Section 8.4. When this option is not feasible, the choice has to be made.

The degree of certainty that the confidence interval contains the population parameter value is represented by the confidence level. It is measured by the percentage of samples that produce satisfactory confidence intervals, where "satisfactory" here means "containing the population parameter value." Thus 68 percent was a confidence level in the example.

In practice, it is more usual to use 90, 95, or 99 percent as confidence levels. In other words, when a confidence interval is calculated, the user wants to be quite sure that the interval actually contains the population parameter value. To obtain these intervals, one simply marks off on either side of the sample average a larger multiple of the SE than is needed for the 68% confidence interval (where the multiple was 1.0). These multiples can be read from the normal probability table, as long as the sample size is large enough to justify the normal approximation to the sampling distribution. (If it is not, see Section 8.5.) You may wish to check that the following do, in fact, come from the normal probability table:

Confidence level (%)	Multiple of SE used
68	1.00
90	1.64
95	1.96
99	2.33

Thus a 99 percent confidence interval for the pulse of the long-distance runners is $52.0 \pm (2.33 \times 1.9)$, which works out to 47.6 to 56.4 beats per minute. Any value in this interval is much less than 72.0. Apparently the sample was large enough to effectively compare the pulse of this group with the norm value 72.0.

This section has described an estimation procedure for estimating a population average based on a random sample from the population. The confidence interval idea incorporates the information about the variability of the estimate into the estimate itself. The sample mean is itself an estimate of the population mean, but without further information about the SE of the estimate, is usually not useful. A procedure that is often used in reporting estimates of population averages is simply to state the sample average and the SE. Those knowledgeable in statistical procedures will feel that the two methods provide the same information, at least for large samples. Others are liable to misunderstand either method; the most common mistake in interpreting the average and SE method is to use the SE as if it were the SD of the sample. The most common misunderstanding of the confidence interval is to think that it describes the range of values where most of the population values lie.

The estimation methods outlined above also apply to an apparently more complicated situation. Suppose that a sample of 50 first-year university students are assessed for the amount of body fat they carry at the beginning and end of the academic year. A method for measuring (approximately) the percentage of body weight that is fat is to measure the "skinfold thickness" using a special pair of calipers. The problem is to estimate the change in this measure over the year. We begin with 100 measurements, two measurements per student, but before we do any calculations of averages or SDs, we take differences for each student. The question concerned the change in body fat, so we should concentrate on the data that measure this change. This seems obvious. Perhaps we should look at the alternative wrong method. For this let us consider just five students' data:

Student	Fall skinfold (mm)	Spring skinfold (mm)	Increase (mm)
1	6	7	1
2	10	11	1
3	5	6	1
4	12	13	1
5	8	9	1
Average	8.2	9.2	1.0

There is a big difference between examining the change in the average skinfold and examining the average skinfold change, even though they both equal 1 mm. The variability of the latter is much less. The fall average is clearly not very well estimated, and neither is the spring average, since there is a lot of variability among students. If we forget that the data in the two measurement times came from the same students, we would be unimpressed by the change in the group. But the change is quite predictable when we use the pairing of the observations, and the data are convincing of a systematic change. The point is that paired data should be treated as such.

There is nothing new about the actual estimation procedure for paired data. If the 50 students had an average increase of 1 mm and a standard deviation of 2 mm, the increase of the skinfold measurement would be estimated by a 68 percent confidence interval as

$$1 \text{ mm} \pm \frac{2}{\sqrt{50}} \text{ mm} \quad \text{or} \quad 1 \text{ mm} \pm 0.3 \text{ mm}$$

Although before-and-after studies are a common source of paired-data sets, the pairing arises in other situations as well. If two different brands of skinfold caliper are to be compared, it would be natural to compare them using pairs of measurements on several subjects at a single time. Or a psychologist might want to compare visual skills in the dominant eye with the opposing eye.

The "paired data" we have been discussing here is different from the "pairs of data" needed for regression and correlation analysis. Paired-data amenable to the treatment in this section is such that to take the difference between the pairs is a natural way to begin the data analysis; this difference has to make sense. In particular, the pairs in paired data will usually be in the same units—this is often not the case in regression or correlation analyses.

EXERCISES

1. A high school alumni association is anticipating its first fund-raising effort. It feels it would be desirable, in its promotional material, to offer choices to contributors about which projects they would like to support. The problem in doing this is that the association

executive has no idea how much money the campaign will raise, so does not know what projects are reasonable. Since the vice-president had taken a statistics course in college, she suggests a sample survey on a small random sample of 20 alumni. Ten alumni would receive appeals mentioning small projects as requiring funding, the other 10 mentioning large projects. After these 20 have responded, the true nature of the pilot survey would be explained, and they may reconsider their gift (or absence of one) in view of the final choice of promotional material. (Otherwise, the 20 could rightly claim that they were not presented with the same choice as the rest of the alumni. Of course, if they had been told of the nature of the pilot study before it was done, this might have biased their response.) The responses from the 20 alumni were recorded as follows:

	Small projects	Large projects
	$ 5.00	$ 7.50
	0.00	10.00
	2.00	0.00
	10.00	25.00
	3.00	5.00
	2.00	2.00
	5.00	20.00
	1.00	5.00
	7.00	10.00
	2.00	5.00
Average	$ 3.70	$ 8.95
SD	2.90	7.48

If the alumni association membership list includes 500 names, estimate the amount that the two types of surveys would probably raise. If the small projects cost about $1000 and the large projects cost about $7000, which promotion should the association choose in order that it be able to fund the prospective projects? (Assume that the SD estimates are close to their population values.)

2. Refer to Exercise 1. Find 99 percent confidence intervals for the contribution per member in each of the two groups.

3. Which of the following statements are true, which are false, and which are ambiguous? Justify your answer.
   (a) Confidence intervals are more informative than point estimates because they give an idea of how much the population average can vary.
   (b) The sample average will vary from sample to sample; its standard deviation can be estimated from a single sample.
   (c) A 95 percent confidence interval states the interval that one can be 95 percent confident will contain the value of the population average.
   (d) The standard error of an average cannot be calculated exactly based on sample information alone.
   (e) A 95 percent confidence interval based on a sample of size 10 will contain the sample mean in 95 percent of the samples of this size that are selected from the same population.

Sec. 8.3   Estimation of Population Percentages   **299**

4. The estimate of the average number of years that certain species of peach tree will grow before bearing its first fruit is 3.5 years. This estimate was based on a sample of 30 trees, and the SE of the estimate is 1.0 years. Estimate what size of sample would yield an estimate with a standard error of 0.5 year.

5. A random sample of 35 cars is selected from the student parking lot at noon on the first day of classes. These cars are appraised to have an average market value of $3500 and the SD of these values is $500. The same cars are again appraised at the end of the school year, to assess the capital loss incurred in the owning of the cars, for students who bring cars to campus. This new appraisal averages $3000 and has an SD of $450. Estimate the students' average capital loss.

8.3 ESTIMATION OF POPULATION PERCENTAGES

In Chapter 5 the special techniques suggested for the description of qualitative variables included the use of contingency tables and percentages. Contingency tables displayed the relationship between two qualitative variables; for quantitative variables it is usual to use correlation and regression to do this. Percentages were used in these contingency tables, whereas averages and SDs were used for quantitative variables. In estimation problems we again have slightly different procedures for qualitative data where percentages are to be estimated than we have used for quantitative data where averages are to be estimated.

Consider the following study. Twenty articles are randomly selected from the recent social science journals; the *Humanities and Social Science Citation Index* is a good list of such articles from which a random sample can be drawn. A panel of experts in several social science disciplines and several professional statisticians assess the adequacy of the research methods used, and they decide whether the methods are acceptable or not. The aim of the study is to estimate the proportion of articles voted by a majority of the experts to be acceptable. The population is all articles in the index. Now the data from this study consist simply of a sequence of notations indicating acceptable (A) or not acceptable (N). That is

$$A, N, N, N, A, N, A, A, N, N, N, A, N, A, N, A, N, N, N, A$$

What is a natural summary of these data? Certainly, it should include the statistic 40 percent acceptable derived from the eight A's out of the 20 papers evaluated. But how close is this statistic to the desired percentage of acceptables in the population of interest? What we need is a standard error for a percentage.

Fortunately, we have already worked this out, as will be shown, since a percentage is actually an average of a sort. Let us first recode our sample of 20 qualitative outcomes as 0's and 1's: let A be coded as 1 and N be coded as 0. The data now appear as

$$1, 0, 0, 0, 1, 0, 1, 1, 0, 0, 0, 1, 0, 1, 0, 1, 0, 0, 0, 1$$

Now the average of this sequence is 0.40, and the SD is 0.48. If we imagine the whole population coded similarly, the estimate of the population average (of 0's and 1's) is $0.40 \pm (0.48/\sqrt{20})$, which is 0.40 ± 0.11. This is a 68 percent confidence interval for the population proportion of acceptable papers, since the average of the coded data *is* the proportion of interest. (We have used the normal approximation.) To express this in terms of percentages is easy—just multiply the proportion by 100: 40 ± 11 percent.

So estimation of proportions and percentages really presents no new challenge at all. In fact, they can be even simpler than for averages because there is a shortcut method of computing the estimates of the population percentages and of the standard error of the sample percentage. This shortcut is based on a simple count of the number of A's in the list and the total sample size.

The average is just the proportion of A's in the list. The standard error of the average is the product of the proportion of A's and the proportion of N's, divided by the sample size, and finally you take the square root. [Keep in mind that we have had to convert qualitative data into numbers (0 and 1) in order to make the terms average and SD meaningful.] Let us see if this gives the same answer as before: 8/20 is 0.40 as before. For the SE of the average we have $(8/20) \times (12/20)$, which is 0.24, divided by 20, which is 0.012, and the square root of 0.012 is 0.11, just as before. For long lists this shortcut is much faster than calculating averages and SDs on a calculator. (Of course, by computer the difference in time would be negligible if only a few such estimates were needed.)

What is a 99 percent confidence interval for the percentage of acceptable papers from the index? The full calculation is shown below—note that we work initially with proportions and convert to percentages at the end:

$$0.40 \pm \left[2.33 \times \sqrt{\frac{(8/20) \times (12/20)}{20}} \right]$$

which is

$$0.40 \pm (2.33 \times 0.11)$$

which is

$$0.40 \pm 0.26$$

In percentage terms this is

$$40 \pm 26 \text{ percent}$$

The interval may also be stated explicitly as (14 percent, 66 percent), meaning an interval from 14 to 66 percent. Now this is quite a large interval, and it would be hard to base a news story on it. In retrospect we can see that a larger sample was needed to get the kind of accuracy we want in this situation. We will deal with this problem in Section 8.4.

Did you notice that the sequence of A's and N's looked very much like the outcome of a binomial experiment? Let's check the assumptions (refer to Section 7.3 for a refresher on this):

1. Yes, the number of "trials" was specified at the outset of the study as 20.
2. Yes, the outcomes were one of two kinds, A or N.
3. Yes, it is reasonable to consider the probability of acceptability as being the same for each article. (The articles were selected at random.)
4. Yes, the acceptability of one article selected at random is probably independent of the acceptability of another article. (Strictly speaking, if some journals have higher standards for than others, the acceptability of two articles in the same journal would be dependent—either both would be acceptable or neither. However, with so many journals being searched for the index, the chance of two selections from the same journal is highly unlikely.)
5. Yes, the thing we are interested in is the number of acceptable articles, since this is just a multiple of the proportion of A's.

In other words, we can use Table A4.2, the table of binomial distribution, to tell us about the sample variability of the number of acceptable articles among the 20. However, we would use this method only if the sample were very small and we were not sure that the normal approximation were good enough. Certainly, for samples of size 20 or more, the normal approximation will be adequate. Usually, it is good enough for practical purposes in samples of size 10 or more. Of course, estimation of a percentage on the basis of a sample of size as small as 10 will not be very precise. In the journal-article example, even a sample size of 20 gave imprecise estimates of a population proportion. But if we must use such a small sample, the normal approximation method is adequate for assessing the precision for samples as small as size 10.

But let us use the binomial table, Table A4.2, to estimate the sampling distribution of the number of A's in the 20. Using $p = 0.40$ (our best point estimate), $n = 20$, the distribution is given in Table A4.2 as

Number of A's	3 or less	4	5	6	7	8	9	10	11	12	13 or more
Proportion of A's	0.15	0.20	0.25	0.30	0.35	0.40	0.45	0.50	0.55	0.60	0.65
Probability	0.015	0.035	0.07	0.12	0.17	0.18	0.16	0.12	0.07	0.035	0.021

Only 1.5 percent of the time is the proportion of A's 0.15 or less and only 2.1 percent of the time is it 0.65 or more. In other words, in $(100 - 1.5 - 2.1)$ percent or just over 96 percent of the sample, proportions from populations having a proportion 0.40 would produce sample proportions between 0.20 and 0.60 (including 0.20 and 0.60). This is not very comforting: apparently, it can happen that our estimate 0.40 could be off by quite a bit. The correct use of this table to produce confidence intervals is complicated, but tables exist which do this.[1]

It is a useful exercise to check that the binomial experiment applies to the sampling of population percentages. However, most applications use large enough sample sizes that the normal approximation can be used. This essentially avoids the complications.

We have used the fact that the sampling distribution of the sample proportion can be deduced from the result of the preceding section on the sampling distribution of averages. However, for ease of reference we state the sampling distribution for sample proportion here separately.

The sampling distribution of the sample proportion is approximately normal with an average equal to the population proportion, p, and an SD equal to

$$\sqrt{\frac{p \times (1 - p)}{\text{sample size}}}$$

The comments concerning paired data made at the end of Section 8.2 apply equally to paired qualitative data. See Exercise 4 for the calculation details.

Qualitative data often has more than two categories. If there is one particular one of interest, collapsing the others will reduce the data set to a kind we have already discussed. If we are interested in the distribution across categories, we need a more complicated analysis—see Section 9.5.

EXERCISES

1. At a particular university there are 800 different textbook titles available at the university bookstore. In a random sample of 50 of these texts, 12 percent are softcover. Find a 90 percent confidence interval for the percentage of the 800 titles that are softcover.

2. A characteristic of statistics textbooks is the percentage of pages that include a diagram or picture. If a typical textbook is 400 pages long, how accurately will this percentage be estimated by examining a random sample of 20 pages?

3. If the percentage estimated in Exercise 2 were the same for each of 10 textbooks, how much variation would you expect to see in the estimated percentages for the 10 textbooks?

4. A random sample of size 30 is obtained of all students who had part-time jobs at supermarkets while they were attending college. Each student was asked the question: "Which activity do you prefer: (1) attending lectures, or (2) working as a grocery clerk?" In the September survey, 22 students choose answer (1), but when these same students were questioned again in April, only 10 students chose (1). The 10 were all from among those who had chosen (1) in the fall. Estimate the proportion of all students in the population sampled that changed their mind between surveys. (*Hint:* Consider the 30 students' pairs of responses as a single sample of 30 responses.)

5. Almanacs regularly list the mayors (or city managers if this position performs mayorial functions) of larger North American cities. There are 610 such in an October 1983 list.[2] To examine the proportion of women among this group, a random sample of 40 names was selected. The sample yielded only two women and 38 men. What is the highest believable value that this proportion, the proportion of mayors that are female, could be among the 610? Assume that sampling is with replacement. (*Hint:* A trial-and-error procedure will work fairly quickly.)

8.4 DETERMINATION OF SAMPLE SIZE

Many novice researchers first seek statistical advice when they are confronted with the question: How big should the sample size be? They are usually dismayed when the statistician replies with a question of his or her own: How precise do you want your estimates to be? The connection between the two questions is the subject of this section.

8.4.1 Sample Size for Estimating Population Percentages

First, let us consider a limitation of all statistical sampling studies. It is certainly possible for a random sample to be unrepresentative of its population. Unlucky draws are possible. In a survey of opinions of academics about economic strategy for the nation, it is possible that of the population of 5000 sampled, the sample of 25 consists of all 25 Marxists from the 5000. The only simple random sample guaranteed to represent the Marxists in the correct proportion is the sample that includes the whole population. In practice, however, we have to be less demanding.

A more reasonable requirement of samples is that they have a high probability of providing estimates that are close enough for practical purposes. But to work with this prescription, we have to say what we mean by "close enough for practical purposes." For example, if the survey of economic strategy included a question about the advisability of rent controls, it might be considered that, if the percentage of academics in favor of rent controls could be estimated to within 5 percentage points, this would be adequate precision in the estimate. This specification must be arrived at from subjective consideration of the eventual use of the estimate, and has nothing to do with the survey data itself.

So the question may be more specifically put: How big does a sample have to be in order that the estimate of a population percentage is within 5 percentage points of the population percentage itself? In terms of proportions, we are requiring that the estimate be within 0.05. When we were calculating confidence intervals for a percentage in Section 8.3, the plus-or-minus figure attached to the point estimate was the amount that we had to move away from the sample percentage to be quite sure that we included the population percentage. The sample size is determined by insisting that this plus-or-minus figure is 0.05. The confidence level will then be the probability that our estimate will be within 0.05 of the population percentage.

Why does the specification of the plus-or-minus number in a confidence interval determine the sample size? Recall that this plus-or-minus number is computed from

$$(\text{multiple from Table A4.1}) \times \frac{\text{SD}}{\sqrt{\text{sample size}}}$$

The multiple depends on the confidence level. The SD for a percentage depended on the population percentage; we seem to be really stuck here because when we are determining a sample size, we do not even have a sample to estimate the population percentage. We would like to know this SD, so we could equate the expression for the plus-or-minus number to 0.05 and infer how big the sample size should be. Before we can proceed with this, we have to find a way around our ignorance of the SD before the sample is collected.

The key to this comes from looking at the expression for the SD of a percentage, or equivalently, of a proportion:

$$\text{SD of a proportion} = \sqrt{\frac{p \times (1 - p)}{\text{sample size}}}$$

where p is an abbreviation for "population proportion." The factor we do not know, $p \times (1 - p)$, can be shown to be not more than $1/4$. It is equal to this value when the population proportion is $1/2$. For population proportions closer to 0 or 1 the factor is less: for example, $0.10 \times 0.90 = 0.09$, which is less than 0.25, as claimed. So we can say for sure that the SD of the proportion in our example is not more than

$$\sqrt{\frac{\frac{1}{4}}{\text{sample size}}}$$

If our plus-or-minus number (0.05 in our example), indicating the precision required of the estimate, is larger than

$$(\text{multiple from Table A4.1}) \times \sqrt{\frac{\frac{1}{4}}{\text{sample size}}}$$

the estimate will be precise enough for practical purposes. If the confidence level we are using is 95 percent, the multiple is 1.96, and we need

$$1.96 \times \sqrt{\frac{\frac{1}{4}}{\text{sample size}}}$$

to be less than 0.05. This is true if

$$1.96^2 \times \frac{\frac{1}{4}}{\text{sample size}}$$

is less than 0.05^2. Or approximately, (1/sample size) is less than 0.0025. Or the sample size is greater than $(1/0.0025) = 400$.

At last, we have an answer to the question concerning estimation of population percentages. When we need the estimated percentage to be within 5 percentage points of the population percentage, 95 percent of the time, a sample size of 400 is big enough.

Table 8.1 is a labor-saving table for other degrees of precision, when the confidence level is specified as 95 percent. The calculation worked through above assumed that the population was much bigger than the sample, so that a finite population correction factor was not needed. Perhaps an overall conclusion from this table is that sample sizes of less than 100 are not very useful for estimating percentages. To

TABLE 8.1 SAMPLE SIZE NECESSARY TO ESTIMATE A POPULATION PERCENTAGE WITH VARIOUS PRECISION SPECIFICATIONS[a]

Precision required (percentage points)	Adequate sample size
1	10,000
5	400
10	100
15	44
20	25
25	16

[a] Sampling is assumed to be with replacement (or population size assumed very large), and the confidence level is 95 percent. The sample size shown would be unnecessarily large when the actual population percentage is close to 0 or 100 percent.

estimate the percentage voting for rent controls to be 50 percent plus or minus 15 percent probably is not good enough.

If the precision limit was one that should be met say two-thirds of the time, rather than 95 percent of the time, the sample size requirement is relaxed somewhat. The procedure above would verify that the sample size need only be one-fourth as large as in Table 8.1 to achieve the same degree of precision. That is, a sample of size 25 would yield percentage estimates within 10 percentage points, with confidence level 67 percent. However this confidence level is seldom adequate for applications.

8.4.2 Sample Size for Estimating Population Averages

There is a new wrinkle to the determination of adequate sample size when the parameter to be estimated is a population average rather than a population percentage. We have no mathematical result to tell us how big the standard error of an average might be. Recall that in the absence of a sample, we were not able to compute even approximately the standard error of the sample proportion, which we needed to relate the desired precision to the sample size. But we were saved in that case by the fact that we could work out the standard error in the worst case (i.e., that the proportion was 1/2) and compute a conservative sample size. But for estimation of averages there is no worst case. About the best we can do is to specify an approximate worst case based on a guess of how big the SD of the population might be.

Let us return to our poll of economists. Suppose that one question is: What will be the price of an ounce of gold one year from today? We would like to estimate the average of all economists' predictions, thinking that this may be of some use for financial planning. We require this estimate to be accurate to within $100, with a high confidence level, say 95 percent.

Let us guess that predictions will vary such that almost all of the economists would predict the price to be between $300 and $700. Certainly, then, the SD must be less than $200 [200 = 1/2 of (700 − 300)]. (It can be shown that the range is usually at least twice the SD.) So the standard error of the average must be less than $(200/\sqrt{\text{sample size}})$. Now the multiple from Table A4.1 is 1.96, so the plus-or-minus part of our estimate will be

$$1.96 \times \frac{200}{\sqrt{\text{sample size}}}$$

This must be less than the $100 specified. The sample size that will just do the trick is 16, since

$$1.96 \times \frac{200}{4} = 100 \text{ approx.}$$

Our survey need poll only 16 economists to achieve the required accuracy in the sample estimate of the population average. In fact, since our guess of the SD was conservative (i.e., probably bigger than the actual SD), we may be able to get away with even a smaller sample.

Is this whole procedure a little shaky because it is based on a guess? Yes and no. We may end up with a sample size that is too small, since it is possible that the SD was larger than we thought possible. However, we will know this by the time we take our sample. Our sample of 16 would be used to estimate the SD formally and compute a proper confidence interval. We would then have firm evidence of just how accurate our estimated average price was. We could use it with due caution. So the guessing is not a careless insertion of subjectivity into the analysis, but rather a useful device aimed at avoiding equivocal results.

EXERCISES

1. An auditor wishes to estimate the average error per account for a client company that has 50,000 accounts. He plans to select a simple random sample of accounts in order to estimate the average account error in the 50,000 accounts to within $1. Past experience with this company indicates that typical account error is in the range −$5 to +$25. How many accounts should be sampled?
2. Redo Exercise 1 replacing the requirement on the accuracy of the estimated average by the requirement that the total error in the 50,000 accounts should be estimated to within $100,000.
3. If the auditor in Exercise 1 were attempting to estimate the proportion of perfectly correct accounts (accounts in which the error is zero) to within 0.05, and if the confidence level of his estimate is to be 90%, how big should his sample size be?

8.5 SMALL-SAMPLE ESTIMATION OF POPULATION AVERAGES

We have covered the estimation of averages of large samples. The important information about the precision of the sample average was estimated by using the sample SD in the formula for the SE of a sample average. With a large sample we could rely on the sample SD to be a good enough estimate that we could consider the SD effectively "known." But for small samples, the sample standard deviation can be quite variable and can give us a misleading idea of the size of the population SD. It is the **population** SD that determines how accurate the sample average is as an estimate of the population average.

Table 8.2 displays the results of a small experiment to demonstrate the variability of estimates of the population SD by the sample SD. This variability is most marked for small samples. The samples in sample-size groups have been ordered for easy viewing. The population SD is 10 in every case, but the sample SDs range widely for the smaller samples.

Only when the sample size gets up to about 30 does the error in the SD estimate get down to about 10 percent of the population value. For the discussion of this section, let us call samples of size less than 30 "small samples," and for these small samples we will have to remember that the SD estimate is not very precise.

The objective in this section is to outline a procedure for the calculation of a confidence interval for an average when the sample size is small. We have two problems to overcome:

1. With small samples we cannot rely on the central limit theorem to tell us the shape of the sampling distribution.
2. With small samples the SE of the average cannot be estimated very precisely.

The first problem we will not be able to avoid completely. However, one situation where we can proceed is when we have reason to believe that the population distribution is normal. In this case the sampling distribution of the average can be shown to be normal, as mentioned in Chapter 7. So in this special case, we would still know the shape of the sampling distribution.

The second problem prevents us from using our previous procedure for the construction of confidence intervals, even in the special case of a normal population. But even though we cannot know the SE of the sample average in this case, we can still calculate confidence intervals that have the coverage property they are supposed to have. The idea is to combine the variability of the SD estimate with the variability of the sample average and to try to determine something in this combination that is known. Then miraculously, the puzzle will be unraveled to produce a confidence interval for the population average. The method will work well when the population is not too different from a normal population.

The small-sample procedure for the calculation of a confidence interval for a population average depends on a sampling distribution called the t distribution. This

TABLE 8.2 VARIABILITY OF THE SAMPLE SD FOR VARIOUS SAMPLE SIZES[a]

Sample[b]	Sample size	Estimated SD
1	3	2.7
2	3	3.5
3	3	7.2
4	3	8.9
5	3	15.7
6	6	6.9
7	6	7.4
8	6	9.5
9	6	11.6
10	6	12.1
11	10	6.6
12	10	9.8
13	10	10.1
14	10	10.2
15	10	10.5
16	20	7.0
17	20	9.2
18	20	10.0
19	20	10.3
20	20	10.6
21	30	9.1
22	30	9.9
23	30	10.6
24	30	10.9
25	30	11.5
26	50	8.6
27	50	9.3
28	50	9.4
29	50	9.9
30	50	10.0

[a] Each sample has been selected from a normal population with SD = 10. The sample SD for small samples is a very unreliable estimate of the population SD.

[b] The samples have been ordered according to the estimated SD, to make the table more readable.

sampling distribution is the distribution of the following quantity, called the *t* **statistic:**

$$\frac{\text{sample average} - \text{population average}}{\text{sample SD}/\sqrt{n-1}}$$

This statistic varies in a known way—all that is needed to specify the sampling distribution of the statistic is n, the sample size.

Before we use this result, the use of the term "statistic" should be justified for a quantity that includes a population parameter value, the population average. Everything else is calculated from the sample, but if we are estimating the population

Sec. 8.5   Small-Sample Estimation of Population Averages

average with a confidence interval, it would certainly be inconvenient to have to know it to calculate the statistic. In some uses of the *t* distribution, such as the hypothesis tests discussed in Chapter 9, the population mean is specified and the statistic is calculable in this case. But the statement is useful for estimation for the following reason: If we can define an interval in which the *t* statistic would lie, we can use this interval to tell us how big the *numerator* of the statistic might be. The denominator just depends on the *sample* SD which is calculable from the sample. Limits on the ratio imply limits on the numerator that can be computed.

Here is the way an *L* percent confidence interval for a population average can be calculated using the *t* distribution:

1. Use the sample to calculate the sample average and SD.
2. Noting the sample size, *n*, look up Table A4.3 for the values of *t* that enclose *L* percent of the distribution, and call them $-t$ and $+t$. Use the row of the table corresponding to df = $n - 1$.
3. Compute

$$(\text{sample average}) \pm t \times \frac{\text{SD}}{\sqrt{n-1}}$$

This is the *L* percent confidence interval for the population average.

For example, consider a sample of size 3 from a large population, where the sample average and SD are 9.5 and 2.0, respectively. What is a 99 percent confidence interval for the population average?

1. The sample average is 9.5 and the sample SD is 2.0.
2. Table A4.3 indicates that $t = 9.92$.
3. The confidence interval is $9.5 \pm 9.92 \times (2.0/\sqrt{2})$, which equals 9.5 ± 14.0. The 99 percent confidence interval is $(-4.5, 23.5)$.

This interval indicates where the population average is likely to be, based on the sample data. It should not be surprising that this interval is very wide—the data were spread over a range of only about 4 (judging from the SD). On the other hand, the population SD might have been 15.0, so our small sample could not be expected to pin down the population average very well. The factor 9.92 allowed for the fact that the SD might have been poorly estimated. The confidence interval still has the coverage property—99 percent of the time the confidence interval is calculated, the interval produced will "cover" the population average.

The *t* table in Appendix A4.3 shows how to relate the values of the *t* statistic to probabilities of certain events associated with this statistic. Specifically, the body of the table gives the values of *t* and the head of each (vertical) column gives a percentage. The *t* statistic will fall in the interval $-t$ to $+t$ with a chance given by the percentage. Each (horizontal) row of the table describes a different *t* distribution. *t* distributions depend on something called **degrees of freedom,** which is the row label.

For the standard calculation of a confidence interval for an average, the degrees of freedom are always one less than the sample size.

More generally, the degrees of freedom of the t statistic are the number of independent estimates of the population SD that are pooled together in the sample SD. The degrees of freedom available for an estimate are the number of independent estimates that we can pool together. Two independent data values provide one estimate of an SD. Clearly, a population SD is estimated by one-half the difference in the two values (see Exercise 4 of Section 4.4). In other words, two values provide one estimate of the population SD. Three values provide two independent estimates: the difference of any pair provides an indication of spread, but given two such differences, the third one does not provide new information since the differences must sum to 0. (The average difference between a list of numbers and their average is always zero.) It is also true, although not quite obvious, that n values provide $n - 1$ independent estimates.

It must be remembered that the t procedure is not simply the "small-sample procedure"; it is the "small-sample procedure when population normality can be assumed." Statistical researchers have shown that the population can depart from normality considerably without invalidating the use of the t procedure. What we have then in the t procedure is a different type of approximation than the central limit theorem. The central limit theorem gives useful results when the sample size is not too small. The t procedure gives useful results when the population is not too different from a normal distribution. When we have a small sample from a very nonnormal population, we have employment for mathematical statisticians.

The t distribution will be used again in Chapter 9.

EXERCISES

1. Calculate a 90 percent confidence interval for the population average, given the following sample from the population:

    $$\{2.4, 13.9, 9.2, 12.3\}$$

2. (a) How often does a t statistic based on a sample of size 5 exceed 2.5 (approximately)?
   (b) For a population that has an average of 10.0, is a sample of size 5, which produces an average of 15 and an SD of 2.0, unusual?

8.6 ESTIMATION OF PROBABILITY DISTRIBUTIONS

In a town of 10,000 households, the manager of a supermarket may wish to know what proportion of the household food expenditures he has captured. To obtain a rough estimate of this, he does a telephone survey. Let us suppose that he contacts a random sample of 25 households, and that he obtains estimated weekly expenditures on foodstore items from each of the 25 households. (This can be a difficult thing to do:

Sec. 8.6 Estimation of Probability Distributions

the problems of obtaining household lists with correct and current telephone numbers, of nonresponse because of absenteeism or noncooperation, are just two of the problems.) Now the techniques for the estimation of the average household food budget has been covered in Section 8.3, so the store owner could combine this information with his own weekly sales to obtain an idea of the proportion of the market he has captured.

Suppose he finds that his share of the market is 40 percent. His strategy for increasing this share will depend on whether all households are buying 40 percent at his store, or whether some smaller precentage of the households are buying 100 percent of their food at his store. Perhaps some additional questions in the survey would have helped to obtain information on this. However, some idea might be gleaned from the distribution of household purchase estimates obtained in the survey. Are these estimates typically small with a few large ones, or are they typically large with a few small ones? In other words, the shape of the distribution may reveal whether the average is larger or smaller than the typical purchase. ("Typical" here is used to refer to amounts that are associated with the highest portions of the histogram, the mode of the distribution.) If it turns out that the skew in the distribution is to the left, or in other words that the distribution contains mostly large values and a few small ones, it might be guessed that the households that shopped at our manager's supermarket bought a high proportion of their weekly food supplies there. In this case, market expansion would require new customers rather than an increased proportion of purchases for present customers. This would be useful information for the manager.

The above is an example of a situation in which the estimation of a whole distribution is required, not merely a single parameter such as the average or SD. A simple approach to this problem is to use the sample histogram as the estimate of the population histogram. If only rough features of the histogram are likely to be important, this is adequate. More sophisticated techniques are available which can provide better estimates, but these are beyond the scope of this text. To appreciate the need for these more advanced techniques, observe that a sample of 25 observations is not going to provide a very smooth histogram, especially if 10 or more intervals are used.

Another feature of population histograms that is often of interest is the presence of more than one mode, a two-humped distribution. In the supermarket example, such a shape would have suggested that there were two kinds of households using the store. Assuming that the community was fairly homogeneous in socioeconomic status, this bimodality (two modes) might be interpreted as a difference in usage pattern by the two types of household: one type shops the bargains at several stores, the other sticks to the one store. Again, this would be useful information for the manager.

Now suppose that the manager is interested in estimating the proportion of households that spend less than $10 at his store. If there were 5 out of the 25 in this category in the sample, 5/25 or 0.20 would be a natural estimate, and the SE would be estimated to be

$$\sqrt{\frac{(5/25) \times (20/25)}{25}}$$

or about 0.08. That is, the 90 percent confidence interval for the proportion is 0.07 to 0.33. This is not very precise, yet the larger sample required for adequate precision may not be feasible. In general, the manager would be stuck with this poor estimate, but there is one situation where he can do better. If the sample histogram looks approximately normal (no obvious skewness and only one mode), he can use the normal approximation to the histogram. Just calculate the average and the SD, and compute the proportion of the population with this average and SD that is less than $10.00. For example, if the average in the sample of 25 were $100 and the SD were $60, $10 is 1.5 SDs below the average, and only about 7 percent (a proportion 0.07) of the households would be estimated to have lesser expenditures than this.

This normal approximation is much more accurate than the other method suggested *if* the population distribution is approximately normal. The use of this method for estimating a whole distribution works well in this case since all the information in the sample is used to fix the average and SD rather than attempting to estimate the height of the population histogram at several values of the variable.

Again, we have described a statistical problem for which the choice of technique depends on judgment. Exercise 1 should tell you whether you have good judgment in this particular situation.

EXERCISES

1. With reference to the example described in this section, which estimate do you think should be used by the manager, 0.20 or 0.07? Give your reason.
2. The distribution of weights from a random sample of the charts from the routine health examination of university applicants is bimodal. Does this finding suggest any interesting theories?
3. Would you use the normal approximation to estimate the percentage of employees in a particular company who have an income below $25,000?

8.7 ANSWERS TO EXERCISES

Section 8.2

1. The SEs for the two samples are $2.90/3.16 = 0.92$ and $7.48/3.16 = 2.37$, respectively. Thus 68 percent confidence intervals for the average donation from the whole association are 3.70 ± 0.92 and 8.95 ± 2.37, respectively. The total contribution would be 500 times these numbers, that is, between $1390 and $2310 for the small projects promotion and between $3290 and $5660 for the large promotions. It looks like the association ought to opt for the small projects, to avoid embarrassment. Alternatively, they could suggest slightly larger projects than were used on the small projects list in the pilot survey.
2. The interval is $3.70 \pm (2.33 \times 0.92)$ for the small projects, and $8.95 \pm (2.33 \times 2.37)$ for the large projects, that is, $1.16 to $5.34 and $3.43 to $14.47, respectively.
3. (a) False, because the population average does not vary.

(b) True, it is estimated by estimating the standard error, which is estimated using the sample SD of the single sample.
(c) Ambiguous, because the phrase "95 percent confident" has not been defined and has no meaning until it is defined. Some texts do define this statement as meaning that the confidence interval includes the population average in 95 percent of samples of the same size, from the same population, but this usage has been avoided in this text. This usage seems to invite misunderstanding.
(d) True, since the standard error of an average depends on the population SD, which is usually unknown. (But it can be estimated.)
(e) False. It contains the *sample* mean 100 percent of the time.

4. The sample size is square-rooted before being used in the denominator of the standard error formula. To get this square root to be twice its former size (so that the SE will be half its former size), one must multiply the same size by 4. So a sample of 120 would be needed.

5. The estimate of a $500 loss is obvious, although one should worry a bit about the population sampled. (First-day noon parkers?) However, the precision of the estimate cannot be properly assessed with the data given. One would need to know the SD of the changes to do this. The precision of the estimate is probably about $100 or less, but this cannot be gleaned from the data given. The unbiasedness of the estimate is in doubt because of the uncertainty that the population is the one intended.

Section 8.3

1. The point estimate of the proportion of softcovers is clearly 0.12. The SE of this proportion is

$$\sqrt{\frac{0.12 \times 0.88}{50}}$$

or about 0.05. The multiple for a 90 percent confidence interval is 1.64. $0.12 \pm (1.64 \times 0.05)$ is the 90 percent confidence interval for the proportion in the 800, and $12\% \pm 8\%$ is the confidence interval for the percentage. Thus the 90 percent confidence interval for the proportion of textbook titles that are softcover is 4 to 20 percent. (The correction factor should be used since sampling would be without replacement, but this does not change the answer much.)

2. The SE of the proportion of pages including diagrams or pictures cannot be determined exactly, but it must be no greater than the square root of $[(1/2) \times (1/2)/20]$, which is 0.11. In other words, the error of the estimate has an SD of 11 percent, and this is a measure of the accuracy.

3. The SE of the percentage, which measures the variation of the estimate, is 11 percent, just as in Exercise 2.

4. Because the data are paired, we can treat the data as a single sample. Of 30 students in the sample, $22 - 10 = 12$ changed their mind, so we have to estimate a population proportion based on a sample proportion $12/30 = 0.40$. A 68 percent confidence interval for the (population) proportion changing their mind is

$$0.40 \pm \sqrt{\frac{0.4 \times 0.6}{30}}$$

which is $(0.31, 0.49)$.

314   Chap. 8   Estimation

5. The proportion is 2/40 = 0.05 women in the sample. What we want is the population proportion that implies that 0.05 is the the first percentile. (The first percentile is my threshhold of belief.) We want the population proportion with the property that our observed 0.05 is 2.33 SEs below it. Try a population proportion of 0.2. The SE of the sample proportion for this population proportion is

$$\sqrt{\frac{0.2 \times 0.8}{40}}$$

or 0.063. Now 2.33 SEs would be 0.147 or about 0.15. But this means that 0.05 is the first percentile of the sampling distribution of the sample proportion when the population proportion is 0.2. So 0.2 is the largest believable proportion of women in the list of 610. (This is the method of "trial." When you know the answer, you can avoid the "errors.") The actual proportion of female mayors in this list of 610 was about 0.03.

Section 8.4

1. The population SD is probably much less than 1/2 the range, that is, probably less than $15. Let us suppose that it is $15. Then the SE of the average is 15/(square root of sample size). 1 SE will be less than $1 if the sample size is 15^2 = 225 or more. If a confidence level of 68 percent is not good enough, we would need a larger sample size. For example, for a 95 percent confidence interval to estimate the population average to within $1 would require 2 × 15/(square root of sample size) to be at most $1. In this case the sample size would have to be 900, since $(2 \times 15)^2$ = 900. A conservative auditor would probably opt for this sample size, 900. (*Note:* The ambiguity in this exercise is typical of many statistical applications. The confidence level is seldom set by any scientific means, but is rather a judgment based on experience and practical considerations.)

2. This new requirement is equivalent to the one that the average be estimated to within $2. The same method as was used in Exercise 1 yields, in this case, a sample size of 151. (Be sure to look up the correct value of the Table A4.1 multiple, 1.64.)

3. Using Table 8.1, the answer is a sample size of 400. The method described there can be used for precisions not included in the table.

Section 8.5

1. Average = 9.5, SD = 4.41. t (3 df) to enclose 90 percent is 2.35. Thus the 90 percent confidence interval is

$$9.5 \pm 2.35 \times \frac{4.41}{\sqrt{3}} = 9.5 \pm 6.0 \quad \text{or} \quad (3.5, 15.5)$$

2. (a) Between 90 and 95 percent of the time. See Table A4.3.
   (b) The t statistic is

$$\frac{15 - 10}{\sqrt{4}} = 2.5$$

So this is a little on the high side, but not unusual. (One would get something more extreme in over 10% of the samples.)

Section 8.6

1. It seems likely that a number of households would spend $0.00 at this store, so there would be marked nonnormality at this end of the distribution. Note also that the SD is large enough that the assumption of normality implies that there would be quite a few negative expenditures at the store. Although this is possible, it is unlikely that it would be admitted in a telephone survey. So the assumption of normality is probably way off. It would be better to stick with the 0.20 in this instance.
2. It should not, since the weight differential between males and females is well known.
3. This would not be a good idea, since income distributions tends to be skewed right, and this would especially be true in a company in which a few executives get much higher salaries than the majority of employees.

8.8 NOTATION AND FORMULAS

A short way to write that the sampling distribution of the sample average is normal with expected value μ and SE $= \sigma/\sqrt{n}$ is

$$\bar{X} \sim N(\mu, \sigma/\sqrt{n})$$

μ is also the population average and σ is the population SD. The sample SD is usually substituted for σ in estimation of the SE.

The confidence level is often expressed as $100 \times (1 - \alpha)$ percent. A 68 percent confidence level corresponds to a "tail" probability of $\alpha = 0.32$, which is the sum of two 0.16 tails here.

A $100 \times (1 - \alpha)$ percent confidence interval for an average is written

$$\bar{X} \pm z_{\alpha/2} \times \frac{SD}{\sqrt{n}}$$

where $SD/\sqrt{n}$ is the estimate of the SE of the average. $z_{\alpha/2}$ is a number which cuts off a proportion $\alpha/2$ of the area under a normal curve. The sampling distribution of a sample percentage $\hat{p}$ is $\hat{p} \sim N(p, \sqrt{p(1-p)/n})$ approximately. In this expression n is the sample size and p is the population proportion. As p is usually unknown in estimation problems, $\hat{p}$ is usually substituted for p in the expression for the SE of $\hat{p}$.

Sample-size problems look much simpler in symbolic form. Let E be the acceptance error, that is, the largest error of estimate one is willing to tolerate, all but $(1 - \alpha)$ percent of the time. Then

$$z_{\alpha/2} \times \frac{SD}{\sqrt{n}} = E$$

must be solved for n, the sample size required. The solution is

$$n = \left(z_{\alpha/2} \times \frac{SD}{E}\right)^2$$

316 Chap. 8 Estimation

For proportions one simply replaces SD by $[p(1-p)]^{1/2}$. The formula becomes

$$n = \left(\frac{z_{\alpha/2}}{E}\right)^2 \times p(1-p)$$

This formula will give the same results as Table 8.1 if $z_{\alpha/2}$ is rounded to 2.0 ($\alpha = 0.05$ in that table) and $p = 1/2$. The t statistic may be written

$$t = \frac{\bar{X} - \mu}{SD/\sqrt{n-1}}$$

It is often written in other texts as

$$t = \frac{\bar{X} - \mu}{S/\sqrt{n}}$$

where

$$S = SD \times \sqrt{\frac{n}{n-1}}$$

The L percent confidence interval is

$$\bar{X} \pm t_{\alpha/2} \times \frac{SD}{\sqrt{n-1}}$$

where $t_{\alpha/2}$ is the tabular value corresponding to 100 $\alpha/2$ percent in the upper tail $\alpha = 1 - (L/100)$.

8.9 SUMMARY AND GLOSSARY

1. Reliable estimation methods depend on the adequacy of the random sampling models for describing the actual sampling process.
2. Random samples provide unbiased estimates of population averages, but the precision of these estimates depends on the sample size and the SD of the population sampled.
3. The sampling distribution of the sample average is approximately normal with average equal to the population average and standard deviation equal to the SE of the average.
4. A confidence interval for a population average, having confidence level 95 percent, is centered at the sample average and extends an amount equal to about 2 SEs on either side of the sample average.
5. The confidence level of a confidence interval is the percentage of the time that the confidence interval constructed would include the population parameter.
6. The estimation of proportion is the same as for averages except that the category to which the proportion applies must be coded as 1 and the complementary category as 0. A shortcut formula is available for the estimation of the SD of a sample proportion based on the sample proportion itself. It is

$$\sqrt{\frac{\hat{p} \times (1 - \hat{p})}{\text{sample size}}}$$

7. The precision with which a percentage may be estimated depends on the percentage itself. The SE of the estimate is largest when the percentage is close to 50 percent and smallest when it is close to 0 or 100 percent.
8. Sample-size determination is performed by deducing the sample size that would produce a confidence interval of acceptable width. Since this confidence interval depends on the population SD, a quantity that is usually unknown, a guess of this quantity will often be required to complete the calculation. In the case of estimating percentages, a worst case of 50 percent may be used in lieu of this guess.
9. When the shape of a population distribution is required, the sample frequency distribution may be used as an estimate. When normality can be correctly assumed for the population distribution, this distribution should be estimated using the intermediate step of estimating the average and SD.
10. The t distribution can be used to calculate confidence intervals for a population average when the population is not too far from being normal, even if the sample size is small.

Glossary

An **L percent confidence interval procedure** for a parameter is a method of calculation that produces intervals from random samples; these intervals include a certain population parameter value for L percent of the samples. This percentage is called the confidence level of the confidence interval procedure. An L percent confidence interval for a population parameter is an interval produced by an L percent confidence interval procedure for that parameter.

t statistic:

$$\frac{\text{sample average} - \text{population average}}{\text{sample SD}/\sqrt{n-1}}$$

(This t statistic is appropriate for one-sample t procedures. For two-sample procedures, see Section 8.5.)

PROBLEMS AND PROJECTS

1. A survey based on a random sample asked the question "Which of candidates A, B, and C do you favor?" It reported the percentage favoring A as 35 percent, the percentage favoring B as 33 percent, and the percentage favoring C as 31 percent. It is expected in this population that voters would actually vote the way they indicated in the survey. The election results may still differ from the survey because of sampling error. Instead of the

candidates being ranked A, B, C in order of popularity, they might be ranked B, A, C or A, C, B. Which of the latter two possibilities is more likely, or are they equally likely? Explain.

2. The presidents of student societies of 900 North American colleges and universities are to be surveyed concerning the responsibilities and financial support of these societies. The budget assigned to the survey allows for a sample of just 100 of these presidents. One aim of the survey is to estimate the proportion of the societies' administrative budgets that are solicited directly from the students rather than through the college administration. What estimation procedure would you use for this?

3. For a psychology research project, the average IQ of students newly accepted to your institution is to be estimated. How large a sample size would you need to estimate the average IQ to within 10 points? (IQs in the general population have average = 100 and SD = 15, for one of the popular tests.) Does it make any difference to your answer how many students are admitted, in total? If so, assume a convenient number. If not, explain why not.

4. How large should be the size of a random sample which is designed to estimate the duration of sleep that adults obtain prior to a working day? The estimate is to be accurate to within 1/4 hour with probability 0.90. (This problem is continued as a computer project in Problem 23.)

5. The times taken for five cyclists to complete a competition circuit are, in minutes: 25.7, 27.1, 31.0, 26.9, and 28.3. These five times were selected at random from 100 competitors who raced the same circuit. Estimate the average time of the 100 cyclists. State any assumptions you need to provide a useful estimate. (This problem is continued as a computer project in Problem 24.)

6. A bank manager suggests the following strategy to detect counterfeit $20 bills. For each package of 50 bills, five are selected at random and the numerical part of the serial number is totaled. There are seven digits in the serial number. It is expected that each digit has an equal chance of being 0, 1, 2, ... , 9 in the legitimate bills, whereas this may not be the case for the counterfeit bills.

   (a) Describe the distribution you would expect for the average serial digit for the legitimate bills.

   (b) Suggest a practical rule for the bank to use to decide whether a given packet of twenties should be examined more closely. Assume that a batch of counterfeit bills would have only a few different serial numbers. (This problem is continued as a computer project in Problem 25.)

7. The standard error of a sample average is the SD of the population divided by the square root of the sample size. Now if the sample were to include the whole population, the standard error, which purports to measure the error in the sample average as an estimate of the population average, should be zero. But the SD divided by the population size could be greater than zero. Explain this apparent contradiction. (This problem is continued as a computer project in Problem 26.)

8. Five thousand people register to compete in the Spuzzum Marathon run. A physical education graduate student is interested in estimating the average value, in these registrants, of a certain index of vital capacity. To do this, the graduate student selects a random sample of 50 names from the registration list, and he is able to measure the index value for all the runners in his sample. The 50 values for the sample runners average 68.0 and have an SD of 15.0.

(a) Find an 80 percent confidence interval for the average value of the index for the 5000 runners.

(b) If a random sample of 100 registrants had been selected, would you expect the SD to be larger, smaller, or about the same value as was obtained in the sample of 50 registrants? Explain.

(c) If a simple random sample of 100 registrants had been selected, would you expect the width of the confidence interval to be larger, smaller, or about the same value as was obtained from the sample of 50 registrants? Explain. (This problem is continued as a computer project in Problem 27.)

9. A graduating class of 1000 is surveyed to ascertain the percentage intending to enter graduate programs. Of 25 students queried, 5 refused to answer, 10 responded that they did intend to enter a graduate program, and 10 responded that they did not intend to enter a graduate program. Provide an estimate of the percentage of the class intending to enter the graduate program, based on this information. Be explicit about whatever assumptions are necessary to do this.

10. A manufacturer of thread tests the breaking strength of one sample from each of 25 batches of thread, the samples having been accumulated over several weeks. The average breaking strength of the 25 thread segments is 10.3 kg and the SD is 1.3 kg.
    (a) Find a 95 percent confidence interval for the average breaking strength.
    (b) What percentage of future thread segments tested, one from each future batch, would have breaking strengths in the confidence interval calculated in part (a)?
    (c) What is the chance that assuming production controls and methods do not change, the testing of 25 segments next year would yield an average that would be contained by the confidence interval from the first year? 95 percent, or more, or less? Explain.

11. A large taxi company owns 1000 taxis. When a taxi will not start at the beginning of the day, it is given a thorough servicing which takes the entire day, resulting in a loss of revenue of about $150. An average of about 10 taxis have this problem each day. Estimate the company's loss of revenue on a given day, indicating the variability of the loss.

12. Indicate whether the following statements are true or false. Explanations must be provided.
    (a) A confidence interval for an average indicates the interval in which a specified percentage of the population averages fall.
    (b) A binomial experiment with a large number of draws (e.g., the tossing of a coin 1000 times) results in an outcome (e.g., the number of heads) that is very close to its expected value.
    (c) Large random samples provide estimates of population averages that are both precise and unbiased.
    (d) The histogram for a sample percentage has an SD that depends on the population percentage; this population percentage is usually not known.
    (e) A random sample of n items, selected with replacement, from a population of two items, can always be modeled as a binomial experiment.
    (f) If a population distribution is not normal, the histogram of the sample average may be very different from a normal curve for small samples.

13. A wholesale distributor of apples agrees to deliver a large shipment of apples to a supermarket. The distributor promises that these apples will have an average weight of 0.20 kg and vary from this average such that 90 percent are between 0.15 and 0.25 kg. The supermarket manager selects a simple random sample of five apples from the shipment. The weights of these five apples are 0.19, 0.21, 0.18, 0.19, and 0.18 kg. Past experience

indicates that the distribution of weights of apples may be assumed to be a normal distribution.
  (a) Compute a 90 percent confidence interval for the average weight of apples in the shipment, based on the sample of five apples.
  (b) With reference to your confidence interval, would you expect it to contain about 90 percent of the apple weights? Explain.
  (c) If the distributor's promise were kept, what is the probability that the average of the sample of size 5 would be at least 0.01 kg away from the average of the shipment? (Note that $0.20 - 0.19 = 0.01$.)
  (d) Of what relevance to the supermarket manager is the probability in part (c)?

14. The *World Almanac and Book of Facts*[3] lists 1833 names under the category "Noted Personalities—Entertainers." Here is a random sample of 45 of these names:

Michael Anderson, Jr.	Nina Foch	Tony Orlando
Bibi Andersson	Jane Fonda	Rick Nelson
Frankie Avalon	Merv Griffin	James Nobel
Harry Belafonte	Mark Hamill	Randi Oakes
Sorrell Booke	Florence Henderson	Merlin Olsen
Timothy Bottoms	Bob Hope	Elenor Parker
Charlie Callas	Trevor Howard	Gregory Peck
Pat Carroll	Ross Hunter	Victoria Principal
Gaby Casadesus	Sammy Kaye	Marjorie Reynolds
Myron Cohen	Frankie Laine	Ann Rutherford
Pat Crawley	Hal Linden	Tom Selleck
Daryl Dragon	Jack Lord	Ben Vereen
Robert Duval	Elaine May	Ray Walston
Tovah Feldshun	Malcolm McDowell	Andre Watts
Peter Firth	Howard Morris	Billy Dee Williams

  (a) Estimate the proportion of the 1833 names that you "know." As a criterion of "knowing," use knowledge of the form of entertainment (film, TV serial, TV occasional, theater, music, etc.).
  (b) Estimate the proportion of the 1833 entertainers that are female.

15. During the 1984 Olympic games, Carl Lewis won a gold medal for his performance in the 100-m run. His time was 9.99 sec. A few days later he competed in the Zurich track-and-field meet and placed first in the 100-m run. His time was 9.99 sec. What, if anything, can be inferred about Carl Lewis from this these facts? Consider the data in Problem 1 of Chapter 6 to be a part of the background information for this problem.

16. The following scheme is proposed to estimate the number of current-year nests of the Great Blue Heron in the Fraser River delta. First the exact boundary of the delta is defined and the region is divided into regions of two types. One type is ideal for nesting, being close to shallow tidal water, and having suitable perches for nests. The other type includes everything else including deep water, industrial areas, and residential areas. The whole delta thus consists of "ideal" and "other" areas. The "ideal" areas are divided into 100 subregions of approximately equal area, and the "other" regions are divided into 500 regions of approximately equal area. A random sample of 10 of the 100 "ideal" subregions and another random sample of 5 of the 500 "other" regions is selected. For each of the

subregions selected, a comprehensive search of current-year nests is to be done. Based on the 15 counts of nests in the subregions, the total number of nests in the delta is to be estimated.

(a) Discuss the reasons for or against the proposed sampling scheme.

(b) If the average number of nests were 15.0 per "ideal" subregion sampled and 0.5 per "other" subregion sampled, estimate the total number of nests in the delta.

(c) The SD of the number of nests per subregion is estimated to be 5.0 in the "ideal" regions and 1.0 in the "other" regions. Explain why the usual estimate of the SE based on a single sample of 15 would not properly estimate the variability of the estimate you have proposed in part (a).

(d) Compute an approximate SE for the estimate you propose in part (a).

17. The ratio of the number of students to the number of faculty in North American colleges and universities can be calculated from data available in the *World Almanac and Book of Facts*. Unfortunately, with about 2300 listings, the arithmetic is considerable. Approximate answers can be obtained as follows:

> *Method 1:* Take a random sample of 25 of the 2300, calculate the ratio for each, and take the average of these ratios.
>
> *Method 2:* Compute the ratio of the sums of numbers of students and faculty for the sample of 25.

(a) Perform both methods on the following:[4]

Institution	Number of students	Number of faculty
Atlantic Christian	1549	100
Arkansas Tech.	3088	150
Beaufort Tech.	1208	89
Duquesne U.	6300	486
Elizabeth City State U.	1560	114
Felican	610	72
Ferris State	11200	600
Goucher	1041	120
Hardin-Simmons U.	1948	123
Hillsborough	12512	405
Illinois Wesleyan U.	1667	136
Jefferson (Watertown)	1578	93
John F. Kennedy U.	1600	400
Lincoln Tech.	737	31
Long Island U.	6851	456
Minneapolis	3000	140
Mobile	946	60
New Hampshire Tech. Inst.	1233	125
New School for Social Res.	25000	1500
Northern Oklahoma	1800	80
Pasco-Hernando	3000	51
Pittsburg, U. of	29315	2333
St. Francis (Loretto)	1564	89
South, U. of the	1278	136
Sullivan Jr. Coll. of Bus.	1325	NA[a]

[a] NA, not available.

322 Chap. 8 Estimation

(b) Which method is likely to be closer to the ratio of the number of students in the 2300 institutions to the number of faculty in the 2300 institutions?
(c) Calculate the standard error of the estimates in parts (a) and (b) if you know how; if not, explain what complications prevent you from proceeding. The data are for 1983. (In some cases the counts include part-time as well as full-time students and faculty.)

18. Five hundred students who are in their last semester at Aeio University are to be surveyed to determine the proportion who intend to enroll in graduate school. For this purpose a simple random sample of 100 students is asked a question. The response is recorded as "yes," "no," "maybe," or "no answer." How accurately will this survey estimate the proportion of the 500 students who would have answered "yes" if they had been asked the same question? (Recall that "accuracy" means precision and lack of bias, and consider both aspects in your answer.)

19. A survey designed to assess the success that employers are having in filling advertised vacant positions is based on a random sample of 50 advertised positions that appear in the local paper over the period of a calendar year. Employer-advertisers are later asked whether they filled the advertised position within one month of first advertising it.
    (a) If 40 out of the 50 sampled answer "yes," estimate the proportion of the annual positions filled within one month. Use a confidence interval and state the confidence level you have used.
    (b) Explain the meaning of the confidence level you have used in part (a) by interpreting it as a "chance," that is, as a long-run percentage. Try to avoid ambiguity in your answer. Your interpretation must relate to the context of survey described here.

20. Ten trees are selected at random from a 1000-tree woodlot. Measurements of the height of a tree and of the circumference of the tree near ground level can be used to estimate the volume of wood in the tree. Assume that this estimate of a tree's volume of wood is imprecise but unbiased.
    (a) From what population may the 10 volume estimates be considered to be a random sample?
    (b) If the average and SD of the 10 numbers are 10 m^3 and 2 m^3, respectively, estimate the volume of wood on the woodlot. To do this, use an interval estimate.
    (c) It is proposed to select a second random sample of 10 trees from the same woodlot, and again volume estimates are to be obtained for each of the trees sampled. If possible, calculate the probability that the average of this second sample lies in the particular confidence interval from part (b). If this is not possible, explain why not.

21. A survey designed to estimate the proportion of students at a large university who have bank loans is proposed by the student association. Funds are available for a survey of 100 students, and the precision of the estimate from the survey is to be such that the estimated proportion having bank loans is within 0.05 of the population proportion with probability 0.90. A representative of a bankers' association has this information already (i.e., knows the proportion having bank loans) but will not release it to the student association. However, the bankers' association representative is willing to reveal that the proportion of students with bank loans is less than 10 percent. Does this information from the bankers' association help the student association to judge the feasibility of the survey? Explain, and include the relevant calculations with your explanation.

22. A statistical terrorist positions himself (or herself, the disguise is so good it is hard to tell the sex) on an entrance path to the university at eight o'clock in the morning. All arrivals are forced to reveal the amount of cash in their pockets, which is noted by the terrorist,

and each arrival is then allowed to proceed (with their cash). At 8:15 hours the terrorist hides away to analyze the data. Thirty data values have been gathered during the 15-minute period. The data have an average of $10 and an SD of $8. (Assume that the data values gathered may be considered to be a random sample from a hypothetical population.)
  (a) True or false? There would probably be only one or two data values greater than $25. Justify your answer.
  (b) True or false? If the prank were repeated under similar circumstances (same day of week, same time and time duration, etc.), the average amount of cash would have more than a 95 percent chance of being within $3 of the average obtained the first time (i.e., between $7 and $13). Justify your answer.

Computer Projects

23. (Continuation of Problem 4.) A sleep-duration study involving a random sample of 10 subjects from an employee roster of 2000 reveals an average sleep duration of 7.7 hr. The SD among the 10 subjects is 1.1 hs. A follow-up study is to be done to estimate the average duration more precisely, this time to within 1/4 hr, in a way that would achieve this precision 90 percent of the time. The problem here is to decide how to use the information that the SD among a sample of 10 was 1.1 hr. This is probably quite imprecise, based on such a small sample, yet it seems that it would be better to use the information than to ignore it.
  (a) Generate several samples of size 10 from a normal distribution with average 7.7 and SD = 1.1 to assess the variability of the sample SD itself.
  (b) From your experience in part (a), choose a "worst-case" value of the SD to determine an appropriate sample size for the follow-up study.

24. (Continuation of Problem 5.) Simulate 100 samples of size 5 from a normal distribution with average 27.8 and SD 1.8, determining in each case the value of the t statistic. Compare these 100 numbers with a t distribution.

25. (Continuation of Problem 6.) If the counterfeit bills all have the number 7369422, what proportion of bills in a packet of 50 would have to be counterfeit in order that the presence of the counterfeit bills could be detected from the sample of five bills? (Use the difference between the population SD and the sample SD as your indicator. When this is large, the presence of the counterfeit bills in the sample is more likely. Try simulations of samples when the proportion counterfeit is 0, 20, 50, and 80 percent.)

26. (Continuation of Problem 7.) For samples of size 25 from the population of integers {1, 2, . . . , 100}, compare the standard deviation of sample averages for several samples with the estimated standard errors of the averages from each of the samples. Comment on the similarity or difference observed between this SD and the estimated SEs.

27. (Continuation of Problem 8.) Repeat the calculation of Problem 8(a) 25 times. Compute the average of the 25 averages. Use these simulations to verify the property that an 80% confidence interval is supposed to have.

Testing Hypotheses with Data

Always do right. This will gratify some people, and astonish the rest.

Mark Twain (1835–1910)

Appearances to the mind are of four kinds. Things either are what they appear to be; or they neither are, nor appear to be; or they are, and do not appear to be; or they are not, and yet appear to be. Rightly to aim in all these cases is the wise man's task.

Epictetus (ca. 50–ca. 120)

In Chapters 4 to 6 methods for summarizing samples were introduced. Chapters 1 to 3 had introduced the need for randomness in our data collection procedures. When the samples are random samples, drawn from populations of interest, the methods of probability are required to describe these populations. These methods were introduced in Chapters 7 and 8. Often a population is of interest because it is suspected that a widely held belief about the population is false. Sometimes the interest in a population is because of the importance to future actions of the truth of a widely held belief about the population. In such cases, a hypothesis about the population needs to be tested against the sample data. Hypothesis tests are the subject of this chapter.

Key words are hypothesis, alternative, P value, critical P value, acceptance error, rejection error, statistical significance, t test, one-sample test, two-sample test, one-sided test, chi-square distribution, dependent samples, contingency table, variance of a population, variance of a statistic, analysis of variance, F distribution, chi-square test of independence, nonparametric test, median test, sign test, and prior knowledge.

9.1 THE LOGIC OF HYPOTHESIS TESTING

The broad setting for hypothesis testing is the following. Someone makes a claim about a population; a random sample is selected from the population, and using the information contained in the sample, some statement about the validity of the claim is made. In other words, a hypothesis is to be tested. Let us first examine some informal hypothesis testing situations that one might encounter in everyday life.

Whenever we observe something that is unusual under ordinary circumstances, we usually suspect that the circumstances are not ordinary. If, for example, you find yourself in unusually heavy traffic, you may wonder whether an accident has blocked the normal flow, or you may realize that it is the Friday before a long weekend. If your neighbor deals himself an unusually large number of aces during a poker game, you may suspect that he is cheating. When three calculators in succession from a single shipment fail to work, you may wonder if the shipment has been subjected to some common shock.

In each of these examples, you are rejecting the initial presumption that you held before you observed the unusual event. Researchers learn in this way also. But when researchers wish to convince others of a new finding, they must counter the explanation that their unusual observation was just a rare concurrence of chance influences of unknown cause. For example, might not the heavy traffic be attributable to a large number of people independently deciding to be at particular intersection at the same time? Might not your gambling neighbor be just plain lucky? The calculators tested might have been the only three in the shipment of several hundred that did not work.

The point of these examples is that when we observe an unusual event, among the many possible explanations that must be entertained is the uninteresting one that it was just a fluke. Tests of hypotheses help us to assess the plausibility of this explanation.

Sherlock Holmes had a slightly different perspective on inference from data: "When you have eliminated the impossible, whatever remains, *however improbable, must be the truth.*" The logic of this is compelling. In practice, however, we find it difficult to decide what is possible and what is impossible. So improbable events are still usefully interpreted as indicators that something is amiss. In the assessment of statistical hypotheses, the observation of an improbable event suggests that the hypothesis is false.

Before we get technical about hypothesis testing, enjoy the cartoons shown in Figure 9.1. They remind us not to leave our common sense behind when we undertake a statistical analysis.

We now address a more formal hypothesis-testing problem. We work through some examples, emphasizing the logic of the method. Then we introduce the standard terminology that will help to reveal the common procedure. A thorough understanding of the logic of these tests is necessary even for the simplest applications.

Let us begin with a trite example. We toss a coin 100 times to determine if it is a fair coin. Suppose we have heard that when coins are biased, they tend to give too many heads, so this is what we are watching for. We observe 58 heads and 42 tails, which suggests that the coin may be biased toward too many heads.

If the coin were fair, how likely would it be that we would get 58 heads? Without going into the details, it can be shown that exactly 58 heads would occur in 2.2 percent of such experiments. It may seem that this calculation suggests that 58 heads is a rare event, but note that even 50 heads occurs in only about 8 percent of experiments. In

326 Chap. 9 Testing Hypotheses with Data

Jumping to Conclusions

Statistics show that most car accidents occur when cars travel at moderate speeds, and that very few accidents occur at speeds of more than 150 kilometers per hour. Does this mean it is safer to drive at higher speeds?

A research study showed that children with big feet could spell better than those with small feet. Does this mean that the size of one's foot is a measure of one's ability to spell?

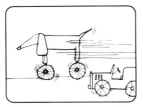

Not at all. Statistical relationships often have nothing to do with cause and effect. Most people drive at moderate speeds, so naturally most accidents occur at moderate speeds.

It does not. The study included *growing* children. All it showed was that older children, who, of course, have bigger feet, spell better than younger ones.

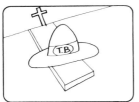

If statistics showed that more people died of tuberculosis in Arizona than in any other state, would that mean that Arizona's climate favored getting TB?

Quite the contrary. Arizona's climate is so *helpful* to TB victims that thousands of them go there. Naturally this would raise the average number of TB deaths.

Figure 9.1 More examples of the failure of common sense.[1]

other words, these calculations just do not seem to help us decide whether or not 58 is a lot of heads for a fair coin.

A quantity that is more helpful in making a decision about the bias of the coin is the chance that 58 *or more* heads are observed, in 100 tosses of a fair coin. This chance turns out to be about 7 percent. In other words, in 93 percent of such experiments with a fair coin, one would get fewer than 58 heads. So if the coin is fair, we have observed a fairly rare event; and this occurred in our one sample of 100 tosses. Do we believe that the single sample of size 100 turned out to be an unusual sample? If not, the only alternative explanation is that the coin is biased toward heads. Maybe we should consider this explanation instead of the rare-event theory. Perhaps we should reject the hypothesis that the coin is fair. Of course, if it were very important that we not make a false accusation, we might still withhold our judgment. In this case we might be forced to continue to assume that the coin is fair until more incriminating evidence accumulated.

This example illustrates how a statistical hypothesis can be tested. We glossed over the crucial calculation. Where did that 7 percent come from? (Recall that the chance that a fair coin produces 58 or more heads in 100 tosses was 7 percent.) To calculate this we need the sampling distribution of a sample percentage. In Section 6.3, this was shown to be approximately normal with average equal to the population percentage and SD equal to the SE of the sample percentage. The calculation is done assuming the coin to be fair, so the population percentage is 50 percent. The SE of the sample percentage is

$$\sqrt{\frac{0.50 \times (1 - 0.50)}{100}} \times 100 \text{ percent}$$

or 5 percent. In other words, if the coin is fair, the sample percentage of heads has a normal distribution with average 50 percent and SD = 5 percent. So 58 heads out of 100 tosses, which is 58 percent heads, is $(58 - 50)/5$ or 1.6 SDs above the average. Looking in the normal table reveals that this value is exceeded about 5.5 percent of the time: so 58 or more heads will occur about 5.5 percent of the time. If we had used the continuity correction (see Section 5.5 for details), we would have reproduced the 7 percent as claimed. That is, $(57.5 - 50)/5$ is 1.5 and the standard normal distribution exceeds this value 7 percent of the time.

Note that the calculation used to assess the fair-coin hypothesis assumed that the coin was fair. This will seem to be self-deceiving unless it is understood that the calculation is a tentative one. *If* the coin is fair, the chance of 58 or more heads is 7 percent. We are not expressing our belief in the fairness of the coin by doing this calculation. We are merely exploring the consequences of the fairness assumption.

What was the population sampled in the coin-tossing experiment? The population sampled was a hypothetical one, and it would be modeled by random sampling with replacement from a population such as {0, 0, 0, . . . , 1, 1, 1, . . .}, with the proportion of 1's indicating the true long-run proportion of heads. This long-run proportion is unknown, and in fact is the object of the investigation. How did we calculate the probability of getting 58 or more heads in 100 tosses if we did not even know the population proportion? We certainly did not use an estimate based on the sample—the only reasonable estimate is 58/100, and the probability of getting 58 or more heads when the population proportion was 58/100 would be about 50 percent. (The average of the binomial is just the population average, and the shape of the distribution is close to normal and is symmetrical; hence the 50 percent.) Any sample from the population would have produced 50 percent this way. The calculation of the unusualness of the sample is done assuming that the hypothesis is true. The hypothesis in this case is that the coin is fair. So the population on which we based our calculation was {0, 1}, so the proportion of 1's in this population *is* known to be 50 percent.

Now we will abandon the coin-tossing paradigm for a slightly more realistic setting. The basic logic remains the same. Suppose that an eminent researcher in education claims that students graduating from high school with first-class honors watch televison an average of 5 hr/week. Now a random sample of 25 students are interviewed and their TV times per week are determined to average 6.0 hr. The sample SD is computed to be 5.0 hr. Is the original claim, that the average is 5.0 hr/week, true or false, or are we unable to decide on the basis of this small sample? What can we say about the claim?

One answer might be that the sample average was 6 hr, and since the sampling method is assumed to be random and unbiased, the 6 hr is a more reasonable belief than 5 hr; so we might be tempted to disbelieve the original claim. But before disputing the claim, should we not evaluate the strength of the evidence against the claim? Perhaps we should give the eminent researcher the benefit of the doubt. Our judgment should really depend on how much doubt there is.

It is usual in hypothesis tests to require strong evidence against a hypothesis before rejecting it; in fact, the hypothesis is specified so that it is appropriate to accept it when the evidence is equivocal. As a first step in evaluating the evidence against the hypothesis in the example, we must determine how well the sample estimates the average TV time for these students. The variability of a sample average was discussed in Chapter 8, and this seems to be what is needed here. The SE of the average is $5.0/\sqrt{25}$ or 1.0, so the 5.0 hr in the claim is only 1.0 SE below the sample average. Even if the population average were 5.0 hr of TV time, the sample average would be greater than 5.0 about 50 percent of the time. The fact that the sample average is greater than 5.0 does not tell us much about the claim. Now the sample average will be greater than the observed average, 6.0 hr, about 16 percent of the time, if the hypothesis is correct (because 6.0 is 1.0 SE greater than 5.0). This means that a sample average of 6.0 hr is not very unusual, *even when the claim is true:* a result more contrary to the claim than the one that occurred would be expected 16 percent of the time, when the claim is true. We have not observed a rare event. In fact, the occurrence of an average of 6 hours in the sample of 25 is quite usual when the hypothesis is true. So perhaps we should not dispute the claim.

Some terminology is needed to discuss hypothesis testing more carefully. In the first place we have to have a **hypothesis** to test. The hypothesis is the statement about a certain population, the credibility of which we want to assess on the basis of the sample. It is less obvious that we will also need to specify an **alternative,** the statement that we will be led to believe if the hypothesis is rejected. Usually, the setting of the problem determines the alternative even before the data are collected, as in the following examples.

Of course, we must have a sample from the population, and we usually boil this down to a value of a **statistic:** a number calculated from the sample values. This statistic value is compared to the **sampling distribution** that the statistic would have if the hypothesis were true. The comparison is summarized into a probability called a **P value:** this is the probability, if the hypothesis were true, that the statistic would be at least as far from its expected value as it was observed to be in the sample. The

Sec. 9.1   The Logic of Hypothesis Testing   **329**

distance from the expected value is measured in the direction of the alternative (this will explained shortly). A **critical P value** is the probability that is set by the person doing the test; it is the threshold for the P value that the tester will use to decide whether the sample is unusual enough, compared to the hypothesized population, to indicate that the hypothesis should be rejected in favor of the alternative. In other words, a **decision** must be made: to reject the hypothesis (and accept the alternative) or to accept the hypothesis. This decision is made based on the size of the P value. When the P value is small (i.e., less than the critical P value), we reject the hypothesis. When it is not small (greater than the critical P value), we accept the hypothesis.

The components of hypothesis testing are listed again for easy reference:

1. The hypothesis
2. The alternative
3. The sample
4. The statistic
5. The sampling distribution of the statistic when the hypothesis is true
6. The P value
7. The critical P value
8. The decision

Let us review the TV-hours example with this checklist.

1. The hypothesis is that the average TV hours, for the population of honors students, is 5.0 hr.
2. The alternative in this situation is not clearly specified. It could be "that the average TV hours are not equal to 5.0 hr" or it could be "that the average TV hours are less than 5.0 hr" or "that the average TV hours are greater than 5.0 hr." The choice depends on what we want to be left with if the hypothesis is rejected. Suppose that a national study by experts in education has recommended that no more than 5.0 hr of TV be watched. Our interest in these sample data, which perhaps have been obtained from students at the local school, is to see if even the best students watch TV for only 5.0 hr. Then the natural choice for the alternative would be the last one: "that the average TV hours are greater than 5.0 hr." (In other words, the alternative is chosen by the person interested in the test, according to what he or she wants to learn from the test.)
3. The sample in this case is a random sample of 25 honors students from a population of honors students (e.g., all honors students graduating from the local school district).
4. The statistic is the sample average.
5. The sampling distribution of the sample average is approximately normal with population average 5.0 and SD equal to the SE of the sample average.
6. The P value is 16 percent. It is calculated by examining the position of the

statistic 6.0 relative to the sampling distribution of the average, when the hypothesis is true. In this example this distribution is approximately normal with average 5.0 and SE approximately $5.0/\sqrt{25}$ or 1.0. The picture is as shown in Figure 9.2.

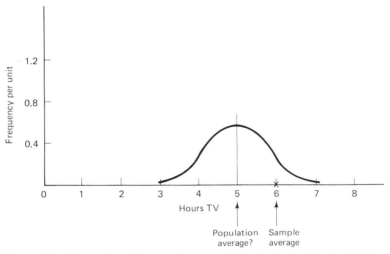

Figure 9.2 Comparison of the sample average 6.0 with the hypothesized distribution of the average (normal with average 5.0 and SD = 1.0).

Using Table A4.1, the shaded area is 0.16, representing a chance of 16 percent that the sample average exceeds 6.0 when the hypothesis is true. Note that we have used the direction indicated by the "alternative" in evaluating how rare a sample outcome is.

The P value ranges from 1.0 to 0.0. As it gets closer to 0.0, it is indicating that the sample is a rare outcome if the population is as hypothesized. The closer the P value is to zero, the stronger the evidence against the hypothesis.

7. Sometimes the hypothesis test stops at step 6. In this example we are left with a feeling that the hypothesis may well be true, since our sample average was fairly typical of values one would expect if the hypothesis were true. There is only very mild evidence against the hypothesis. However, if the aim of the hypothesis test is to make a decision whether to accept or reject the hypothesis, we need a threshold on the size of the P value that will help us to decide one way or the other. This is the critical P value.

A small P value indicates that the hypothesis is false. ("If one observes something that is unusual under ordinary circumstances, this is evidence that the circumstances are not ordinary.") A large P value suggests that the evidence against the hypothesis is too weak to reject it, so the hypothesis is accepted. If the critical P value were 0.10, we would accept the hypothesis in this example. (*Reminder:* The term "P value" refers to a calculated probability which depends on both the sample and the hypothesis. "Critical P value" is the threshold number with which we compare the *P* value to make a decision. It is specified usually before the test of hypothesis is performed, without reference to the sample or the hypothesis but with regard to how rare an event has to be before we will suspect that the hypothesis is false.)

8. The decision is to accept the hypothesis that the average TV time among honors students in this school district is 5.0 hr.

The Epictetus quote at the beginning of the chapter has its analogous expression in hypothesis testing. If we view the hypothesis test as a decision-making method, the aim of the test is to accept or reject the hypothesis. Whatever we decide, the population sampled may or may not be in agreement with the hypothesis. So there are four possible outcomes of a hypothesis test:

1. Hypothesis true, test accepts hypothesis.
2. Hypothesis false, test accepts hypothesis.
3. Hypothesis true, test rejects hypothesis.
4. Hypothesis false, test rejects hypothesis.

Outcomes 1 and 4 mean that the test has led us to a correct conclusion, while 2 and 3 mean that the test has led to an erroneous conclusion. Outcome 2 is a **hypothesis acceptance error,** and outcome 3 is a **hypothesis rejection error.** How do we make these errors as infrequent as possible?

The critical *P* value is actually a specification of the hypothesis rejection error probability (outcome 3) that you are willing to accept. The *P* value will be smaller than this number a proportion of the time equal to the critical *P* value itself if the hypothesis is true. If the critical *P* value is 5% and the hypothesis is true, the *P* value will be less than 5 percent just 5 percent of the time.

The reason that we cannot in practice set the probability of a rejection error arbitrarily small is that there is a trade-off between acceptance errors and rejection errors. By reducing one, you increase the other. By setting the critical *P* value at 1 percent instead of 5 percent, we are going to increase the chance of missing a false hypothesis; that is, accepting it. The only way to reduce both at once is to increase the sample size. This effectively increases the precision of the estimates, which helps in the decision making.

One last tidbit of terminology. When a hypothesis is rejected, the result is said to be "statistically significant," and when the hypothesis is accepted, the result of the test is said to be "not statistically significant." This unfortunate custom has caused as much confusion as the labeling of the "normal" distribution, but it is a widespread custom that one should know. Hypothesis tests are similarly called "tests of significance." The problem with this is that a hypothesis test can be extremely important (one is tempted to say "significant") when the result is to accept the hypothesis. For example, if it were found that smokers did *not* have higher blood pressure than nonsmokers, this would be a very important result. But the test of the hypothesis that there is no difference between the two groups would in this instance have to be called "not statistically significant" according to the customary terminology. More will be said in Section 9.7 about the interpretation of hypothesis tests.

Before you drown in details, take a look at the data plotted in Figure 9.3. The smooth curve is the histogram of the normal distibution having average 5.0 (the hypothesized value in the TV example) and an SD of 5.0. (We are assuming that the sample SD can be used as the population SD.) The data plotted average 6.0 but visually would be credible as a sample from the population described by the normal curve. The *P* value is an objective way to assess this credibility. Based on it, we decide whether we believe the hypothesis (the smooth curve) or reject it. (For the figure, we had to make the assumption that the population distribution was normal. For the numerical test, we did not need to assume this because we could rely on the central limit theorem to guarantee the sampling distribution of the average to be approximately normal.)

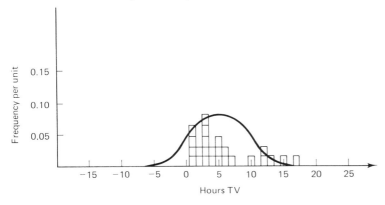

Figure 9.3 Visual test of a hypothesis. The smooth histogram is for a normal distribution with average 5.0 and SD of 5.0 The question is whether the data might be a random sample from a population whose histogram is the one shown. If so, accept the hypothesis. If not, reject the hypothesis.

There are many details to sort out in the execution of a hypothesis test, but to see if you have the overall strategy, you should try Exercise 1 at this point.

The rest of this chapter explains how to execute this hypothesis-testing procedure in a variety of situations. The criteria that distinguish between the procedures are:

1. The statistic used to summarize the sample. This in turn depends on the hypothesis to be tested.
2. The size of the sample.
3. Knowledge about the population before sampling is done.
4. The number of populations sampled.
5. The number of variables observed.

We will refer back to these criteria in the introduction of the various tests.

All the tests in this chapter assume that the sample is a random sample. In practice it is important to determine if this randomness assumption is valid. We have discussed in Chapter 7 the requirements of this assumption. The practical problem with nonrandom samples is not only that they tend to be biased, but that we have no basis for calculating any probabilities with such samples. Nonrandom samples can produce estimates, but the estimates have unknown accuracy. When it is the population rather than the sample that we are interested in, we must have random samples on which to base our estimates and hypothesis tests.

The tests discussed in this text are only the most commonly used tests. Several tests that are applicable to regression models have been omitted.

EXERCISES

1. A statistics professor wishes to demonstrate the logic of tests of hypothesis using a simple classroom experiment. He tosses a coin in the usual way and reports the result, "heads" or "tails," to the class. When he calls out the sixth head in the first six tosses, a student runs up to the front of the lecture hall near the professor, to watch the next toss more closely. Explain the behavior of the student and speculate on an explanation of the behavior of the professor.
2. A man takes his car in to a service station to have the brakes checked. When he picks the car up at the end of the day, he notices a thumping sound, which turns out to be a loose wheel. The customer claims that the mechanic must have forgotten to tighten the wheel after checking the brakes since he cannot believe that the wheel had been loosened by some vandal (or gremlin) after the mechanic had finished with the brake job. Does this experience have the elements of a formal hypothesis test? Discuss with reference to the eight points outlined in the text.
3. Redo the TV-hours example described in this section with the sample average equal to 7.3 hr instead of 6.0 hr.
4. In the TV example, if rejecting the hypothesis falsely were to be especially avoided, in view of the eminence of the researcher, would you set your critical P value to be very small (e.g., 0.001) or relatively large (e.g., 0.10)?
5. At a day care center, a small child pulled a high chair down onto his head while playing. After a few stitches, the child seemed fine but the next morning had his first epileptic seizure. The doctor claimed that the two events were unrelated, just a coincidence. Is there any reason to doubt this opinion?

9.2 TESTS CONCERNING AVERAGES

9.2.1 Normal Tests for an Average

The two tests described in this section use the identical computational procedure but apply to two different situations. Both tests examine a hypothesis about the value of a single population average, but one is restricted to use with a large sample.

The normal population test for an average. The normal population test has two special requirements:

1. That the population be normal.
2. That the population SD be known, or else that a good estimate of the population SD be known or estimable from the sample. By "good" here we mean that the SD should be estimated by a sample size of at least 20, or else the SD should be known to at least this same precision from other sources. (For simplicity we will refer to this requirement as "SD known.")

If the population distribution is normal, the sampling distribution of the sample average will also be normal. This is true for samples of any size. (It is obviously true for samples of size 1.) This property is useful for making probability statements about the possible values of the sample average; but when can the assumption of normality *of the population* be reasonably assumed? One case is when the investigator has some experience with populations thought to be similar to the one of interest, and knows that normal distributions commonly occur. Another case is when the sample is large enough that one can simply look at the sample itself to see if normality is present. Or the mechanism generating the sample data may involve some averaging, and although the components cannot be observed, it may be reasonable to assume that the central limit theorem produces approximate normality of the population in these situations. In other situations the normality of the population should not be assumed.

Some people have the impression that "normal" is "good" and they reason that since they have been careful collecting the data, the data should be "normal." This is faulty reasoning. The normal distribution is not the "usual" distribution, notwithstanding its deceiving name. Life is too complicated for there to be one "usual" distribution.

When the population can reasonably be assumed to be normal, we may proceed. We refer to the eight steps outlined in Section 9.1 and also to the TV-hours example described there.

1. The hypothesis states the value, or range of values, that the population average is claimed to have. In the example it is that the average weekly TV hours among honors students is 5.0 hr.
2. The alternative, as previously explained, is that the TV hours are greater than 5.0 hr.
3. The sample consists of the 25 honor students.
4. The statistic is the average TV hours in this sample, 6.0 hr.
5. Because the population is normal, the sampling distribution of the sample average is also normal. This sampling distribution has average 5.0 (under the hypothesis) and SE estimated to be the sample SD divided by the square root of the sample size, 5.0/5 or 1.0 in the example. (The sample size for this estimate was 25, so condition 2 is satisfied.)
6. The P value is 16 percent.
7. If the critical P value is 10 percent, we would
8. Decide to accept the hypothesis.

When we have a normal population but cannot satisfy condition 2 concerning the estimate of the population SD, we may use the procedures of Section 9.2.2 (the t procedure).

The large-sample normal test for an average. When it cannot be assumed that the population has a normal distribution, we can still use the test procedure just described if the sample is large. The central limit theorem tells us that the sampling distribution of a sample average is approximately normal no matter what the population distribution is. This approximation improves as the sample size used to calculate the sample average increases. A rule of thumb is that samples of size 20 or more can be considered to be large enough for practical purposes to allow this approximation to work well. If the population is not too non-normal, even smaller samples will give quite normal sampling distributions.

The procedure is the same as the one just described for the normal population test of an average since the sampling distributions turn out to be the same.

What do we do for small samples from a nonnormal population when we want to test the population average? The nonparametric tests, which are described in Section 9.6, may be used in this case.

9.2.2 The t Test for an Average

Just as in the small-sample estimation of averages (see Section 8.5), the small-sample testing of averages requires the t distribution.

The t test is a valid test of a population average whenever the population is normal. However, when the population SD is known, the normal population test is better. We will explain the mechanics of the t test shortly, but first let us consider in more detail when we might use this test. The condition needed for the "normal population test," that the population SD should be known, is unnecessary for the t test, because the t test takes into account the effect of the variability of the sample SD as an estimate of the population SD. However, it is still necessary that the population be normal, or at least approximately normal, for the t test to be valid.

If you review the conditions under which the normal tests can be done, you will see that the only situation in which the t test is applicable and the normal tests are not applicable is when all of the following are true:

1. The population is known to be normal.
2. The population SD is unknown.
3. The sample size is small (say fewer than 20).

This situation arises only in certain fairly specialized application areas, such as routine testing in industrial quality control or environmental inspection. It is in such situations where experience with the shape of the distribution justifies the assumption of normality, even when samples are too small to check the assumption. In spite of this, the t test is widely used for tests on means in small samples. It is known that the t test is fairly "robust" to lack of normality in the population, so in practice it is not necessary to have exact normality of the population. Nevertheless, it is a risky procedure because in small samples there is no opportunity to check with the sample data whether the normality assumption is reasonable. (A histogram based on 20 or fewer values does not produce a reliable impression of the shape of the population histogram.)

To illustrate the t test procedure, consider the TV-hours example again, but this time suppose that the sample size is 4. Suppose that we observe a sample average of 8.8 hr and a sample SD of 5.0 hr, we want to test the hypothesis that the population average is 5.0 hr. Does the following procedure seem correct?

The SE of the average is estimated to be $(5.0/\sqrt{4}) = 2.5$ and the observed value 8.8 is at $(8.8 - 5.0)/2.5 = 1.52$ standard units, so the P value is 0.064. Since the critical value is 0.10, we would reject the hypothesis. However, there is one problem with this calculation, and that is that the SD, 5.0, is based on a very small sample and is unreliable. For the normal test procedure, the SD is supposed to be known and not subject to large sample-to-sample variability. The correct procedure is as follows.

The statistic for the t test, which we will call the t **statistic,** is calculated as shown in Chapter 8:

$$t = \frac{\text{sample average} - \text{population average}}{\text{SD}/\sqrt{n-1}}$$

In the example, the t statistic is

$$\frac{8.8 - 5.0}{5.0/\sqrt{4}} = 1.32$$

For the t test, the P value is obtained from the t table, Table A4.3. The appropriate row in the table is the one for the degrees of freedom equal to 3, the sample size less 1. Reference to Table A4.3 shows that the P value is greater than 0.10. That is, the probability of getting a value for the sample average that is at least this much larger than the hypothesized average of 5.0 is more than 0.10.

Compare this with the value 0.064 obtained using the normal distribution. The t procedure is the correct one in this situation. The difference between the two procedures can be considerable for very small samples. But of course this is exactly the situation in which the normality of the population distribution would be most in doubt. This presents a dilemma in many practical situations.

What happens if the population distribution is not normal and we still try to use the t procedure? We get the wrong P values, and sometimes they are very wrong. Here is an extreme example. Suppose that the population is {0, 1} and we are considering samples (with replacement) of size 2. Possible samples are:

0, 0 with average 0 and SD 0

0, 1 with average 0.5 and SD 0.5

1, 0 with average 0.5 and SD 0.5

1, 1 with average 1.0 and SD 0

These possibilities are equally likely, each having probability 0.25 of occurring.

The value of the *t* statistic for each possible sample, using the true population average of 0.5, is infinite, 0, 0, and infinite, respectively. Thus the sampling distribution of the *t* statistic is as follows:

Value of *t* statistic	Probability
0	0.5
Infinite	0.5

Compare this with the tabulated distribution of the *t* statistic for 1 degree of freedom (Table A4.3). The probability of a rejection error would be 50 percent, no matter what critical *P* value were used. In practice, the inadequacy of the *t* procedure when the population is not normal is not nearly this marked, but we should not ignore completely the normality assumption that underlies the procedure.

EXERCISES

1. In a countrywide survey of food prices, a random sample of 25 stores is selected. The price of a standard basket of goods on the day of the survey is calculated for each of these 25 stores. The average for this sample is $85 and the SD for the sample is $15. A particular store manager notes that the same food basket at his store costs $75. Are this store's prices lower than in the rest of the country, or would a difference of this magnitude be best explained as sampling variation? (That is, is it plausible that the food stores in the population of food stores sampled have an average cost for the food basket of $75?)

2. A lottery claims to have sold all 10,000 of its tickets. These tickets bear serial numbers 0000, 0001, 0002, . . . , 9999. Five winning numbers are announced; their serial numbers average 4350, and have an SD of 2450. Do you think that all the tickets really were sold? (Assume that unsold ones have the high serial numbers.)

3. The staff of a carpet cleaning company periodically makes telephone calls to homes in a certain suburban area to offer the company's services. Each such blitz involves calling thousands of homes. To monitor the bottom-line success of the several callers, the boss, Mr. Big, records the length of time (called the waiting time) that the caller takes to establish an appointment with a customer. (Several customers will usually refuse to make an appointment, so the time taken to call these is a part of the "waiting time" to make an appointment.) At the end of each evening, Mr. Big records the average waiting time for each caller. Miss Glib asks the boss for a raise after her first day (a 7-hr day), since her waiting time was quite low, on average. All the callers receive the same hourly rate so far, since no one has dared approach Mr. Big. The boss decides to look at the numbers, and he comes up with the following:

    - *Miss Glib:* average waiting time = 15.6 min, SD = 3.9 min
    - *Other 23 callers:* average waiting time = 22.2 min, SD = 4.5 min

    Is Miss Glib's performance really better than average? (*Hint:* The problem does contain enough information to do a useful calculation.)

4. The concentration of mercury in a certain lake has been measured many times over a period of time. The measurements reveal an average concentration of 1.20 mg/m^3 with an SD of 0.15 mg/m^3. Following an accident at a smelter on the shores of the lake, nine more measurements are taken. These have an average mercury concentration of 1.45 mg/m^3. Making whatever assumptions are necessary, but stating what they are, test the hypothesis that the concentration of mercury has not increased. Use a critical P value of 0.01. (Several assumptions are needed here.)

9.3 TESTS CONCERNING PERCENTAGES

In Chapter 8 we explained how the sampling distribution of a percentage was in a sense just the same as the sampling distribution of an average: we just used an appropriate population of zeros and ones in the sampling model. Of course, in a practical situation, it is not necessary to work through this process each time a percentage is to be estimated; formulas are available which depend only on the sample size and the proportion of 1's. The same is true of tests concerning percentages. The relationship of tests for proportions to tests for averages is helpful in sorting out what assumptions must be met.

The normal population test for averages required, among other things, a normal population. Any population containing only 0's and 1's is clearly not normal, so we can forget about customizing this one to percentages. The large-sample normal test can be so customized, however, since large samples produce averages that have a normal sampling distribution, approximately, no matter what the population distribution. This is the test described in Section 9.3.1. In Section 9.3.2 we have a test that will work for small samples. It is not just an application of the t test, however, since, again, we do not have a normal population when we are working with percentages. This small-sample test is the binomial test.

9.3.1 The Normal Test for a Percentage

A municipal government that supports a number of local libraries is asked to expand several of the libraries because of the need for more shelf space. The municipality decides that libraries should have no more than 10 percent of their books with publication dates 10 or more years old. Each library is supposed to ship its old books to a central reference library, to minimize the need for additional shelf space at each of the local libraries. The libraries claim that they have already done this and that they still need to expand. The municipal government decides to see for itself.

A random sample of 30 books is selected from the entire current holdings of a local library. Five of these 30 books are more than 10 years old. Should the local head librarian be drawn and quartered?

We have to test the hypothesis that the percentage of old books in the entire collection is 10 percent, against the alternative that it is greater than 10 percent. The statistic we must use is the sample percentage: in this example it is (5/30) × 100 percent, or 16.7 percent. The distribution of the sample proportion, under the hypothesis, is approximately normal with average = 0.10 and SD = $\sqrt{(0.10 \times 0.90)/30}$ = 0.055. That is, the sample average has expected value 0.10 and SE 0.055 if the hypothesis is true. (The calculation of the SE of a proportion was discussed in Section 6.3.) By examining the normal probability that an observation exceeds $(0.167 - 0.10)/0.055 = 1.22$, the P value for the observed proportion 0.167 is obtained. This P value is found from Table A4.1 to be approximately 0.11. The surveyor from the municipality is not sure whether to reject or to accept the hypothesis, because she forgot to specify the critical P value. In a serious matter in which the tester wishes to give the hypothesis the benefit of the doubt, the critical P value is usually set at 0.01 (i.e., 1 percent) or even less. The decision for the municipality to make in this case should probably be to accept the hypothesis that the percentage of old books in the entire collection is 10 percent. The local head librarian explains this to the survey team, since he knows the statistical theory well. This knowledge has saved the librarian's hide. The answer to the original question is "no."

Were you surprised that 5 out of 30 was not convincingly greater than 10 percent of books in the library? If so, you should be all the more convinced of the necessity of formal tests of hypothesis.

This example illustrates that the normal test for a percentage is done the same way as the large-sample normal test for an average, with one exception: the SE can be calculated exactly since the sample size and the hypothesized population proportion are known. (In the test for averages, it had to be estimated or known from previous experience.)

9.3.2 The Binomial Test for a Percentage

When the sample is so small that normal approximation to the distribution of the sample percentage is poor, we need another strategy for the test of hypothesis of a percentage. Since the percentage times the sample size is the number of times a certain outcome has occurred in a binomial experiment, we can use the binomial distribution to do tests in this situation. Of course, we could use the binomial distribution for large samples, too, but the calculations can be very time consuming and tables are not available to cover all possible situations.

Let us redo the library survey example of Section 9.3.1. The procedure we will use is applicable to smaller samples and does not assume any normality. We have a very large population of books and the hypothesis of 10 percent old ones is to be assumed for the calculation. We have a binomial experiment with 30 independent trials and the outcomes are of two kinds: old and not old. Each book selected has a 10 percent chance of being old, according to the hypothesis. So the binomial tables (A4.2) give the chance that 0, 1, 2, 3, or 4 books are old. This is (4.2% + 14.1% + 22.8% + 23.6% + 17.7%) or 82.4 percent. Thus the chance that 5 or more are old is 17.6 percent. Because this 17.6 percent represents an event that is not unusual, we infer that 5 old books in 30 is not a rare event when the population percentage is 10 percent. Again we would accept the hypothesis.

Why did we get such a different P value for this method? The normal approximation was 0.11 and the binomial, which is the exact calculation, gives 0.176. The difference would mostly disappear if we had used the continuity correction method with the normal approximation (Section 7.5.2): the 5/30 expressed in standard units was $(0.167 - 0.10)/0.055 = 1.22$, but with the continuity correction it is $(4.5/30 - 0.10)/0.055 = 0.91$ and the normal table gives the P value to be 0.181, which is very close to 0.176. So the normal approximation, done properly, does give a good approximation for this fairly large sample.

EXERCISES

1. Redo the library example described in this section assuming that six old books were found among the 30 sampled. Use both the normal approximation and the binomial test, and comment on the closeness of the two.
2. A coin is tossed, and nine heads are observed in 10 tosses. Is the coin fair?

9.4 TWO-SAMPLE TESTS

All the tests we have described so far concern a single population. However, many important questions involve comparisons between two populations; often we wish to test if the two populations are the same. Naturally, such tests are based on two samples, one from each population. Such tests are usually called **two-sample tests**.

9.4.1 Normal and t Tests for Averages

Here is an example of a situation requiring a two-sample hypothesis test. This one involves a comparison of two population averages.

Students at AOK College are of two types: those that live in residence on campus and those that live off-campus. A sociologist wishes to know whether there is any marked difference in the socioeconomic backgrounds of these two groups. A survey of a sample of students from each group is collected to document the parents' incomes for each student. The higher parental income is considered an indicator of socioeconomic status, and this is to be the focus of the analysis. The sample of residence students is a random sample of size 20 from the entire list of residence students; a second random sampling procedure is performed to obtain 15 students from the population of off-campus students. The question is whether or not the average incomes in the two groups are the same.

The sample data consist of two sets of incomes, which we can summarize by:

- *On-campus:* 20 incomes with average $40,000 and SD = $10,000
- *Off-campus:* 15 incomes with average $38,000 and SD = $12,000

The sampling distributions of the two sample averages are based on large enough samples to provide an adequate approximation to normality. (The central limit theorem states that for large enough samples, averages are approximately normally distributed). One procedure for evaluating an average is to reexpress it in standard units and use the normal approximation, but this is valid only if the sample SD, which is usually substituted for the population SD in these sampling distributions, is based on a large enough sample that this substitution does not affect anything. With samples of 15 and 20, the estimates of the population SDs are a bit shaky, but we will need to combine them in this two-sample test, as will be shown, so we can be a little easier on the sample-size requirements for this two-sample test than we could in the one-sample test. A rule of thumb for the two-sample test is that each of the samples should be at least 15. When this size requirement is satisfied, the test procedure shall be called the normal two-sample test for averages. When one of the samples is less than 15, we can use a two-sample t test (Section 9.4.2) or a nonparametric test (Section 9.6.2).

We will develop the procedure for the normal two-sample test for averages according to the usual steps. The test will be described in terms of the incomes example.

1. The hypothesis is that the two populations have identical average incomes.
2. The alternative has not been clearly specified by the wording of the problem. Let us assume that the alternative of interest is that the off-campus average income is less than the on-campus average. Later we will see the effect of varying this supposition.
3. The two samples are not given, but the summary statistics shown will be adequate for the test procedure.
4. The statistic we will use must clearly include the difference in the two-sample averages: $40,000 - 38,000 = $2000. But to assess the statistical meaning of this difference, we need to know the SD of the difference. We can deduce this from the result in Section 7.4 about the SD of a sum. The difference $A - B$ can be written as a sum $A + (-B)$. This implies that the *SD of the difference between two variables is the sum of the SDs of the variables, as long as the variables are independent*. The independence of the sample averages is guaranteed here by the independence of the samples. Consequently, the difference in the two sample averages has

$$SD = \sqrt{\frac{10,000^2}{20} + \frac{12,000^2}{15}}$$

which is about $3820. In standard units the difference is therefore 2000/3820, which is about 0.5.

5. The statistic calculated as in step 4 will have a normal distribution, approximately. If the hypothesis is true, the average of this sampling distribution will be zero and its SD will be 1.0.

342 Chap. 9 Testing Hypotheses with Data

6. The *P* value is the probability associated with values of the statistic as far or farther away from the hypothesized average than was observed in this instance, in the direction of the alternative. The alternative is that the off-campus average is less than the on-campus average, and the statistic tells us that it is observed to be 0.5 standard unit less. The *P* value for more extreme departures from the hypothesis is 0.31. (See Table A4.1).

7–8. The critical *P* value has not been specified. If a decision is necessary here, we would have to accept the hypothesis, since the evidence against it is very weak. (The chance of getting a larger difference in the sample averages is 31 percent.) Otherwise, we would simply conclude that the evidence against the hypothesis is weak. This hypothesis test would result in the conclusion that the income levels of the parents of the two groups of college students are about the same. (The strict conclusion is that the income averages for the two groups of college students have not been shown to be different.)

Now let us consider how the test would have been modified if we were interested in detecting a difference in the two averages, without a particular interest in which direction the difference went. The alternative changes: it would be that the two population averages are different. (Compare this with the alternative stated above.) In this case "the direction of the alternative" is really two directions, and when we are calculating the *P* value, we should also consider the possible sample outcomes in which the on-campus sample average is much less than the off-campus sample average. These outcomes would also be evidence against the hypothesis in this case. The statistic still comes out to be 0.5, but the *P* value is twice as large as before since we must add the area in the two tails of the normal histogram. The *P* value in this case would be 0.62 instead of 0.31. Our conclusion would be the same in this example (see Figure 9.4). When the alternative indicates two directions of deviation from the hypothesis, the test is called a **two-sided test.**

When the samples are very small, that is, if at least one sample is less than 15, our estimate of the population SD will be too poor to use it as if it were known exactly (and this is what we are doing if we assume that the difference in sample averages

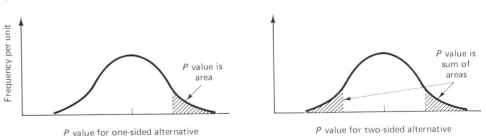

Figure 9.4 *P* values for one- and two-sided alternatives. For the two-sided alternative, very large or very small values of the statistic indicate rejection of the hypothesis, and we have to take account of this even though, for a given sample, it would be impossible for a statistic to be both very large and very small.

divided by the calculated SD has a normal sampling distribution). Furthermore, we may not be able to count on the central limit theorem to produce approximate normality even if the SDs were known. However, if both populations are approximately normal and the sample sizes are small, we can use a t procedure to assess the hypothesis. A hitch with this is that one additional assumption must be satisfied: the SDs of the two populations have to be equal.

To summarize, the assumptions for the two-sample t procedure are:

1. The population distributions are normal.
2. The population SDs are identical.

The technique will be used only when the sample sizes are small, since the normal test is adequate with larger sample sizes and requires fewer assumptions. But small sample size is not a requirement of the t procedure; the t procedure is valid for large samples as long as the extra assumptions above (1 and 2) are satisfied. The t procedure gives P values identical to the normal procedure for large samples.

The differences in the t procedure from the normal procedure outlined above are:

4'. The denominator of the t statistic for the difference of two averages is calculated differently; the sample SDs for the two groups are replaced by a single "pooled" SD. We will denote this pooled SD as SD_p. This difference in procedure from the normal one is due in part to the new assumption about the equality of the population SDs.

SD_p is a particular combination of the two sample SDs. If one sample SD is written SD_1 and the other is SD_2, and the corresponding sample sizes are n_1 and n_2, then

$$SD_p = \sqrt{\frac{(n_1 \times SD_1^2) + (n_2 \times SD_2^2)}{n_1 + n_2 - 2}}$$

5'. The t statistic for the difference in two sample averages described in step 4' will have a t distribution with $n_1 + n_2 - 2$ degrees of freedom.

The other steps are similar to the normal procedure. If we apply steps 4' and 5' to the example, we get

$$SD_p = \sqrt{\frac{(20 \times 10{,}000^2) + (15 \times 12{,}000^2)}{20 + 15 - 2}}$$

which is approximately $11,200. The denominator of the statistic is therefore

$$\sqrt{\frac{11{,}200^2}{20} + \frac{11{,}200^2}{15}}$$

which is about 3826. So the 2000/3826 is again about 0.5. If the sample SDs had been very different, the two procedures for calculating the denominator of the statistic would have produced more different results.

Proceeding with the incomes example, the distribution against which to compare the 0.5 value is the t distribution with 33 degrees of freedom. From Table A4.3 we can see that the P value is only slightly less than 0.31. The normal and the t procedure give about the same P value in this case. The difference in the two P values would be larger if the sample sizes were smaller.

If the assumptions of the t test for two averages are satisfied, it is the correct test to use. Unfortunately, the difference between this t test and the normal procedure is greatest in just that situation in which the assumptions are hard to check: very small samples. Again, the t procedure is quite robust to nonnormality of the population distributions, but the assumption of equal variances is more crucial. This must be taken into account before relying on the results of this two-sample procedure.

9.4.2 2 by 2 Contingency Table Tests

When data are qualitative, the feature of the population of interest is often a percentage (or a proportion). When we have two samples of qualitative data, we usually want to compare the population percentages. Consider, for example, an experiment in which 50 patients who have passed a general health examination are randomly allocated into two groups of 25. One group is issued a one-month supply of high-dose vitamin tablets to take each day; the other group is issued an apparently similar supply of tablets that are harmless but do not contain any vitamins (i.e., a placebo). At the end of the month the 50 subjects are evaluated and labeled as "passed" or "failed" by a doctor who does not know which regimen the patients have been on. In other words, the data consist of two lists of length 25 where each position in the list consists of a "pass" or "fail." The question of interest is: Are the patients on the vitamin regimen more likely to pass the final health exam than those on the placebo?

Let us suppose that 20 of 25 pass in the vitamin group, but only 10 of 25 pass in the placebo group. It seems that the vitamin is having a beneficial effect, but we must check whether the apparent effect might be reasonably attributed to chance. This is why we do a test of the hypothesis that there is no difference in the population percentage of "pass" patients due to the vitamin regimen. The alternative in this case is that the vitamin regimen leads to a higher pass percentage. The samples have been described. A statistic that is calculated for both samples is the sample proportion: 0.80 in the vitamin group "passed" and 0.40 in the placebo group. However, a statistic that focuses on the difference between the proportions is the difference itself divided by the standard error of the difference. As long as the samples are large enough that we can assume the sample proportions to be normally distributed, the difference statistic will also be normally distributed. We show below that the SE of the difference in the two proportions is 0.14. The statistic for the hypothesis test is therefore $(0.80 - 0.40)/0.14$, or about 2.9. The P value is 0.002, from Table A4.1. We would certainly reject the equality hypothesis in favor of the alternative that the vitamin group has a higher pass rate.

The calculation above follows a familiar pattern. The reason this test required special consideration is that the calculation of the SE is a little different from what you might expect based on the SE calculation for the difference between two averages. The reason for this difference is that the hypothesis about the equality of the proportions implies that the two population SDs must also be equal. Recall that the SD of a population containing a proportion p of 1's and a proportion $(1 - p)$ of 0's is just $\sqrt{p \times (1 - p)}$. This p is unknown, and we estimate it using *both* samples. The two samples of 20 of 25 passes and 10 of 25 passes are combined to give $(20 + 10)$ of $(25 + 25)$ passes, or 30/50, which is 0.60. Now we use this "pooled" estimate to calculate the SEs of the individual sample proportions, and combine these two SEs as we did with the averages.

$$\text{SE of } (0.80 - 0.40) = \sqrt{\frac{0.60 \times 0.40}{25} + \frac{0.60 \times 0.40}{25}}$$

$$= 0.14$$

This is the figure that was used to complete the hypothesis test as described above.

If one of the samples had been much larger than the other, the test described above becomes very close to a one-sample test of a population proportion (Section 9.3.1). Let us redo the example above assuming that 200 of 250 in the vitamin group pass and, again, 10 of 25 in the placebo group pass. The test statistic in this case turns out to be

$$\frac{0.80 - 0.40}{\sqrt{[(210/275) \times (65/275)/250] + [(210/275) \times (65/275)/25]}}$$

which works out to about 4.5. The P value would be very small.

The comparable one-sample test that the sample proportion based on the placebo sample of 25 is equal to 210/275, the test statistic is

$$\frac{0.40 - (210/275)}{\sqrt{(210/275)(65/275)/25}}$$

which is equal to 4.2, almost the same as the two-sample test. The P value and the conclusion would be almost identical to the two-sample test. This should not be too surprising. When one of the two samples is large, the estimates the large sample provides are so precise that they may be treated as known population values; this is what was done to do the one-sample test. Note the importance in this procedure of the assumption that the two population SDs are equal.

We have shown how to compare two samples in order to test whether the associated population proportions are equal. We required the samples to be large enough that we could appeal to the normal distribution as an approximate sampling distribution for the statistic of the test. If the samples are quite small, say one is less than 15, we cannot count on this approximation. The extension of the binomial test to two samples is complex, and in any case the small-sample estimates of proportions are quite imprecise. We are stuck in such a situation. Note, however, that if one sample is small and the other is very large, we can use the one-sample binomial test. The large sample is treated as providing exact information, and is the hypothesized proportion in the hypothesis.

There is an alternative but equivalent approach to the testing of the equality of two proportions. It will be useful to explore this as an introduction to Section 9.5.2. We begin this new approach by using what is called a contingency table to summarize the information in the two samples of qualitative data. See Table 9.1.

TABLE 9.1 DATA FROM HYPOTHETICAL HEALTH EXAMINATION OF TWO EXPERIMENTAL GROUPS

	Vitamin	Placebo	Total
Passed	20	10	30
Failed	5	15	20
Total	25	25	50

In examining this table, what do you look at to see if the pass-fail split is the same in the two groups? If the two proportions 20/25 and 10/25 are from populations with the same underlying proportion, surely 30/50 is a reasonable estimate of this common proportion. But what does this common-proportion hypothesis imply about the "expected" proportions in the body of the table? The word "expected" here is used in the sense of "expected value": the long-run average value in repetitions of the whole experiment. The best estimate of the common proportion is 30/50 or 0.60. If we apply this 0.60 to the group totals, which in this case are both 25, the "expected" split is 15–10 in both groups. The "expected" frequencies may be summarized in a similar table (see Table 9.2). One way to view the comparison of the two proportions is to see how different these two tables are. If the entries in the actual data table (Table 9.1) are called "observed" and those in Table 9.2 are called "expected," we want to examine (observed − expected) for each of the four cells.

These differences, (observed − expected), will be large in magnitude when the

TABLE 9.2 EXPECTED FREQUENCIES, UNDER HYPOTHESIS OF EQUAL PROPORTIONS, FOR THE DATA IN TABLE 9.1

	Vitamin	Placebo	Total
Passed	15	15	30
Failed	10	10	20
Total	25	25	50

two sample proportions are far apart (i.e., when the vitamin-group pass rate is much different from the placebo-group pass rate). The sample proportions will tend to be far apart when the hypothesis of equal population proportions is false. In other words, if we have a way of evaluating the chances for differences (observed − expected) of the size observed, we would also be able to use it to test the hypothesis of equal population proportions.

A mathematical result related to the central limit theorem (see Section 7.5.1) gives us a way to decide whether the differences (observed − expected) are unusually large. It involves a new probability distribution called the **chi-square distribution**, which is tabulated in Appendix A4.4. The result is that for large samples, the statistic formed by adding up

$$\frac{(\text{observed frequency} - \text{expected frequency})^2}{\text{expected frequency}}$$

for each of the four cells in the contingency table has a chi-square distribution with 1 degree of freedom. (The table of the chi-square distribution is like the t distribution in that several distributions are partially tabulated, with the number of degrees of freedom defining the different distributions.)

The use of the chi-square statistic for comparing two proportions is based on the central limit theorem and works best for large samples. A rule of thumb to decide whether the sample is large enough to warrant use of the chi-square approach is that the expected cell frequencies should be at least five. This requirement is roughly the same as the requirement that each sample be based on a sample of at least size 15. For smaller samples a more complex analysis is required. However, proportions are seldom estimated with such small samples, since the estimates would be so poor, and this is not a common problem in practice.

Let us apply this chi-square procedure to the same data set as we used with the normal approximation. For the original example in which we compared the vitamin and placebo regimens in two groups of 25 each, the observed and expected frequencies are as shown in Tables 9.1 and 9.2. The quantities based on (observed − expected) for each cell are as shown in Table 9.3. The sum of these four quantities is the value of our **chi-square statistic.** In this example it is 8.33, and the chi-square table shows

TABLE 9.3 COMPONENTS OF CHI-SQUARE CALCULATION RELATING TO DATA IN TABLE 9.1

	Vitamin	Placebo
Passed	$\dfrac{(20 - 15)^2}{15}$	$\dfrac{(10 - 15)^2}{15}$
Failed	$\dfrac{(5 - 10)^2}{10}$	$\dfrac{(15 - 10)^2}{10}$

that the associated P value is less than 0.005. We would reject the hypothesis in this case, just as with the normal test.

We have described two ways of examining the equality of two sample percentages. Let us call the first one the normal approximation test for two sample proportions, and the second one the chi-square test for a 2 by 2 contingency table. They are both large-sample tests and require the same assumptions. Do they always produce the same results? A short answer is "yes," but the way the two test procedures depend on the alternative is slightly different.

A "one-sided" alternative was actually used for the normal procedure; the alternative was that the vitamin group had a *higher* pass rate than the placebo group. We have stated that the P value is the probability of obtaining a result as far or farther from the hypothesis than was observed *in the direction of the alternative*. So the specification of the alternative actually affects the calculations.

Our P value in the normal test where we used a one-sided alternative (it was appropriate for the original description of the example) was 0.002. Had the alternative been two-sided, the P value would have been $2 \times 0.002 = 0.004$. When we discussed the chi-square test, the probability from the chi-square table (Table A4.4) was not given exactly, only that it was less than 0.005. Had the table been more complete, it would have also given this P value as 0.004. It gives P values for two-sided alternatives because in the formula for the chi-square statistic, the differences (observed − expected) are squared; the direction of departure from the hypothesis is ignored. To get the proper P value for a one-sided alternative using the chi-square method, we would have to halve the 0.004 from the table to get the 0.002. Of course, since the table does not give the P value accurately for the two-sided alternative, this halving cannot be done accurately. This difficulty can be avoided by using the normal test for one-sided alternatives.

In most practical situations, the question of whether the alternative is one-sided or two-sided will not make any appreciable difference. The slight difference in the application of the two tests (normal versus chi-square test of the difference in two proportions) would be irrelevant in such a setting. The normal test is a bit easier to do the table look-up for, since such tables are more complete. Also, the calculations for the normal test are slightly simpler do to. When a computer program is computing the P value, even these distinctions disappear. For hand calculations and the use of textbook tables, the normal test is simplest.

The real advantage in introducing the chi-square procedure is that it can be extended to take care of hypothesis tests involving many populations and samples, not just two. This is done in Section 9.5.2.

A final detail should be mentioned in case other texts or computer programs give slightly different results from those obtained using the methods just described. We have ignored the "continuity correction" that some authors include in their descriptions of the normal test and the chi-square test. A correction similar to that used to improve the normal approximation to the binomial can be used with the two tests, but in the large samples for which the tests are applicable, these adjustments are minor, and we have chosen to avoid this further complication.

9.4.3 Dependent Samples

We have considered two-sample tests for averages and proportions when the two samples are selected independently: the values observed in one sample do not help to predict the values in the other sample. A common situation when the samples do not have this independence property is when the samples are paired: the first observation in one sample is linked in some way to the first observation in the second sample, and the other observations are similarly paired. Analysis of this type of data was discussed in Chapter 8, but we repeat it here to point out the implications for testing hypotheses.

Paired data commonly occur as before and after measurements. Twenty people may have their weight measured before and after a weight-loss program. Another kind of pairing occurs in certain medical studies in which two patients who are found to be similar with respect to diagnostic category, age, sex, occupation, and perhaps many other things, are allocated to two drug treatments that are to be compared. If the treatment is judged to be successful (S) or failing (F) after a certain period of time, the data consist of paired data such as the rows in the following list.

Pair number	Drug A	Drug B
1	S	S
2	S	F
3	S	S
4	F	S
.	.	.
.	.	.
.	.	.

In such studies, one counts the number of pairs in which A fails and B succeeds, and compares this with the number of pairs in which A succeeds and B fails. This comparison will be shown to reduce to a one-sample binomial test of a proportion.

Paired samples are certainly dependent (not independent) samples. We should not compare the proportion successful with drug A with the proportion successful with drug B. This comparison is not illogical; it is simply not the best way to do the comparison. A difference in the effectiveness of the two drugs could be missed by ignoring the pairing.

The paired samples of before and after weights should be analyzed by looking at the weight change; the two samples really reduce to a single sample of differences. The question is whether these differences average out to zero in the population from which the subjects are selected. The two-sample test again reduces to a one-sample test of an average.

The general strategy for analyzing paired samples is to reduce the data to an appropriate single sample and apply the appropriate one-sample test.

Suppose that the weight-reduction clinic selects five patients at random from its files of hundreds of patients. The weights for these five patients are as shown in Table 9.4.

TABLE 9.4 ANALYSIS OF PAIRED QUANTITATIVE DATA BY USING DIFFERENCES

Patient	"Before" weight (kg)	"After" weight (kg)	Weight change (kg)
1	93	90	−3
2	105	106	+1
3	80	76	−4
4	123	110	−13
5	86	81	−5
Average	97.4	92.6	−4.8
SD	15.3	13.4	4.6

By scanning the data you should get the impression that the weight-reduction program seems to have had a small but reasonably consistent effect. And yet you can easily verify that the two-sample t test of whether the "after" average is less than the "before" average will accept the no-change hypothesis. Let us do the calculations for the one-sample t test.

To proceed with this problem, we must assume that the distribution of weight changes is normal. The sampling distribution of the average weight change is estimated to have SD $= 4.6/\sqrt{5} = 2.1$. The observed average weight change is 4.8 kg, which is 2.3 standard units above the hypothesized value. The appropriate t distribution with which to compare this is the one with $5 - 1 = 4$ degrees of freedom. The P value for the t statistic of 2.3 (the alternative in this case is one sided—we are interested in whether or not the average weight change is positive) is between 0.025 and 0.05. The evidence against the hypothesis is quite strong. The program does seem to result in a reduction of weight, on average. This conclusion is in contrast to the test that ignored the pairing.

Let us return to the example of paired qualitative data concerning the success rate for two drugs. The data have not yet been stated completely: Table 9.5 gives a complete summary. In other words, 63 pairs of patients were in the experiment. Forty-five of these pairs gave no evidence of any difference between the two drugs. Of the 18 others, 6 indicated that A was better, and 12 indicated that B was better. Under the hypothesis that there is in fact no difference in the success rate of the two drugs, one would expect these 18 pairs to be allocated to the two categories, just as a fair coin tossed 18 times would result in heads and tails. So we can test the hypothesis by calculating the probability of obtaining as unbalanced a split as 6–12 when the expected values are 9–9. Equivalently, we test the hypothesis that the population proportion is 0.50, having observed a sample proportion of 6/18, 0.33, in a sample of size 18. The binomial distribution gives these probabilities. The probability of 0, 1, 2, 3, 4, 5, or 6 out of 18 is 0.12, so the P value for this test is 0.24. We would accept the hypothesis that the two drugs have the same success rate.

For quantitative data, the paired-sample test of averages is reduced to a one-

TABLE 9.5 SUMMARY OF PAIRED
QUALITATIVE DATA BY A CONTINGENCY TABLE

		Drug A	
		Succeeds	Fails
Drug B	Succeeds	25	12
	Fails	6	20

sample test. The choice betweeen the t test and a normal test is made just as for a one-sample test. For qualitative data, the paired-sample test of equality of proportions is reduced to a one-sample binomial test. Whether the normal approximation to the binomial can be used will depend on the sample size.

EXERCISES

1. A flower seed distributor wishes to study the effect of storage temperature on the germination rate for a certain flower species. The only way to obtain this information is to take packages of seeds and plant them all in the same way that a customer would do it. Because of the expense of this, only a small number of packages is to be tested. Also, since the higher-temperature storage is more expensive, this treatment receives a smaller number of packages than does the low-temperature treatment. The data are shown below. It may be assumed that the two treatments have been allocated randomly to the 15 seed packages.

Treatment	Number of packages	Average proportion of seeds germinating	SD of proportion germinating
Warm	5	0.65	0.10
Cool	10	0.57	0.07

Past experience with this kind of experiment indicates that the proportion germinating in each package is approximately normally distributed. Test whether the warm treatment produces a higher germination rate than the cool treatment. Use a critical P value of 0.05.

2. A large metropolitan government wishes to compare the awareness among its public servants of the methods of cardiopulmonary resuscitation (CPR). Part of this survey compares the firefighters and the police. A random sample of 25 from each group is given an oral examination to determine, for each individual, whether he or she is or is not adequately trained in CPR. The results are

   - *Firefighters:* 5 not adequately trained, 20 adequately trained
   - *Police:* 10 not adequately trained, 15 adequately trained

   If all the 700 firefighters and 2000 police were similarly questioned, can we predict with confidence which group would have the greater proportion adequately trained?

352   Chap. 9   Testing Hypotheses with Data

3. Verify that the two-sample t test for independent samples, if applied to the weight-loss example in Section 9.4.3, would accept the hypothesis that the average weight loss is zero; in other words, show that this test, which is the wrong one for this situation, would give an answer contrary to the paired-sample t test worked out in Section 9.4.3.
4. Use a chi-square procedure to solve Exercise 2.
5. A study of the academic performance of athletes at Bowling Green University during 1970–1980 showed that 34.2 percent of athletes (based on 1457 athletes) graduated within five years, whereas 46.8 percent of the general student population (based on 4188 students) graduated within five years.[2] All students entering university between 1970 and 1975 were included in the study.
   (a) Are the percentages graduating in five years "statistically significantly different"?
   (b) The same study showed that 33 percent of 1270 males and 41 percent of 189 females graduated within five years. The P value for this difference was reported as being less than 0.001. What does this mean?
6. In a study of the possible factors that influence the frequency of birds being hit by aircraft (which, ironically, is viewed as a hazard to the aircraft), the noise levels of various jets were measured just seconds after their wheels left the ground.[3] The jets were described as wide-bodied (DC10, 747, L1011) or narrow-bodied (727, 707). Twenty-two wide-bodied jets had noise levels averaging 106.4 decibels (dB) and an SD of 3.3 dB, while 10 narrow-bodied jets had a noise level averaging 114.0 dB with an SD of 2.0.
   (a) Test whether the difference in the average noise levels, on repeated and prolonged testing, would be zero or greater than zero. (Assume that the normal approximation is adequate.)
   (b) Does the inference imply a causal link?

9.5 MANY-SAMPLE TESTS

The tests of Section 9.4 can be modified to apply to three or more populations: that is, each of several populations is sampled, and the hypothesis of the equality of the population averages is assessed in light of the samples. This extension involves some new conceptual issues as well as new procedures. The extension of the t test for independent samples is called the analysis of variance. The most natural extension of the 2 by 2 contingency table test is called the chi-square test of independence.

9.5.1 Analysis of Variance

In an agricultural experiment, 30 small plots of land are allocated randomly to three varieties of wheat, 10 plots to each variety. The yields, in kilograms, for each of the 30 plots are given in Table 9.6. The averages and SDs are also shown. The yield for variety A seems to be a bit greater than that for B or C, but the variation in yield is quite large even among the plots planted with the same variety. The question is whether there is enough evidence to conclude that variety A is the highest-yielding variety. Let us first address a simpler question: Is there any difference in the average yield that would be produced by these three varieties?

TABLE 9.6 YIELD OF THREE VARIETIES OF WHEAT, EACH VARIETY GROWN IN 10 PLOTS

	A	B	C
	6.7	3.6	7.1
	6.2	5.9	5.8
	6.2	5.6	5.9
	7.4	5.1	4.7
	4.2	5.5	6.5
	5.2	4.6	6.5
	8.8	3.5	6.9
	5.7	7.3	6.0
	7.8	5.1	4.7
	7.5	5.2	6.6
Average	6.57	5.14	6.07
SD	1.29	1.05	0.79

One approach to this question is to perform two-sample t tests on each pair of samples. For example, the t statistic for testing whether or not varieties A and B have different average yields is

$$\frac{6.57 - 5.14}{\sqrt{SD_p^2/10 + SD_p^2/10}}$$

where

$$SD_p = \sqrt{\frac{10 \times 1.29^2 + 10 \times 1.05^2}{10 + 10 - 2}}$$

So $SD_p = 1.24$ and thus the t statistic is $1.43/0.55 = 2.60$. The t table for 18 degrees of freedom shows that this difference has a P value between 0.02 and 0.05 (for a two-sided test). The evidence for a nonzero population difference is quite strong. Similar t tests for the other two pairs of varieties produce a t statistic of 1.00 (P value greater than 0.20) for the A–C comparison and 2.12 (P value between 0.05 and 0.10) for the B–C comparison. Thus there seems to be quite convincing evidence of a difference between the average yield of the A variety and the average yield of the B and C varieties.

The pairwise procedure just described is incorrect. The inadequacy of it is subtle. The performance of many hypothesis tests on every possible pair of varieties will lead us to detect too many group differences. To see this, imagine that we have 100 varieties and that we perform t tests on each pair of varieties. If our critical P value were 0.05, we might expect about 5 percent of these tests to reject the equal-population-averages hypothesis, *even when all the population averages are equal*. (This expectation is not quite correct, because the tests are not independent, but it is close enough for the point being made here.) There are about 4950 such pairs, so about 250 hypothesis tests among these 4950 will reject the hypothesis incorrectly. Clearly, there is something wrong with this procedure. We would like to have a fair chance of concluding that all the population averages are equal when this is the case. We return now to our three-group example to explore another way of dealing with the many-population comparison of averages.

Recall that hypothesis tests calculate a probability, the P value, on the assumption that the hypothesis is correct. If we contemplate doing this for our three-group example, we are going to assume, for the calculations, that all three population averages are the same. We also assume that the samples are independent. An additional assumption that we shall make, which is not always a valid one in practice, is that the three populations all have normal distributions with the same SD. The combined assumption, then, is that the populations all have the same normal distribution, with unknown average and variance. If the hypothesis is correct and the assumptions are satisfied, the three samples should look like three independent random samples from the same normal population. Note that these assumptions are direct extensions to three samples of the assumptions we made for the two-sample t test in Section 9.4.

Now if we did have three samples from the same population, there would be two independent ways to estimate the common population SD:

1. Each sample provides an estimate based on 10 observations, and we can pool these estimates to provide a *within-group estimate* of the population SD.
2. Since the population $SD/\sqrt{n}$ is an estimate of the SD of the sampling distribution of the sample average, and since we also have three averages whose SD we can compute directly, the latter SD times $\sqrt{n}$ gives the second estimate of the population SD. This is called a *between-group estimate*.

In the example, for method 1 we have 1.29, 1.05, and 0.79 as the three sample estimates. This suggests that $(1.29 + 1.05 + 0.79)/3$ or 1.04 would be a reasonable estimate of the population SD. Let us tentatively accept this. Method 2 works as follows. We have three averages, 6.57, 5.14, and 6.07, whose SD is 0.59. This is an estimate of (population SD)$/\sqrt{10}$, so the estimate of the population SD is $\sqrt{10}$ times 0.59, which is 1.87.

So the two estimates of the common population SD are:

Within-group estimate: 1.04

Between-group estimate: 1.87

The population SD estimate based on the variability in the group averages is greater than one would expect from the variability within groups. A reasonable explanation for this is that the group averages varied partly because the three population averages were different, not only because of the common population variability.

The problem with the foregoing analysis is that a method of subjecting the two SD estimates to a formal hypothesis test is unknown. However, a very closely related test is known. The logic of the analysis is the same, but the variability is measured by "variance" rather than by SD. This apparently minor change is a big step toward simplification of the analysis.

First a word on estimates of variance. The (sample SD)^2 is not an unbiased estimate of variance; however,

$$\frac{n}{n-1} \times (\text{sample SD})^2$$

is an unbiased estimate of variance. We combine this fact with the logic proposed in the SD analysis above to outline the development of a procedure for testing the equality of several group averages; it is called **analysis of variance,** since it is based on variability comparisons that use variance to measure variability.

We begin again with three samples of size 10 from our population, and if the population has a common variance, each of the samples will provide an independent estimate of this common variance. The three sample estimates of variance are $(10/9) \times (1.29)^2$, $(10/9) \times (1.05)^2$, and $(10/9) \times (0.79)^2$ (i.e., 1.85, 1.23, and 0.69). Now to pool these together, we simply average them:

$$\frac{1.85 + 1.23 + 0.69}{3} = 1.26$$

So 1.26 is the "within" variance estimate.

Now for the "between" variance estimate. The three averages are 6.57, 5.14, and 6.07 and the variance estimate based on these three is

$$\frac{3}{2} \times (\text{SD}^2 \text{ of } 6.57, 5.14, \text{ and } 6.07) = \frac{3}{2} \times 0.35 = 0.53$$

This is an estimate of the variance of the sampling distribution of the averages of samples of size 10. But since

variance of independent sum = sum of variances

we know that the variance of an average is $1/n$ times the variance of the population sampled. In other words, our second estimate of the population variance, the "between" estimate, is 10×0.53 or 5.3.

So now we have two estimates of the population variance; 1.26 based on observed variation within the samples and 5.3 based on the variation between samples. The sampling distribution of the ratio of the two estimates obtained in this way is known. Specifically, when the equal averages hypothesis is correct, the ratio

$$\frac{\text{"between" variance estimate}}{\text{"within" variance estimate}}$$

is known to have a distribution called the **F distribution.** As with the t and chi-square distribution, there is more than one F distribution, depending on the degrees of freedom. The F distribution depends on the degrees of freedom used for the numerator estimate and on the degrees of freedom used for the denominator estimate.

The ratio in our example is $5.3/1.26 = 4.21$. The degrees of freedom for the numerator and denominator are 2 and 27 (to be explained shortly), so we should compare the value of the F statistic, 4.21, with the F distribution having 2 and 27 degrees of freedom. The F distribution is tabulated in Appendix A4.5. This table gives the value of the F statistic that exceeds a certain percentage of the distribution. An excerpt from the table is as follows:

Value of F statistic	Tail probability
2.5	0.10
3.3	0.05
4.2	0.025
5.5	0.010
8.8	0.001

The P value for the test of the equality of variance estimates is just over 0.025. This indicates that the equality of variances is unlikely. We reject that hypothesis. But the hypothesis that we were originally interested in testing was that the three population means were equal. However, unequal population means will cause the "between" variance to be large relative to the "within" variance, so we have actually tested this hypothesis. The conclusion of the analysis of variance test is that the three populations averages are not all equal.

Notice that by analysis of variance we have tested the equality of averages. This explains the seemingly paradoxical term "analysis of variance." One might have expected the analysis of equality of population averages to be called "analysis of averages"—not so.

The analysis-of-variance test has not told us whether wheat variety A has higher yields than those of variety B or variety C. It has only told us that the varieties do not all have equal yields. Of course, a glance at the averages and standard deviations for the three varieties does suggest that the variety A yield is higher. But the question of which varieties are different from which is a different question requiring a slightly more complicated analysis; the techniques required for this are called *multiple-comparison tests*, and are beyond the scope of this book. We will be content to perform this step informally by reexamining the data to determine the relationships between the averages, in the case that the analysis of variance rejects the equal-averages hypothesis.

A word of explanation about the "degrees of freedom" claimed for the two variance estimates. The degrees of freedom is equal to the number of independent estimates that we have pooled together to form the combined estimate, an estimate of a variance in this case. When we start with 30 independent observations, we certainly cannot have more than 30 independent estimates of a parameter based on these 30 observations. When we are estimating a variance, we have fewer than 30. To explain this is a bit tricky without more mathematical machinery, but here is an intuitive explanation.

To estimate a variance with two observations, we will look at the two differences between the observations and their average. These two differences have to be the same size and are certainly not independent. So the two observations are producing one estimate of variability. For example, if the two observations are 1 and 2, the two deviations from the average are $-1/2$ and $+1/2$. The sample SD would be $1/2$. If we are estimating a variance based on these, it would be $n/(n-1)$ times the SD^2 or $(2/1) \times (1/2)^2$, which equals $1/2$. This is how two observations produce one estimate of variance.

Now suppose that we had three observations, 1, 2, and 3. How many variance estimates can we construct now? We have three deviations: 1–2.0, 2–2.0, and 3–2.0. But they are not independent; any two will determine the third because the deviations will always sum to 0.0. This is because the average is estimated by adding all the observations. But we do have more than one independent estimate; one deviation does not determine the other two. So we have two independent estimates, each using one of the three deviations as its basic information about variability of the population. That is, three observations produce a combined estimate of variance having two degrees of freedom. Similarly (here is the intuitive leap), n observations produce $(n-1)$ degrees of freedom in the variance estimate.

Now we had three "observations"; they were group averages, to estimate the "between" variance, so the "between" variance has 2 degrees of freedom. For the "within" variance, we started with 30 observations, but there are only 27 independent estimates of variance available. We used the three groups of 10 observations each to produce a variance estimate, and each group had 9 degrees of freedom. We lose 1 degree of freedom every time we have to calculate an average with which to compute deviations.

The explanation of analysis of variance required the sample sizes to be identical. This limitation is easily avoided, but the explanation of the analysis-of-variance strategy is a bit more complicated. However, the definition of the problem and the interpretation of the result are the same whether one is analyzing samples of equal size or samples of unequal size. The computations are slightly different but may be entrusted to a reputable package of statistical computer programs.

You should keep firmly in mind that the groups being compared by analysis of variance are qualitatively defined groups. There is no natural ordering of the varieties of wheat before the results on yield are known. The following is an example of an interesting study that should not have used analysis of variance for its analysis.

A research paper appeared recently in the *Journal of Research in Music Education* with the title "Effects of music loudness on task performance and self-report of college-aged students."[4] The study tested the ability of students to solve arithmetic problems while listening to selected theme music from popular movies. Four experimental groups were formed and subjected to the music at 0, 65, 75, and 85 decibels. The number of problems correctly solved was one of the measures of performance used. The results are shown in Table 9.7.

There is a hint that performance improves with increased background noise. But note that the same analysis of variance would have resulted if someone had inadvertently mixed up the group ordering. So analysis of variance just does not capture this information at all. The fact that the F statistic was fairly small and the P value large does not justify the conclusion that the music had no statistically demonstrable effect.

TABLE 9.7 ARITHMETIC PERFORMANCE UNDER VARIOUS LOUDNESS LEVELS OF BACKGROUND MUSIC[4,a]

Loudness group (dB)	Number in group	Number of correct problems	
		Average	SD
0	50	19.30	5.02
65	50	20.20	5.46
75	50	21.82	4.43
85	50	20.20	4.79

[a] F statistic 2.259, to be compared with the F distribution on 3 and 196 degrees of freedom.

These data should have been subjected to a kind of regression analysis, with the performance as the dependent variable and the loudness as the independent variable. Unfortunately, we have not been able to cover in this text the extension of regression methods to the hypothesis-testing context, but this would be a reasonable approach (see Exercise 5 at the end of this section).

One often sees reference to "one-way" analysis of variance or "two-way" analysis of variance. We have described a one-way analysis of variance: the variance of the 30 observations was partitioned into one category that was to explain some of the variation (the various populations) and a residual category which we call unexplained error. In two-way analysis of variance, we might have tried to detect an effect on yield of soil fertility as well as of wheat variety. Our table of average yields might look like:

		Variety		
		A	B	C
Soil	Good	7.5	6.1	7.0
	Poor	5.7	4.1	5.2

The term "two-way" is clear from this table, a "two-way" table. However, we shall not detail the two-way analysis of variance in this text.

9.5.2 Chi-Square Tests

In discussing the comparison of two sample proportions, we showed how the data could be represented as a 2 by 2 contingency table. The example used there was the health exam pass rates for patients on one of two regimens, vitamin or placebo. Let us now consider the same experiment, but here we include the data for a second vitamin regimen (see Table 9.8). A common hypothesis for such a table is that the proportions passing in each group (i.e., in each population) are equal. The alternative is that they are not equal.

The "expected values" are as indicated in Table 9.9. They are calculated by

TABLE 9.8 RESULTS OF A THREE-TREATMENT EXPERIMENT SUMMARIZED BY A 2 BY 3 CONTINGENCY TABLE

	Vitamin 1	Vitamin 2	Placebo	
Pass	20	25	10	55
Fail	5	15	15	35
	25	40	25	90

TABLE 9.9 EXPECTED VALUES FOR THE CONTINGENCY TABLE OF TABLE 9.8 ASSUMING THAT THE HYPOTHESIS OF EQUAL PROPORTIONS IS TRUE

	Vitamin 1	Vitamin 2	Placebo	
Pass	15.3	24.4	15.3	55
Fail	9.7	15.6	9.7	35
	25	40	25	90

allocating the vertical column totals to the pass and fail categories in proportion to the horizontal row totals. For example, $(55/90) \times (25) = 15.3$ gives the expected value for the upper left-hand cell.

For each cell (observed − expected)^2/expected is calculated and the sum of these over all cells measures the departure of the sample data from their expected values under the hypothesis. This statistic is again called the chi-square statistic, but this time the degrees of freedom are 2: the number of populations less one. (The number of degrees of freedom is greater than this if there are more than two rows in the table; see the general formula, page 360.) In this example the chi-square statistic is computed as:

$$\frac{(20-15.3)^2}{15.3} + \frac{(25-24.4)^2}{24.4} + \frac{(10-15.3)^2}{15.3}$$
$$+ \frac{(5-9.7)^2}{9.7} + \frac{(15-15.6)^2}{15.6} + \frac{(15-9.7)^2}{9.7} = 8.49$$

The P value associated with this statistic is found from Table A4.4 to be between 0.01 and 0.025. There is strong evidence that the hypothesis is false. We would conclude that there are differences in the proportions passing the health exam among the three treatment groups.

The many-sample chi-square test is a large-sample test. If any of the expected values of the cells are less than 5, the P values of the test will not be correct.

What if the health exam had more than two categories of result: not merely "pass" and "fail" but "pass," "fail," "did not show," and "outcome pending laboratory tests"? We would then have the 4 by 3 contingency table, shown as Table 9.10.

TABLE 9.10 A 4 BY 3 CONTINGENCY TABLE, SHOWING ROW AND COLUMN TOTALS ON WHICH EXPECTED VALUES ARE BASED

	Vitamin 1	Vitamin 2	Placebo	All
Pass	10	20	8	38
Fail	5	10	10	25
No show	7	5	4	16
Pending	3	5	3	11
All	25	40	25	90

The calculation of the chi-square statistic follows the same pattern as before. Its value can be shown to be 5.47. The distribution of the chi-square statistic, under the null hypothesis that the outcome proportions are the same for each regimen, is chi-square with 6 degrees of freedom. Where does the 6 come from? For a contingency table with r horizontal rows and c vertical columns, the degrees of freedom for the chi-square statistic are $(r - 1) \times (c - 1)$. Here the degrees of freedom are $(4 - 1) \times (3 - 1) = 6$. The value 5.47 of the chi-square statistic has a P value between 0.10 and 0.90, and in this case we would accept the hypothesis of equal proportions of outcomes for the three regimens.

The use of the chi-square distribution for the chi-square statistic has been proposed without any qualification about the population distribution except that it be a distribution of qualitative data. However, there is a restriction as to the sample size. The chi-square distribution is an approximate one for the statistic described in the same way that the normal approximation applies to the comparison of two proportions. It is another application of the central limit theorem, although a somewhat disguised version. A practical rule of thumb is that the expected values of the cells in the contingency table should be at least 5, although this rule may be relaxed for tables with several rows or columns.

The test just described has another interpretation: it is sometimes called the **chi-square test of independence.** The categorizations indicated by columns and by the rows are said to be independent categorizations if the knowledge of a row category for an individual does not help with the prediction of the column category, and vice versa. This is the same as saying that the proportions by row are the same for each row and that the proportions by column are the same for each column. The row categorization and the column categorization are said to be *independent*. Another way of stating the hypothesis is that the population relative frequencies in each cell are the same as would be given by the expected values.

A final word about the applicability of the chi-square contingency-table test. It is intended for purely qualitative (categorical) data. It can fail to detect fairly clear trends across categories when the categories have a natural order. For example, the responses to a student society opinion poll about the abolition of examinations might be to record the status of the respondent (first year, second year, third year, fourth year) and the response to the question itself: "disagree" or "agree." If it is true that the agreement with the proposal declines as students struggle through the system (with exams), one would expect the proportion agreeing to decrease as the year increases. This would certainly suggest a relationship between the agreement variable and the year variable. But chi-square does not look for such trends. It examines only whether all the year categories have proportions consistent with a common proportion, in the population, over all categories. Thus the chi-square test may reject the equal-proportions hypothesis, but we will be unable to conclude why: namely, that there is a linear trend across years. Or the test may accept the hypothesis, even when there is a clear linear trend across categories, because the trend does not make the category proportions different enough for the chi-square statistic to detect.

When the variation across categories is slight but consistent (65 percent, 62 percent, 59 percent, 55 percent), the hypothesis of equal population percentages will be accepted by the chi-square procedure in samples of moderate size. The test that takes note of the ordering of the categories is beyond the scope of this book (in the literature, look for "chi-square tests for ordered contingency tables"). Note that the chi-square statistic does not change one bit if the order of the categories is rearranged, so obviously it cannot measure "trends" across categories.

EXERCISES

1. A random sample of 100 potatoes from a farmer's field are classified by grade and by size. Grades are A, B, C, and D and sizes are recorded as L, M, or S. To answer the question "Does grade depend on size?" a chi-square contingency-table test is proposed. Is this appropriate? Explain.
2. Redo the vitamin–health exam chi-square test after elimination of the data corresponding to the category "pending." (The table will then consist of data from 79 persons.)
3. Refer to Table 9.6. Eliminate variety C from the data set and perform an analysis-of-variance test to determine whether varieties A and B have the same yield. Also do a two-sample t test for this reduced data set. Compare your P values and the statistics on which they are based. Comment on this comparison.
4. A psychology study has a panel of judges classify a random sample of 100 students according to two criteria: aggressiveness and academic style. Are the two criteria independent in this group of students? The sample data are:

		Aggressiveness	
		Aggressive	Not aggressive
Academic style	Organized, many diversions	25	5
	Binge studier	15	5
	Organized, few diversions	5	25
	Does not study	10	10

5. (a) Verify the calculation of the F statistic shown in Table 9.7 concerning the music–performance experiment.
   (b) Compute the P value for the analysis-of-variance test.
   (c) Explain why analysis of variance is inappropriate for this study.

6. The following table gives the results of a study on the relationship between "racism" and "sexism" in a population of 505 professionals.[5] (The categories were formed by dividing the range of a questionnaire-based index into high and low values.) Test the independence of these two variables in the population sampled.

		Racism	
		Low	High
Sexism	Low	39.0%	18.6%
	High	15.9%	26.5%

9.6 NONPARAMETRIC TESTS: AN INTRODUCTION

The tests we have described that require the population to have a normal distribution are parametric tests. Normality is defined by a formula relating the histogram shape to the parameters of the normal distribution (see Chapter 4). Tests that do not require the specification of a parametric formula for the population distribution are called **nonparametric tests.** Because these tests do not begin with a specific population distribution, they also often avoid other assumptions needed for the parametric tests. (For example, certain nonparametric tests aimed at comparing the centers of two populations would not require the equality of SDs as is required by the two-sample t test.) Nonparametric tests are used when the validity of the usual assumptions of the parametric tests are in doubt.

Actually, we have already covered a number of nonparametric tests. Large-sample tests, which used the normality of sample averages no matter what the population distribution, do not require parametric assumptions. The large-sample normal tests for averages (Sections 9.2.1 and 9.4.1), the normal test for a percentage (Section 9.3.1), and the chi-square tests (Sections 9.4.2 and 9.5.2) are nonparametric tests. Some nonparametric tests do not use normality at all. Not only are the population distributions not required to be normal, but the tests work for small samples where the normal approximation would be invalid. A few such tests are presented in this section.

9.6.1 One-Sample Median Test

The median was introduced in Chapter 3 as an alternative to the average for indicating the center of a data set; it is the middle-ranked value in the data set. Medians can also be calculated for populations; the median of a population is the middle-ranked value in the population. When the population is described by its probability distribution, the median is the value that exceeds 50 percent of the population values and is less than 50 percent of the population values as well.

The normal distribution is an example of a symmetrical distribution: symmetrical distributions have histograms that are the same shape on either side of the median. For symmetrical distributions, the average and the median are equal. Tests on medians of symmetrical distributions are automatically tests of averages. But these same tests of medians can also be applied to skewed distributions where normal distribution assumptions are not appropriate.

Here is an example where the median test is ideally suited. A computer software company complains to its supplier of diskettes that a recent customer survey showed that the diskettes had a lifetime of only about 100 uses. This 100 was the median number of uses from the time the diskette was first used until the time that defects were apparent. To check this claim, the diskette supplier asks to perform a similar survey on a random sample of 25 recent customers of the software company. It is expected that the distribution of the number of uses until defects are noted is skewed to the right (i.e., with a long tail of the histogram extending to the larger values). The data from the 25 show that 8 have failed before 100 uses, while for the other 17, 7 were still in use after 100 uses and 10 failed after 100 uses. Does this refute the claim of the software company?

The hypothesis for the test of the claim is that the median is 100. The alternative is that it is greater than 100. The test statistic in this case is just the number of diskettes among the 25 that have a number of uses greater than 100. This is clearly 17. (Note that we did not have to wait until all the diskettes had failed to produce our data set; the method of analysis is able to accommodate such incomplete data.) The sampling distribution of the statistic, if the hypothesis is correct, is binomial with $n = 25$ and $p = 0.5$. (The 100 was claimed to be the median, the middle-ranked value. One-half of the diskettes lasted longer than 100 uses; this is where the 0.5 comes from.) Table A4.2 tells us the probability that 17 or more out of 25 would outlast the claimed median of 100; this probability is just over 0.05. The claim is in doubt, although perhaps not clearly refuted. The P value of 0.05 is often used as a critical P value, so this would be a borderline case.

The procedure above is an example of the **median test.** Note that the actual distribution of the number of uses before failure was never assumed to be of a particular shape, except there was a suggestion that assuming normality would be a bad idea. The sample size of 25 was large enough for large-sample results to be used, but the procedure did not make use of any large-sample results. The median test is valid for small samples. In fact, we would not have been able to use a large-sample result in this case since the average number of uses was not hypothesized (and large-sample approximations by the normal distributions apply to averages, not medians) nor was the sample average calculable from the incomplete data available to the surveyor.

9.6.2 Two-Sample Tests: The Median Test and the Sign Test

Further simple applications of the binomial distribution provide tests for comparing two samples. The median test can be used for comparing two independent samples, the sign test for comparing two paired samples.

Two judges rank 13 gymnasts on their performance at a national gymnastics meet. Their scores are shown in Table 9.11. Does one judge give higher scores than the other?

TABLE 9.11 TWO JUDGES' SCORES ON 15 GYMNASTS' PERFORMANCES

Gymnast	Judge 1	Judge 2
1	8.4	7.4
2	7.3	7.5
3	9.4	8.5
4	6.8	7.2
5	7.8	7.3
6	7.4	7.3
7	7.9	8.0
8	7.3	7.3
9	9.3	9.1
10	8.9	7.9
11	8.4	7.4
12	8.8	8.5
13	7.9	7.7

A normal test of equality of averages might be used here, but the assumption of normality may well be in doubt. Note that the two samples are paired. A simple way to see if one judge is more generous than the other is to look at the *sign* of the difference in the two ratings for each gymnast. If we subtract judge 2's score from that of judge 1, we get nine minus signs, three plus signs, and one tie.

The hypothesis here is that the median difference is zero: in other words, that the sign of the difference has a 50 percent chance of being plus. When this hypothesis is true, we would be satisfied that one judge gives about the same level of scores as the other judge. When we are focusing on the signs of the differences, we ignore ties because they do not indicate either judge as having higher scores. (There is also a 50 percent chance of a minus among the 12 gymnasts if the hypothesis is true.) The alternative is that the chance of a plus is different from 50 percent. So we need to do a two-sided test. We again use the binomial ($n = 12$, $p = 0.50$) to calculate the probability that a result as far or farther from the expected value of six pluses would be observed. That is, the probability of 0, 1, 2, 3, 9, 10, 11, or 12 plus signs is, from Table A4.2, 0.14. This *P* value is not very small, so we accept the hypothesis in this case. There is no consistent difference in the evaluations of the two judges.

The test just described is called the **sign test.** It did not assume any particular form for the distribution of judges' scores, yet one could calculate the *P* values exactly even though the sample size was small.

Before leaving this example, we should consider the following question: What population is being described in this example? One could argue that whether one judge was more or less generous than the other may be answered completely by looking at their ratings of the 13 gymnasts. In nine cases judge 1 was more generous, and in three judge 2 was, and in one case they were the same. Period. The problem with this is that even judges whose median scores over their judging careers are identical would be expected to differ somewhat on any particular group of gymnasts. Given another 13 gymnasts, the judges' relative scores might swing the other way. We imagine a hypothetical population of scores from which the 13 we see are a random sample. The issue of hypothetical populations was discussed in Chapter 7.

As an application of the two-sample median test, let us reuse the same data as with the sign test, but ignore the pairing. For this to be a reasonable procedure, we will now assume that the gymnasts rated by judge 1 are a random sample of gymnasts selected independently from the gymnasts rated by judge 2. The hypothesis is that the two judges have the same median scores (in the hypothetical population). The alternative is that they are different. The median of the combined sample is 7.85.

We can construct a contingency table summarizing the relationship of each sample to the combined median:

	Less than 7.85	Greater than 7.85
Judge 1	5	8
Judge 2	8	5

If the judges have different median scores, we would expect the ratios in the two column categories to be different. The way to test whether the sample evidence is convincing is to form a chi-square statistic from this table in exactly the same way as has been done in Sections 9.4 and 9.5. In this case we have

$$\frac{(5 - 6.5)^2}{6.5} + \frac{(8 - 6.5)^2}{6.5}$$
$$+ \frac{(8 - 6.5)^2}{6.5} + \frac{(5 - 6.5)^2}{6.5} = 1.4$$

The P value obtained by comparing 1.4 with the chi-square distribution (Table A4.4) with 1 degree of freedom is much greater than 0.10, so we would accept the hypothesis of equal medians in this case. We have to accept the hypothesis of no consistent difference between the judges.

Thus we have compared two paired samples with the sign test, and two independent samples with the two-sample median test. Neither test made assumptions about the parametric form of the population distributions. Note that these two tests throw away some of the sample information before proceeding with the test; the sign test used the sign of the differences but ignored the size of the difference; the median test used the ranks of the observations but not the actual sizes of the observations. Tests that use some of this additional information go by the names of Wilcoxon rank sum test (sometimes called the Mann–Whitney U test) and Wilcoxon signed rank test. We will not cover these tests.

EXERCISES

1. A simple random sample of 20 cars of a certain model called the XS is selected from the annual production of 1000 of this model. The original purchasers of the 20 cars in the sample are contacted to ask if they were satisfied with the car's performance over the first six months of use. The owners were asked to indicate their satisfaction on a subjective scale of 0 to 10, 10 indicating complete satisfaction. A similar survey was done on the preceding year's model, but all owners were contacted in that survey. The median score obtained in the large survey was 6.0. The 20 scores obtained this year are

    8, 7, 7, 5, 4, 7, 6, 6, 7, 5, 8, 8, 9, 2, 4, 7, 8, 6, 8, 6

    Are this year's owners of the XS more satisfied than last year's?

2. (Continuation of Exercise 1.) A competitor does the same survey on a comparable model, called the XL. The result in this survey of only 10 owners is

    3, 8, 7, 5, 9, 7, 9, 5, 8, 8

    Do the two surveys indicate that the XL gives more satisfaction to its owners?

3. A study of "nature" versus "nurture" is aimed at determining the genetic contribution to artistic skill compared to the environmental contribution. Artistic skill is measured by a battery of tests upon which an index of artistic skill is constructed. The index has been newly devised and its distribution is unknown. Ten pairs of genetically identical twins are selected for the study; one of each pair has been exposed to parental encouragement of artistic development, while the other has received no special encouragement in this area. The index values for the 20 subjects are shown as follows:

Twins	Artistic nurture	No artistic nurture
1	79	63
2	84	73
3	68	79
4	57	58
5	92	79
6	48	53
7	60	51
8	85	68
9	94	88
10	77	80

    Does artistic nurture improve the artistic skill? (There are no genetic differences between these twins.) Use a critical P value of 0.10.

9.7 THE LOGIC OF HYPOTHESIS TESTING, REVISITED

If you have worked your way through the tests introduced in this chapter, you will have a good grasp of the basic strategy of hypothesis testing. However, there are a few issues of practical importance that have not yet been discussed. They point out that application of statistical methods is not merely a matter of ritual.

We will consider the following issues in this section:

1. When is a hypothesis test required?
2. What determines the choice of the critical P value?
3. How is our "prior" knowledge about a population combined with information in the sample?
4. How does the sample size affect the conclusion of a hypothesis test?

1. When is a hypothesis test required? First, if the sample itself is all we are interested in, the "sample" is really the "population." In this case descriptive methods are all that are required. But if we want to infer something from the sample about the concrete population from which the sample was selected, or about the hypothetical population that describes the mechanism generating the sample, we need to consider how well the sample can help us with this inference. This requires the knowledge of the method of sampling, sampling distributions for sample statistics, and the relationship between the sample statistics and certain population parameters. The one big question then is: When do we need an estimate (i.e., a confidence interval, for example), and when do we need to perform a test of hypothesis?

The main criterion is whether the investigator, or his or her peers, have any preconceived ideas about what the population should be, or at least what the value of certain parameters should be. If not, there is no hypothesis to test. When inference about a population is required, yet there is no hypothesis to test, estimation procedures are likely to be useful.

There is a close connection between hypothesis tests and confidence intervals, although their purposes are quite different. Suppose, for example, than an auditor is sampling accounts and works out the error, for each account, between the reported income and the actual income. Let us suppose that the accounts are sufficiently complicated that a nonzero "error" is not too incriminating, but that the average error should be close to zero for a conscientious company accountant. Now, we might use a hypothesis test to test whether the average error might reasonably be zero. But if we were to work out a confidence interval for this average, we would find that it would include the value zero in exactly those instances in which the hypothesis test accepted the hypothesis of zero average error. Both methods are based on the same sampling distribution, so perhaps this is not too surprising. In practice, the calculation of confidence intervals (logically, an estimation technique) may be substituted for a test of hypothesis when this is convenient. However, the confidence level for the confidence interval (e.g., 95 percent) must be equal to 100 percent minus the critical P value of the hypothesis test (e.g., 100 percent $-$ 5 percent).

2. What determines the choice of the critical P value? In the first place, it should be pointed out that many hypothesis tests are reasonably approached without specification of a critical P value. The outcome of the test may be the P value itself. We may not wish to be forced to make a decision as to whether to accept or reject the hypothesis. If the evidence is equivocal, we may wish to withhold any decision. For example, if a test is based on the hypothesis that a vitamin supplement has the same effect as a placebo regimen, we may feel that a P value of 0.15 suggests that we should neither pronounce the vitamins as worthless nor promote their use by others. We have a P value in the "twilight zone." Our conclusion should be that our knowledge about the effect of the vitamin supplement has not changed. We are still ignorant about it.

However, there are a variety of situations in which a decision one way or the other is necessary. One such situation is the testing by drug companies of a large number of compounds on experimental animals (rats, for example). For example, a compound might be tested for an allergic reaction by a skin test. Some rats would react to a particular compound and some would not. A decision would have to be made whether to perform further tests on larger numbers of experimental animals, or to discontinue testing of the compound. Following up all compounds is not feasible. A decision has to be made.

Another situation in which a decision about the acceptability of the hypothesis is important is when the cost of not making a decision is great. If an oil exploration company holds drilling rights on a property, and sample cores are used to estimate the quantity of oil that might be recovered, the failure to decide whether or not to drill further will cost the company money—the interest on their investment. They must proceed as if there is enough oil to recover, or sell their rights. A decision must be made.

When a decision has to be made, the ideal is to have the risks of a wrong decision balanced according to the costs of each possible mistake. That is, if the property does contain commercially profitable reserves of oil, a decision not to proceed with exploration will cost that lost profit. On the other hand, if the oil is insufficient to justify recovery, the decision to proceed will cost the futile future drilling expenses. If the former mistaken decision is more costly than the latter, we should adjust our decision procedure accordingly. However, this process requires some statistical theory that we have not covered. So we use a simplified procedure: we adjust the critical P value. When the biggest cost arises by incorrectly accepting the hypothesis, we would use a relatively large critical P value (say, 0.15). When mistakenly rejecting the hypothesis is the most costly error, we choose a small critical P value (0.001, say). However, the choice of critical P value within these bounds is admittedly a subjective choice. The reason for accepting this subjectivity in an otherwise objective procedure is the desire to keep the procedure simple and widely applicable.

There is a difference in the application of hypothesis tests depending on whether the study is exploratory or confirmatory in nature. Exploratory studies tend to use large critical P values; confirmatory studies use small critical P values.

3. How is our prior knowledge of a population combined with the information in a sample? In statistics, "prior" knowledge is knowledge that we have before we examine the sample at hand. For example, several of the techniques that have been discussed in this book have assumed that the population distribution is normal. Knowledge that a population distribution is normal must derive from past experience with data populations such as the one under study. So this is one way that we make use of prior information. Another example of the use of prior knowledge is the knowledge that a sample has been generated by a mechanism such as a binomial experiment; this has been used in some of our analyses.

A different sort of prior knowledge is our knowledge of what constitutes reasonable values for an unknown parameter. The average age of eighth-grade children in a North American school might be as low as 12.0 or as high as 15.0, but outside these limits is unlikely. This type of knowledge is known before we look at any sample data. If a sample had an average value outside this range, we might select another sample. This is not in accord with our principles of sampling, but it would keep us from wasting our time with an unlucky sample.

One school of statistical inference views the sample as simply updating our knowledge about a population. The prior knowledge about the parameter is thus a probability distribution. We have a prior distribution for a population parameter, and this in combination with the sample produces another distribution for the parameter. This new distribution is called the *posterior distribution* for the parameter. This approach to statistical inference is called *Bayesian*, after Thomas Bayes (1702–1761). Note that in this text we have stayed with the classical theory of estimation, in which a parameter is considered to have a fixed value rather than a probability distribution.

Another example of the use of prior knowledge is the use of the knowledge of a population SD. This knowledge would allow us to use a normal estimation procedure instead of the t procedure. The resulting confidence intervals for the normal procedure would be narrower and more informative.

Most statistical methods do not formally include the effect of our prior assumptions on the inference in question. This is a shortcoming of the foundations of the subject at the present time.

Nonparametric methods are most useful when our prior knowledge is minimal. One such method that has been recently proposed is called the "bootstrap" because it lets the sample specify the shape of the population distribution. (The sample provides information usually thought of as "prior" information, as well as providing the sample estimates in the usual way. This double duty for the sample suggests the metaphor of a person "pulling himself up by his own bootstraps.") The idea is to select samples, with replacement, from the numbers that constitute the original sample. If one were to calculate the value of the statistic for each of these resampled samples, one would be able to estimate a sampling distribution for the statistic. This technique is not yet commonly used but may be soon.

4. How does the sample size affect the conclusion of a hypothesis test? When a sample is very large, a hypothesis will tend to be rejected even when it is very nearly correct. If we use a sample of 500 children to test whether the average age in grade eight is 14.0 years, then even if the population average is 14.1 years, the hypothesis will tend to be rejected. Large samples reject hypotheses more than we would wish for practical purposes.

Small samples tend to suggest acceptance of false hypothesis, even when they are far from correct. If the average age in grade eight is 13.0, and a sample of size

5 has a sample average of 13.0, this sample would probably not reject the hypothesis that the average is 14.0. The small sample does not produce a very precise estimate of a population average.

When a large sample rejects a hypothesis, or when a small sample accepts a hypothesis, the conclusion is logically appropriate but in practice we reserve judgment.

Hypothesis tests are often called *tests of significance*. When a hypothesis is rejected, the result is said to be **statistically significant.** The confusion among non-statisticians between "significance" in this technical sense and "importance" in a practical sense has led some statisticians to avoid the term. There does not seem to be any particular merit to the "significance" terminology.

These issues will give some idea of the looseness of statistical theory. The important thing to remember is that any procedure that does not seem reasonable in a given situation should not be relied upon, no matter how respectable the technique among statistical experts. To use a statistical technique wisely, the user must understand the strategy on which the technique is based. See Figure 9.5.

EXERCISES

1. Canadian packaged goods companies distributed 3.14 billion coupons in 1983, up 12 percent from the 2.8 billion distributed in 1982.[6] Is this a significant change?
2. A sample of five books that have been returned to the library on a certain day indicates that the books were last checked out an average of 6.6 days; the SD of the five usage times is 2.0 days. Experience in the past has indicated that the average usage is 5.0 days with an SD of 1.5 days. Has the usage changed?

9.8 ANSWERS TO EXERCISES

Section 9.1

1. The professor is probably reporting a head on every throw regardless of the outcome of the toss. He wants to see if any of the students will doubt his reports because of the rare event of six heads in six tosses. If so, the students will realize that they have used the same logic as formal hypothesis tests. If not, they will have a simple experiment with which to remember the basic strategy of hypothesis tests.

    The student observes that a rare event has occurred. He says to himself that it is rare *if* the experiment is actually as the professor has portrayed it. The student reasons that the occurrence of the six heads suggests that the experiment is not being conducted as it has been portrayed by the professor. Consequently, he runs to the front to see if the professor is reporting the results as they actually happen, and also to make sure the coin does not have two heads.

Hempel's Ravens

   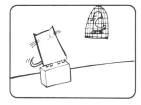    

Figure 9.5 Some scientists think they know all about the scientific confirmation of hypotheses.[7]

2. Yes.
   (1) The hypothesis is that the mechanic did tighten the wheel after the brake job.
   (2) The alternative is that he did not.
   (3) The sample is the one incident at hand.
   (4) The statistic is the state of the wheel when the car is picked up: tight or loose.
   (5) The state of the wheel under the hypothesis [see (1)] is that the wheel will almost certainly be tight at the time it is driven away. (In other words, the statistic is only roughly specific here.)
   (6) The P value is very, very small.
   (7) A critical P value of 0.10 would still be giving the mechanic the benefit of the doubt. So choose it = 0.10.
   (8) Decision. Reject hypothesis. (Sue the mechanic.)
3. The only change is that 7.3 is $(7.3 - 5.0)/5.0$ or 2.6 in standard units, and the P value is about 0.005. (See Table A4.1) The hypothesis would be rejected in this instance. In other words, it is concluded that the honor students actually watch more than an average of 5.0 hr of TV per week.
4. Small. The critical P value is the probability of a rejection error, and we want this to be small.
5. Yes! The "coincidence" is incredible.

Section 9.2

1. Hypothesis: population average = \$75.00. Alternative: Population average is more than \$75. The sampling distribution of average is approximately normal with average = \$75 and SE = (15/5) = 3.0. The observed average value = \$85, which is 1.67 standard units above the hypothesized average. The P value is 0.05, indicating that an observed average this far from the hypothesized average would only occur 5 percent of the time, when the hypothesis is true. This suggests that the hypothesis is false. The manager's prices do seem lower than elsewhere, even allowing for sampling variation.
2. You cannot answer this question with the techniques you have learned in this book—the population is nonnormal and the sample is small. Also, some of the evidence relating to the question is summarized by the smallness of the SD. So far, we have only described ways to assess averages. You should know enough from Sections 7.1 and 7.2 to know that you need help with this question.
3. In the 7 hr, Ms. Glib must have wangled $(7 \times 60/15.6)$ appointments (i.e., 27 appointments). If we assume that these 27 are like a random sample from the population of waiting times experienced by the other callers (an assumption of a situation in which Ms. Glib would certainly not merit extra pay), we can treat the 15.6 as an average from a population whose hypothesized average is 22.2 min and whose SD is estimated to be 3.9 min. The normal test gives a z value of $(15.6 - 22.2)/(3.9/5.2) = 8.27$. The P value for this is very small, so Ms. Glib has demonstrated her superiority, at least relative to the average of the others. (Note that the average 22.2 was treated as a known number; it was based on a very large sample, so its precision was likely very good. For this reason the SD of the 23 was ignored.)

4. It must be assumed that the nine measurements are obtained using a procedure similar to the procedure that resulted in an average of 1.20 and an SD of 0.15. Furthermore, it must be assumed that the measurements are done independently, so that the SD in the past is a reasonable measure of the SD in the sample of nine and so that it is reasonable to treat these nine measurements as a random sample from a population comparable to the one with average 1.20 and SD 0.15. (Presumably the measurements are done in the same part of the lake.) The question posed will be addressed by testing whether the average of the random sample of nine new measurements could reasonably be from the population with average 1.20 and SD 0.15. Because the sample is small, we will have to assume that the population is normal to guarantee that the sampling distribution of the sample average is also normal. (If we were given the raw data, it would have been possible to use a nonparametric test and avoid the normality assumption.)

With all these assumptions, we can proceed to do the test. The hypothesis is that the sample is from a population with average 1.20. The alternative is that the sample is from a population with average greater than 1.20. The t statistic is $(1.45 - 1.20)/(0.15/3) = 5.0$. The P value for this value is less than 0.005. The hypothesis is rejected because the critical P value is 0.01.

Section 9.3

1. The chance of 0, 1, ..., 5 is 91.6 percent, so the new P value from the binomial is 0.084. The normal approximation, with the continuity correction, is found from $(5.5/30 - 0.10)/0.055$, which is 1.52, which gives a P value of 0.064. The evidence against the hypothesis is a bit stronger. If the critical P value were 0.10, both tests would reject the hypothesis. If it were 0.05, both would accept it. The normal approximation is good enough for practical decision making here.

2. The hypothesis is that the proportion of heads in many tosses would be 0.50. The alternative is that it is greater or less than 0.50. The proportion observed in the sample is the statistic; its value here is 0.90. For this small sample we use the binomial test. The probability of 9 or 10 heads in 10 tosses is $0.010 + 0.001$ or 0.011 (Table A4.2). The events 0 heads or 1 head (which also have probability 0.011) would also have been considered as evidence against the hypothesis (see the form of the alternative), so the P value is $0.011 + 0.011 = 0.022$. This P value is quite small and most people would be inclined to reject the hypothesis in this case. In other words, they would conclude that the coin is biased.

Section 9.4

1. We do a t test of the hypothesis that the two treatments have equal germination rates, on average. The alternative is that the warm treatment yields a higher rate. (Since this treatment is more expensive, we are not interested in it as an equivalent to the cool treatment; to be useful the warm treatment must have a higher germination rate.) We have two small samples here, and the statistic we will use is the one described for assessing the difference in averages in two small samples. The normality of the population is justified by the past experience mentioned in the problem.

The statistic is the difference in averages $(0.65 - 0.57)$ divided by a quantity similar to an estimate of the SE of the difference: that is,

$$\sqrt{\frac{SD_p^2}{n_1} + \frac{SD_p^2}{n_2}}$$

374   Chap. 9   Testing Hypotheses with Data

which in this case is the square root of $(0.09^2/5 + 0.09^2/10)$, or 0.05. In other words, the t statistic is $0.08/0.05 = 1.6$. The t distribution to compare this with, to get a P value, is the t distribution with $5 + 10 - 2 = 13$ degrees of freedom. Table A4.3 shows that the P value for the one-sided alternative is between 0.05 and 0.10, and is certainly greater than the critical P value that was specified to be 0.05. Thus we would accept the hypothesis in this case. The warm treatment has not been demonstrated to be superior.

2. The question is asking us to do a hypothesis test on the equality of two sample proportions. The suggestion is that the alternative should be two-sided. The two sample proportions are 0.80 and 0.60, and the SE of the difference between these is

$$\sqrt{\frac{0.70 \times 0.30}{25} + \frac{0.70 \times 0.30}{25}}$$

which equals 0.13. (0.70 is the estimate of the common proportion if the two proportions are the same as hypothesized.) The statistic is therefore $(0.80 - 0.60)/0.13 = 1.5$ approximately. The P value associated with this is 2×0.14 or 0.28 (normal table). The proportions might well be equal in the two groups of public servants. No, we cannot predict with confidence which proportion would be greater.

3. The difference in the averages is $97.4 - 92.6 = 4.8$ kg. The SE of this difference is estimated to be

$$\sqrt{\frac{(5 \times 15.3^2) + (5 \times 13.4^2)}{5 + 5 - 2}} \times \sqrt{\frac{1}{5} + \frac{1}{5}}$$

which is 8.3. The t statistic for the difference between the two averages is $(4.8/8.3) = 0.58$. The t distribution with 8 degrees of freedom will have a value more extreme than 0.58 with high probability. This verifies that the test would accept the hypothesis of no weight loss.

4. The observed data may be displayed as follows:

	Adequately trained	Not adequately trained	Total
Firefighters	20 (17.5)	5 (7.5)	25
Police	15 (17.5)	10 (7.5)	25
Total	35	15	50

The expected values are shown in parentheses, assuming that the pooled common proportion applies to both groups. The chi-square statistic is

$$\frac{(20 - 17.5)^2}{17.5} + \frac{(15 - 17.5)^2}{17.5} + \frac{(5 - 7.5)^2}{7.5} + \frac{(10 - 7.5)^2}{7.5} = 2.38$$

The P value for the chi-square statistic is greater than 0.10, so we would accept the hyopthesis. (This checks with our calculations for Exercise 2.)

5. (a) Since the entire population "sampled" is included in the sample, the question does not make sense. What you see (in the data) is what you get. (The authors of the paper apparently realized this because no hypothesis test was done.)
   (b) Normally, it would mean that two population percentages are different. But what

populations are being used here? These male and female athletes are not a random sample from any easily describable population, concrete or hypothetical. So the P value does not really tell us anything in this case. Of course, the fact that the proportions graduating is different among male and female athletes *in the study* is indisputable. Generalization to a larger population, such as male and female athletes at all North American universities, may lead to a correct conclusion, but there is no evidence in the study to support this.

6. (a)
$$SD_p = \sqrt{\frac{22 \times 3.32 + 10 \times 2.02}{22 + 10 - 2}}$$

which is 3.05. Thus the SE of the difference in the averages is $3.05 \times (1/22 + 1/10)$ or 1.16. The difference $(114.0 - 106.4)$ is $7.6/1.16$ or 6.6 SEs greater than 0.0. Comparing this with the normal distribution shows that P is tiny, and we conclude therefore that the "population" difference in noise level is greater than zero. That is, there does seem to be a noise-level difference between narrow- and wide-bodied jets.

(b) No, the available planes were not allocated at random to "narrow-bodied" and "wide-bodied." An alternative explanation is that the wide-bodied planes were quieter because they were newer or because they were better maintained. In fact, it is very unlikely that the noise level is due to the body width per se.

Section 9.5

1. The categories have a natural order. It would be a good idea to consult a statistician about an appropriate test, especially if the chi-square test accepts the null hypothesis when the contingency table seems to indicate a dependence across categories.

2. Expected values for the table are as follows:

    10.6   16.8   10.6
    7.0   11.1   7.0
    4.5   7.1   4.5

    and the chi-square statistic is 4.80. This should be compared to the chi-square distribution on $(3 - 1) \times (3 - 1) = 4$ degrees of freedom. The P value is somewhere between 0.10 and 0.90 and suggests that the hypothesis should be accepted. Health outcome does not depend on treatment.

3. The averages 6.57 and 5.14 have SD = 0.715. (The fact that $6.57 - 5.14 = 1.43$ reveals a formula that works for two groups; the SD of two numbers is one-half their difference.) The "between" estimate of variance is $10 \times (2/1) \times 0.715^2 = 10.225$. The "within" estimate of variance is

$$\left(\frac{1.29^2 + 1.05^2}{2}\right) \times \left(\frac{10}{9}\right) = 1.537$$

Thus the F statistic is $10.225/1.537 = 6.65$. The P value (from the F distribution with 1 and 18 degrees of freedom) is between 0.01 and 0.025.

376   Chap. 9   Testing Hypotheses with Data

The value of the t-statistic is $1.43/\sqrt{1.24^2/10 + 1.24^2/10} = 2.579$. From Table A4.3, the P value is just greater than 0.02 for the two-sided alternative (see Section 9.5.1).

The analysis-of-variance test statistic for two groups, which in the example worked out to 6.65, is exactly the square of the two-sample t statistic:

$$F = 6.65 = 2.579^2 = t^2$$

The ranges determined for the P values were F: 0.01 to 0.025 and t: 0.01 to 0.02. These ranges suggest that more detailed tables would give identical P values for the two tests. (It can be deduced that this indication is correct.)

4. The chi-square statistic is

$$\frac{(25-16.5)^2}{16.5} + \frac{(15-11)^2}{11} + \frac{(5-16.5)^2}{16.5} + \frac{(10-11)^2}{11} + \frac{(5-13.5)^2}{13.5}$$

$$+ \frac{(5-9)^2}{9} + \frac{(25-13.5)^2}{13.5} + \frac{(10-9)^2}{9} = 23.0$$

From Table A4.4, the P value is less than 0.005. The conclusion is that the two criteria are not independent among the population of students concerned.

5. (a) The sum of the squares of the SDs (times 50/49, since we are estimating a variance) is 97.60. Dividing by 4 gives 24.40 as the "within" estimate of population variance. The SD of the four averages is 7.38; we square this and multiply by 50/49 to get the "between" estimate of population variance = 55.08. The F statistic is therefore 55.08/24.40 or 2.26, as stated in Table 9.7.
   (b) It is a bit greater than 0.05.
   (c) See the explanation in the text.

6. The first thing to notice is that the contingency table quotes percentages instead of frequencies. The chi-square statistic must be applied to frequencies. (The strength of the evidence for a relationship obviously must depend on the frequencies.)

Multiplying the percentage times the total of 505 gives:

		Racism	
		Low	High
Sexism	Low	197	94
	High	80	134

The expected values are:

		Racism	
		Low	High
Sexism	Low	159.6	131.4
	High	117.4	96.6

and the chi-square statistic is 45.8. So the *P* value is less than 0.001, and the dependence of sexism and racism in the population sampled is the resulting conclusion (i.e., reject the hypothesis of independence).

Section 9.6

1. If the median score in the entire population of customers during the current year is 6 (the hypothesis to be tested), the chance that any one of the scores in our sample is greater than 6 is approximately the same as the chance that it is less than 6. So if we ignore the 6's in the sample, the scores 7, 8, 9, and 10 would occur with a chance of 50 percent. But we have 11 such scores against only 5 in the range 0, 1, 2, 3, 4, 5. The *P* value associated with this outcome is found from the binomial Table A4.2. The company presumably is looking for evidence of improvement in satisfaction, so a one-sided alternative should be used (i.e., that the median satisfaction index has increased). The *P* value is the probability that the binomial with $p = 0.5$ and $n = 16$ would have 11, 12, 13, 14, 15, or 16 in the "above 6" category. This probability is read from the table by adding the probabilities for these outcomes, and is 0.10. Thus there is weak evidence of an increase in the median index value. This year's customers do seem to be more satisfied, but a larger sample would have to be taken for a definitive answer to the question.

2. Here we have two samples and the two-sample median test could be used to see if the second sample provides enough evidence that the XL gives more satisfaction than the XS. We need the combined sample median; to obtain this we order the 30 sample values:

    2, 3, 4, 4, 5, 5, 5, 5, 6, 6, 6, 6, 7, 7, 7, 7, 7, 7, 7, 8, 8, 8, 8, 8, 8, 8, 8, 9, 9, 9

    The median of these is 7. In the first sample three values are less than the median, and five are greater. In the second sample, nine are less and six are greater. The contingency table for these results is

	Less than median	Greater than median
The XS	3	5
The XL	9	6

    The chi-square statistic is

    $$\frac{(3 - 4.1)^2}{4.1} + \frac{(5 - 3.8)^2}{3.8} + \frac{(9 - 7.9)^2}{7.9} + \frac{(6 - 7.2)^2}{7.2} = 1.02$$

    Comparing with the chi-square distribution with 1 degree of freedom, the *P* value is much larger than 0.10, so we accept the hypothesis of equal medians. There is inadequate evidence that the satisfaction of the owners of the competing models differ in their opinions of their respective cars.

3. Because the data are paired and the population distributions are unknown, we shall use the sign test. The hypothesis is that the chance that the nurtured twin has a higher score is 50 percent. There are six of 10 pairs in which the artistic nurture is associated with the higher index score. The *P* value associated with the one-sided alternative (that artistic nurture produces higher scores) is found from the binomial distribution for $p = 0.5$ and $n = 10$. It

378   Chap. 9   Testing Hypotheses with Data

is almost 0.4, so there is no reason to doubt the hypothesis. This test suggests that we should accept the genetic basis of artistic skill.

If you feel uneasy about the conclusion reached by the sign test in view of the data, you may have been looking not only at the sign of the differences but at their size as well. The Wilcoxon signed rank test is a more sensitive test which uses this information. Since we have not described this Wilcoxon procedure, we indicate the improved sensitivity by using the t test for paired data. This may be risky if the population distribution is unknown, but we proceed anyway to illustrate a point.

The differences in scores for the 10 twins have average = 5.4 and SD = 9.5. The t statistic for testing whether this sample average could have been from a population with average = 0 is $(5.4 - 0)/3.2 = 1.7$, where 3.2 is $SD/\sqrt{9}$. This should be compared with the t distribution with 9 degrees of freedom. The P value in this case is abut 0.05. We should reject the hypothesis in this case. The reason that the two tests give such different results is that the t test has made better use than the sign test of the sample information. The positive differences tended to be larger than the negative differences. This lent support to the genetic basis of artistic skill which the sign test ignored. The Wilcoxon signed rank test would have given a result similar to the paired t test.

Section 9.7

1. The change may well have been important to marketing people reading this fact, but it is not a "statistically significant" change; there is no reasonable sampling model for these data.
2. One has to make a normality assumption to do anything with this small sample, since we only have the summary statistics to work with. The only real question is whether we will use the SD of 2.0 based on our sample, or the 1.5 based on past experience. The sample is so small that a t test comparing 6.6 with a supposed population value of 5.0 would not be as good as a normal test assuming the SD to be 1.5. This is an example of the use of prior information to improve the analysis.

$$\frac{6.6 - 5.0}{1.5/\sqrt{5}} = 2.39$$

P is about 0.02 for a two-sided test, so we would conclude that the average usage is not like it has been in the past.

9.9 NOTATION AND FORMULAS

When a population is sampled, we may let X represent any possible sample value. If the population is normal and has average μ and SD = σ, we write $X \sim N(\mu, \sigma)$. If μ_0 is a particular value of the average of the population that we wish to test using the sample, we write $H: \mu = \mu_0$. In words, the hypothesis is that the population average is equal to the value μ_0. The alternative is written symbolically as $A: \mu \neq \mu_0$. When $X \sim N(\mu, \sigma)$, then $\bar{X} \sim N(\mu, \sigma/\sqrt{n})$, where n is the size of the sample. The normal population test is based on the sampling distribution of $\bar{X}$. When the sample size is large, the sampling distribution of $\bar{X}$ is approximately equal to the sampling distribution of $\bar{X}$ when the population is normal, with the approximation improving as n gets larger.

When the population is normal and the sample is small, the population SD, σ, is not known. However, the distribution of $(\bar{X} - \mu_0)/(SD/\sqrt{n-1})$, when $\mu = \mu_0$, is known to be the t distribution with $n - 1$ degrees of freedom. A test of hypothesis $H: \mu = \mu_0$ would be based on this sampling distribution in this case. In hypothesis tests, the P value is usually represented simply by the symbol P, and the critical P value traditionally has the symbol α. The report of a hypothesis-test calculation might be $P < \alpha$ or $P > \alpha$. If a critical P value is not specified, we write $P = 0.07$ (or whatever) or if tables do not allow an accurate determination, something like $0.01 < P < 0.025$. The symbol for a sample proportion is $\hat{p}$. When we multiply $\hat{p}$ by the sample size n, we have a quantity $= X$, say, such that $X \sim \text{Bin}(n, p)$, where p is the population proportion. Thus the sampling distribution of a sample count is binomial with average np and

$$SD = \sqrt{n \times p \times (1 - p)}$$

A simple approximation to this sampling distribution can be expressed most simply in terms of the sample proportion $\hat{p}$. In fact, $\hat{p} \sim N(p, \sqrt{p(1-p)/n})$.

Two-sample tests for averages involve two samples with averages $\bar{X}_1$ and $\bar{X}_2$ and sample SDs SD_1 and SD_2. The two-sample test of a nonzero difference in population averages is based on one of two statistics:

1. $t = \dfrac{\bar{X}_1 - \bar{X}_2}{SD_p(1/n_1 + 1/n_2)^{1/2}}$

   where

   $$SD_p = \sqrt{\dfrac{(n_1 \times SD_1^2) + (n_2 \times SD_2^2)}{n_1 + n_2 - 2}}$$

2. $z = \dfrac{\bar{X}_1 - \bar{X}_2}{[(SD_1^2/n_1) + (SD_2^2/n_2)]^{1/2}}$

The t statistic for this two-sample case has a t distribution with $n_1 + n_2 - 2$ degrees of freedom. It is valid for any sample sizes $n_1 > 1$ and $n_2 > 1$. This statement requires not only normality of the population distributions but also equality of the population standard deviations ($\sigma_1 = \sigma_2$). The t test is usually thought of as a small-sample test. This is because for large samples we can use the normal approximation and avoid these two assumptions. For large samples we use the statistic z in statistic 2 above, which has, approximately, a normal distribution.

The tests that use the chi-square statistic all have the same formula:

$$\chi^2 = \sum \dfrac{(o - e)^2}{e}$$

where o and e are observed and expected frequencies, respectively, and the sum is over all cells of the contingency table. The chi-square statistic has a chi-square distribution with $(r - 1) \times (c - 1)$ degrees of freedom; r and c are the number of rows and columns in a contingency table. The χ in the formula has nothing to do with the X used for tests on averages. It is simply the usual symbol used in the definition of the chi-square statistic.

The symbolism for one- and two-sided alternatives are indicated by the following examples:

1. A one-sided alternative is written $A: p > 0.5$ (or $A: p < 0.5$).
2. A two-sided alternative is written $A: p \neq 0.5$.

9.10 SUMMARY AND GLOSSARY

1. Statistical hypotheses are tested by calculating a probability, called a P value, on the assumption that the hypothesis is true. The P value depends on the sample used for the test, and the statistic used to summarize the information in the sample about the parameter being tested. The statistic may be close to its expected value under the hypothesis, in which case the hypothesis is accepted, or it may be far from its expected value under the hypothesis, in which case the hypothesis is rejected. The P value defines what is "near" and "far"; it is the probability that the value of the statistic is as far or farther from its expected value under the hypothesis, in the direction of the alternative, than is the observed value of the statistic in the sample at hand. When this P value is close to zero, the value of the statistic is judged to be unlikely to be from a sample from the hypothesized population, and the hypothesis is rejected.

2. The hypothesis is specified to be the statement that one wishes to continue to believe unless strong evidence from the sample refutes it. The alternative is the statement one would make if the hypothesis is rejected. Both hypothesis and alternative are determined before the data are examined.

3. The calculation of a P value depends on the sampling distribution of the statistic, the value of the statistic in the sample at hand, and the direction of the alternative.

4. Sample averages and percentages based on large samples have an approximately normal sampling distribution, no matter what the population distributions. Small-sample averages have a normal distribution only if the population distribution is also normal. Small-sample percentages have a distribution closely related to the binomial distribution. [If the percentage is (count/sample size) $\times$ 100 percent, the count has a binomial distribution.]

5. Probabilities for a normal distribution depend on the average and SD. When these are unknown, they must be estimated. The expression

$$\frac{\text{sample average} - \text{population average}}{\text{SD}/\sqrt{n - 1}}$$

has a t distribution. For small samples this is much more dispersed than the standard normal distribution.

6. Two-sample tests for averages and proportions require estimates of the standard error of a difference; this involves the combination of the individual sample (or population) SDs. The large- and small-sample procedures for this are slightly different; the latter uses a pooled estimate of a population SD. This population SD must be assumed to be common to both populations.
7. The number of degrees of freedom in a sample estimate is the maximum number of independent sample estimates that have been combined to form the estimate.
8. Qualitative data are summarized either by percentages or by contingency tables. Tests for such data sometimes require the use of the chi-square statistic and its distribution.
9. Two-sample test procedures depend on whether the samples are independent. Dependent samples are more difficult except when the dependence is due to a natural pairing of the samples. In this case the two-sample test can usually be reduced to a one-sample test, by contrasting the pairs before summarizing the data.
10. Many-sample tests cannot be done by simply combining two-sample tests. The analysis of variance and the chi-square contingency table tests are many-sample tests.
11. When small-sample tests are to be performed which would involve questionable parametric assumptions, nonparametric tests are available. The median tests and the sign test are nonparametric tests.
12. Hypothesis tests are for making decisions about population parameter values based on sample evidence. The critical P value for a hypothesis test is determined by the costs involved of possible wrong decisions.
13. The sample size of a test can influence the meaning of the conclusion. Very large samples reject hypotheses that are almost true, and very small samples accept hypotheses that are quite false. Accepting a hypothesis with a large sample, or rejecting a hypothesis with a small sample, are the most definitive conclusions.

Glossary

A **hypothesis** in statistical testing is a statement, about a certain population, whose credibility we want to assess on the basis of the sample. (9.1)

An **alternative** in statistical testing is a statement that we will be led to believe if the hypothesis is rejected. (9.1)

The ***P*** **value** is the probability, if the hypothesis were true, that the statistic would be at least as far from its expected value as it was observed to be in the sample. The distance from the expected value is measured in the direction of the alternative. (9.1)

A **critical *P* value** is the probability that is set by the person doing the test; it is the threshold for the *P* value that the tester will use to decide if the sample is unusual enough, compared to the hypothesized population, to indicate that the hypothesis should be rejected in favor of the alternative. (9.1)

An **acceptance error** is the error incurred when a hypothesis is false, but the hypothesis test leads us to accept the hypothesis. (An "acceptance error" is a hypothesis acceptance error.) (9.1)

A **rejection error** is the error incurred when a hypothesis is true, but the hypothesis test leads us to reject the hypothesis. (A "rejection error" is a hypothesis rejection error.) (9.1)

When a hypothesis is rejected, the result is said to be **statistically significant,** and when the hypothesis is accepted, the result is said to be **not statistically significant.** (9.1)

A *t* **test** is a hypothesis test concerning averages which uses a *t* statistic to compare the sample with the hypothesized population. (9.2)

A **one-sample test** compares a single sample with a hypothesized population. (9.4)

A **two-sample test** compares two populations by examination of two samples, one sample from each population. (9.4)

A **one-sided test** is a hypothesis test in which the alternative is one-sided. Similarly for a **two-sided test.** (9.4)

A **chi-square** distribution is the probability distribution of a chi-square statistic; the chi-square statistic measures the departure of frequencies in a contingency table from their expected value under certain assumptions. (9.4)

Dependent samples are samples for which the relative values of one sample, in some order, are predictive of the relative values of the other sample. (9.4)

A **contingency table** is a table involving two categorical variables which records the frequency of each combination of levels of the two variables. (9.4)

The **variance of a population** is the square of the population SD. (9.5)

The **variance of a statistic** is the square of the SD of the sampling distribution of the statistic. (9.5)

Analysis of variance is a sample-based technique for detecting inequalities of population means. It does this by comparing within- and between-group variances. (9.5)

The *F* **distribution** is the probability distribution that models the sampling distribution of the ratio of two independent estimates of the same population variance. (9.5)

The **chi-square test of independence** tests the independence of two categorical variables. The two variables are independent if the frequency distribution of one variable is the same for all categories of the other variable. (9.5)

Tests that do not require the specification of a parametric formula for the population distribution are called **nonparametric** tests (e.g., **median test** and **sign test**). (9.6)

Prior knowledge in a statistical testing context is the knowledge about the population that one has before the sample information is available. (9.7)

PROBLEMS AND PROJECTS

1. A survey of 10,000 welfare recipients is done to assess their utilization of hospital services. A simple random sample of 200 welfare recipients is asked the question: "Have you been hospitalized for one or more days in the last 12 months?" All those sampled respond, and 30% answer "yes" while 70 percent answer "no." The persons sampled were also asked the question: "Have you had any formal education up to the grade 12 level?" Again all those sampled responded to this question, 30 percent answering "yes" and 70 percent answering "no." Only 3 percent of respondents answered both questions "yes." What can you conclude about the relationship between hospitalization and educational background, as measured by the questions posed?

2. Mr. Nicely is a politician who claims that he has the support of a majority of the electorate. A sample survey in which a simple random sample of size 100 is selected from the electorate indicates that 45 of the 100 support Mr. Nicely, while 55 support other candidates. Does this evidence conclusively refute Mr. Nicely's claim? (This problem is continued as a computer project in Problem 33.)

3. Research productivity of Canadian university professors is to be compared with those in the United States by examining the average number of papers per faculty member published in refereed journals. Of several thousand faculty members at Canadian universities, a random sample of 500 faculty members is selected, and for these 500, the number of research papers published in 1984 averages 2.1 with an SD of 1.0. A similar study in the United States based on a random sample of 5000 faculty members yielded an average of 2.2 with an SD of 1.0. The difference between those two sample averages is determined to be highly statistically significant, with a P value of less than 0.01. The Canadian Minister of State for Science and Technology is worried, especially since the difference is deemed to be highly statistically significant. What principles has the minister forgotten from his first statistics course? (The data are hypothetical.)

4. A factory produces batches of 1010 firecrackers. As a quality control procedure, a random sample of 10 firecrackers is selected from each batch of 1010 firecrackers. The sampled firecrackers are ignited one at a time and an instrument measures the loudness of the explosion. The data for one batch, in decibels, are

$$93, 76, 91, 68, 0, 75, 89, 0, 95, 77$$

The manufacturer wants to guarantee that the average loudness for the 1000 firecrackers sent to a customer is at least 75 decibels. Would a t test be an appropriate means of deciding whether or not to release the batch? If so, perform the test. If not, explain why not. (This problem is continued as a computer project in Problem 34.)

5. A special roulette wheel has 38 possible spin outcomes: 2 green, 18 red, and 18 black. If the wheel is properly balanced, the 38 outcomes are all equally likely.
   (a) What is the chance that the first spin results in a green?
   (b) If the wheel is spun six times, what is the chance that exactly four reds result?
   (c) The wheel is spun 3800 times in a balance test. The results are: 225 greens, 1732 reds, and 1843 blacks. Use the chi-square test to test whether or not the wheel is balanced.

6. Two diets, A and B, are to be compared in a population of obese patients. The investigator thinks that although the diets have equal caloric content, diet A should allow a greater proportion of obese people to lose 20 lb (or more) in three months than does diet B. One

hundred volunteer obese subjects are allocated to diets A and B using the flip of a coin to indicate the allocation. The results are indicated by the following frequency table:

	Diet A	Diet B
Lost at least 20 lb	12	8
Did not lose 20 lb	35	45

   (a) Show that the chi-square statistic for comparing the success of the diets is 1.70, and determine the P value.
   (b) What is the exact question that the calculation in part (a) is attempting to answer concerning the diets?
   (c) What logical design defect(s) does the proposed study contain?
   (d) What is your conclusion, assuming that the defect in part (c) can be ignored without error?
   (e) The investigator ignores the test and instead makes the following inference: "12/47 is greater than 8/53. I think I will do a further study comparing diet A with the most successful diet reported in the literature to date." Discuss the merit and folly of this inference.

7. A biology student has a laboratory project in which she must distinguish between three types of cells from a large population of cells. The student knows that if she cares for the population properly and classifies them with care, she should have approximately 25 percent cell type 1, 50 percent cell type 2, and 25 percent cell type 3. The project requires that the student select a random sample of 50 cells and report the type of each one. However, the project deadline is 10 minutes away and the project would take about 2 hours. You are a close friend, and being an expert in statistics, you agree to advise your friend on how to construct artificial results (just this once). You are aware that even the famous geneticist Gregor Mendel has been caught by a statistical evaluation to have obtained results too close to the theoretical ones to be experimentally complete. Yet results too far from the theory will receive a low mark because of assumed carelessness with the project. Help your friend in need by suggesting reasonable type-frequencies to report.

8. A king is presented by a statistician with a special gift: a perfectly fair coin. The king is suspicious, however, and demands that the statistician prove that the coin is fair (or else "off with his head"). "Proof" to the king means tossing the coin several times and having exactly half the tosses come up heads. He is willing to let the statistician decide between 4 and 400 tosses. (He was a good king.) Which option should the statistician choose in order to prolong his miserable life? Explain with calculations.

9. (a) What sort of question can an analysis-of-variance test answer that a t test cannot answer?
   (b) Describe the rationale for the use of a ratio of variance estimates (i.e., an F statistic) to determine the answer to the sort of question that you state in your answer to part (a).

10. Seventy students choose either section 1 or section 2 of a course. Given the frequencies in the following table, do you think there is convincing evidence that this choice is associated with the students' declared major area? Perform the appropriate calculations to help make your decision.

	Arts major	Science major
Section 1	15	25
Section 2	15	15

11. A random sample of five students from a statistics class of 500 students is selected from those who took the final examination. Midterm test scores and final examination scores for these students are as follows:

Student	Midterm	Final exam
1	50	60
2	73	69
3	37	51
4	83	85
5	61	73

(a) Assess the evidence for the claim that the average score on the midterm is lower than on the final for students who wrote the exam.
(b) If the third student claims that he was ill for the midterm and you wish to delete this one mark (the 37), what would you do with the nine other data values to answer the same question as that posed in part (a)? Give some rationale for your choice.

12. Eighty percent of a random sample of 10 Saskatchewan farmers report that type A, a new fertilizer, produces higher yields than type B. Is this convincing evidence that a majority of Saskatchewan farmers would report that A had higher yields?

13. A survey of a random sample of 25 people from a voters' list of 3000 names yields the following results:

- 10 favor a tax increase.
- 15 are against a tax increase.

To assist in speculating on the value of the proportion favoring a tax increase among the 3000 on the voters' list, the following table is constructed:

Proportion favoring a tax increase among the 3000	Probability that the observed sample (10F, 15A) would be selected
0.0	0.0
0.1	0.0001
0.2	0.01
0.3	0.09
0.4	0.16
0.5	0.10
0.6	0.02
0.7	0.001
0.8	0.0
0.9	0.0
1.0	0.0

(a) Is this table a probability distribution?
(b) On the basis of this table only, suggest reasonable values for the proportion favoring a tax increase in the voters' list of 3000 names.
(c) Perform a test to determine if the proportion favoring a tax increase is greater than 0.6. Compare with your answer to part (b).

14. A simple random sample of 60 ears of corn from a farmer's field shows that 33 percent of these are withered and that 25 percent of the withered ears are from tall plants, whereas 60 percent of the nonwithered ears are from tall plants. Is this withered condition associated with growing on tall plants, at least for ears of corn in this farmer's field?

15. A test is to be made of two brands of golf ball, X and Y. A machine is used to simulate the drives an experienced golfer would execute under ideal conditions. It is expected that the distances traveled by the ball before coming to rest would be about the same on successive drives if the balls were identical. The distances traveled on successive hits may be assumed to be independent. The data obtained from 30 hits of each of brand X and brand Y are summarized by:

Brand X: average distance = 275 yards, SD = 20 yards
Brand Y: average distance = 265 yards, SD = 18 yards

(a) Can you conclude from these data that brand X tends to travel farther than brand Y? Justify your answer.
(b) If 10,000 balls of each brand were subjected to a similar comparative test, how small do you think the difference in the average distances traveled might be? (It was 10 yards in the small experiment, but what is the smallest plausible value of this difference in the large experiment, based on the data given?)

16. Each morning, John the weatherman looks at the sky at 7:00 a.m. and guesses the amount of rain that will fall that day. He then compares this with the amount recorded subsequently by a rain gauge. The recorded data for a recent week are as follows:

Day	Guessed rainfall (cm)	Actual rainfall (cm)	Error
1	2.5	1.0	1.5
2	0.0	0.0	0.0
3	1.0	1.7	−0.7
4	4.0	3.0	1.0
5	2.5	0.3	2.2
6	0.2	0.0	0.2
7	0.0	1.0	−1.0

If possible, test whether or not John's guesses are unbiased. If it is not possible, explain why not.

17. A librarian is attempting to determine the amount of new shelving required for the next year's acquisitions. A random sample of 10 weeks out of the last eight years of acquisitions has yielded the following number of yards of shelving required for acquisitions in each of the 10 weeks: 7, 4, 6, 9, 3, 8, 5, 6, 6, 7. The capital budget has allowed for an average of 5 yards of shelving per week over the next 52 weeks. If the acquisition rate is the same over the next 52 weeks as it was over the last eight years, will the budgeted amount of shelving be enough?

18. A geneticist is studying the relationship between fur length and eye color of laboratory mice. The fur length is noted to be short or long, and the eye color is noted to be red or brown. Two hundred mice with parents of the same genetic strain are sorted according to these characteristics, with the following results:

		Hair length	
		Short	Long
Eye color	Red	25	45
	Brown	52	78

   (a) Are hair length and eye color independent characteristics?
   (b) A genetic hypothesis suggests that the frequencies should occur with the following frequency distribution:

Red-short	11.1%
Red-long	22.2%
Brown-short	22.2%
Brown-long	44.4%
Total	100% (except for rounding)

   Test this hypothesis.
   (c) If the genetic hypothesis is correct, are hair length and eye color dependent or independent? Explain.

19. An automobile manufacturer produces several thousand model XXX cars each year. The manufacturer claims that the cars have an average lifetime of 10 years and that 10 percent of the cars actually last 15 years or more.
   (a) Assuming that the distribution of lifetimes is normal and that the claim is correct, what proportion of cars have a lifetime of 20 years of more? Is this assumption of normality reasonable?
   (b) A simple random sample of 30 model XXX cars manufactured in 1960 reveals that all 30 cars have "died" and that their lifetimes average 8.5 years and have an SD of 4 years. Is the manufacturer's claim believable in the light of these data?
   (c) A simple random sample of 30 1970 model XXX cars reveals that only 33 percent of the cars are still on the road in 1980. Using these data alone, estimate the proportion of 1970 cars of this model that last 10 years. Use an interval estimate.

20. Refer to Exercise 5 of Section 9.5 and Table 9.7.
   (a) Draw the four group averages and SDs on a scattergram. (This is a new kind of diagram.)
   (b) Compute a regression line from the four points. Might this be misleading, as suggested in Sections 5.2 and 5.3? (See the "correlation-of-averages" problem on p. 148.)
   (c) Is loudness causally related to performance? Use whatever reasoning you can invent. (The standard method has not been covered in this book.)

21. A theater tries the following innovative scheme to attract customers. For each ticket that is purchased at the regular price of $5.25, the ticket salesperson tosses a fair coin. If it turns

up heads, the purchaser gets a refund of $5, so the effective cost of the ticket is only 25 cents; if the coin turns up tails, the purchaser receives no refund.
(a) What is the expected value of the effective cost of a ticket?
(b) What is the SD of the cost of a ticket?
(c) Describe the histogram of the revenue derived from the sale of 100 tickets.
(d) The manager of the theater checks with the salesperson after the first 100 tickets are sold and is told that the revenue so far is $225. If you were the manager, what would you do next? Explain.
(e) The first day this coin-tossing scheme is used, 1000 tickets are purchased. What can be said about the average ticket price for the day? What can be said about the total revenue for the day?

22. Indicate whether the following statements are true (T) or false (F). Indicate your reasoning.
(a) The 90th percentile of the chi-square distribution with 10 degrees of freedom is 4.86.
(b) A 95 percent confidence interval indicates the interval of values in which about 95 percent of the data values can be expected to fall.
(c) Cluster sampling is used instead of random sampling to increase the accuracy of estimates of population parameters.
(d) The sampling distribution of an average of a random sample of size 5 is the t distribution.
(e) The square of the sample SD is an unbiased estimate of the variance of the population sampled.
(f) The sampling distribution of a sample proportion has a binomial distribution.

23. A random sample of 100 cows is selected from a herd of 500 cows for a study of the effect of the distance covered during a day by each cow on the quality of its milk. The milk is tested at the end of the day and is assigned a quality: High or Low. The distance traveled by the 35 Low cows was 1.5 km with an SD of 0.5 km, while the distance traveled by the 65 High cows was 0.7 km with an SD of 0.4 km. Discuss and assess the hypothesis: Milk quality of a cow is affected by the distance traveled by the cow.

24. A small survey was done based on a random sample of five union members to determine whether a strike should be called. Of the five questioned, not one indicated that they favored a strike. The executive of the union is contemplating a larger survey, of 1000 members. What can be said about the executive's hope that 50 percent or more of the union members will favor the strike?

25. The following data were collected concerning the systolic blood pressures (SBP) of five patients before and after treatment for hypertension:

Patient	SBP before	SBP after
1	190	150
2	200	145
3	150	110
4	250	210
5	180	140

The sample was extended to 50 patients; the "before" average and SD were 200 and 50, respectively, and the "after" readings averaged 150 and also had an SD of 50. Perform a test to determine whether there would have been a change in the average SBP of a large number of patients of which this sample of 50 is a random sample. Is your conclusion (based on the data set of 50 cases) in agreement with the trend seen in the five cases shown? Explain why or why not.

26. A downhill ski race results in the following times for the five contestants: 120.11, 121.13, 119.96, 120.53, and 121.44 sec. Estimate the average time that these skiers would accomplish (as a group) under similar conditions in a subsequent run. Use an interval estimate. State whatever assumptions are necessary to provide the desired estimate.

27. The interpretation of significance tests depends to an extent on sample size. Very large or very small samples can produce P values which indicate a conclusion that is not justified from a practical standpoint. Discuss with examples of your own invention.

28. A mailing list contains 10,000 names of candidates to subscribe to a trade magazine. Ninety percent of the addresses on the list are supposed to be valid. Before going to the expense of sending your advertising material to all 10,000 prospects, you select a random sample of 100 names from the list and check the addresses by telephone and other means. Of these 100, 19 are incorrect. Is the claim of 90 percent correct addresses in the 10,000 believable?

29. An ecologist predicts that the number of three species of plant in a meadow should be in the proportions 60, 30 and 10 percent. By sampling plants in the meadow using a scheme equivalent to simple random sampling, the ecologist collects numbers of plants 550, 290, and 160 (comparable with the 60, 30 and 10 percent). Are these findings in accord with the ecologist's prediction, allowing for sampling error?

30. At a track-and-field meet, two independent timing instruments are used to time the 100-m run. The meet is unusual in that runners are timed in separate "races." That is, each runner is the only runner in the race in which he is being timed. The two timing instruments are of the same brand and model and they are both calibrated to eliminate bias. The average of the two times recorded for each runner (in a single race) is used to decide on the ranking of the runners. The times for the two best runnners is shown in the table below.

Runner 2 claims that the timers are too imprecise to justify the claim that his run was slower than the run of runner 1; he feels that the race should be declared a tie. Do some calculations that would help to decide the merit of this claim of runner 2.

	Runner 1	Runner 2
Timer 1	10.02	10.18
Timer 2	10.10	10.24

31. A marketing survey is designed to determine if there has been a change in the proportion of households that use certain brands of detergent. A random sample of households from the marketing area is contacted and the usual detergent used is determined for every household in the random sample. The data are summarized as the number of households using brand A, brand B, or "other." The previous survey was a very large scale survey including 10,000 households and the responses were: 1000 use brand A; 2000 use brand B; and 7000 use "other." The current survey is comparatively small scale, with only 100 households surveyed, with the following results: 16 use brand A; 24 use brand B; and 60 use "other."

(a) Does this current survey provide strong evidence of a change in brand use, or is the difference from the large-scale survey reasonably attributable to sampling variation? Justify your answer.

(b) Assess the strength of the evidence for the claim that the proportion of households using "other" has increased since the large survey.

(c) Comment on the relationship of your answers to parts (a) and (b).

32. A pet food wholesaler does a huge survey to determine the proportions of households that own a dog, a cat, or both. Ten thousand households are contacted and the proportions from the survey are:

Dog only	31.1%
Cat only	29.6%
Both dog and cat	15.2%
Neither dog nor cat	24.1%

(a) Is the proportion of "dog only" households significantly different from the proportion of "cat only" households? Perform a test at the 5% level of significance. (That is, use 5 percent as the critical P value.)

(b) What is the significance to the pet food wholesaler of the conclusion in part (a)? Discuss the general statistical principle relating to this question.

Computer Projects

33. (Continuation of Problem 2.) Simulate the number of heads in 100 tosses of a fair coin. Repeat this experiment enough times to estimate the sampling distribution of the number of heads fairly well, at least near the center of the distribution. Compare this with the binomial table in Appendix A4.2. Compare the probability of 45 heads, as estimated by your simulation, with the normal approximation that is based on the correct average and SD. Comment on the applicability of your findings to Problem 2.

34. (Continuation of Problem 4.) To study the t test proposal in Problem 4, we would like to have an appropriate population from which to select many samples of size 10 so that we can see what sort of sampling distribution the t statistic has. One idea that has been proposed in the statistical literature is to assume that the original sample of 10 (see Problem 4) faithfully describes the population distribution. We can construct our artificial samples by selecting a random sample of size 10, with replacement, from the 10 numbers in the original sample. The result is that each artificial sample will contain some of the original values more than once and will miss some values (typically). For each artificial sample, we calculate the value of the t statistic. The histogram of these t statistic values will be an estimate of the sampling distribution of the t statistic. Note that this method does not make any assumption about the normality of the population. This procedure is called "bootstrapping."

Determine the bootstrap distribution of the t statistic. Is this a t distribution? Is it close to a t distribution? (You may be able to answer this question with only a few artificial samples, say 15 or 20. Of course, if the programming is no problem, a larger number of artificial samples is advisable.) Use your findings here to answer Problem 4.

Appendix 1

Use of Computers for Statistical Tasks

Computers are not an essential feature of an elementary statistics course. Most of the calculations described in this text are simple to do by hand or with a calculator, and it may reasonably be questioned whether the effort of learning to use the computer in a first course in statistics is worthwhile.

The most obvious reason for using computers in a course in statistics is that computers are now the usual means of doing statistical calculations in applications. Students who will be analyzing data will be several steps ahead if they have experience with computer-based calculation methods. Even students who would only be evaluating the analysis of others should know the special hazards of automated calculations. Moreover, it is a more stimulating exercise to learn how to use computer packages of statistical programs than it is to execute various arithmetical procedures by hand. Hand calculation may help to familiarize students with a formula, but there is a limit to how much of this is necessary.

Another reason for using computers in this course is that students interested in an experimental or inductive approach to statistics can use the repetition facility of the computer to try many variations of analysis on a single data set, or to explore the effect of a single analysis on many similar data sets. These repetitions tend not to be feasible for hand calculations, but are very instructive with only a little investment of time when the computer is used.

It may certainly be argued that the application of computers to statistics could be left to a second or subsequent course. As long as students are required to take a subsequent course involving computers in statistics, this is a reasonable argument.

However, for many students in the social sciences, computing skills are never required. For these students the real nature of modern applied statistics may never be learned.

To those brave souls who would venture into the world of statistical computing, this appendix is a starting point. It is assumed that the reader knows nothing about computers and very little about statistics. After reading this appendix you will have to do a bit of work on the documentation of the statistical package that is available at your institution. This appendix tells you what to look for in such documentation that is directly related to this text.

A1.1 INTRODUCTION TO THE USE OF PACKAGES OF STATISTICAL COMPUTER PROGRAMS

The difficult part in learning to use statistical computer programs is to find out how to get the data into, and the results out of, the computer. Once this is mastered, the rest is quite easy. Not that it is easy to know what analysis to do; that is the subject of this text. But once you know what you want to do, and the data to analyze have been input to the computer, the designation to the computer of the desired calculation procedure is quite simple. After that is done, the job of getting at the answer is all that is left.

As a first step to data handling, let us consider what a data set looks like. Table A1.1 gives an example of a typical data set. This data set has six **cases** (1 through 6) and four **variables** (STUDNO, CREDITS, CREDSREG, and GPA). The 24 positions at which data values appear are arranged in a format called a **matrix.** A matrix has a certain number of rows (six in the example) and a certain number of columns (four in the example). There is no compelling reason to display the data this particular way, but this is the convention used in most packages. The columns are always vertical, and one column lists the values of a particular variable for all the cases. The rows are always horizontal, and one row lists all the variable values for a particular case. The matrix below would usually be referred to as a "six by four" matrix (not a "four by six" matrix). It is necessary to know the matrix language because it is often used in the documentation of the programs.

TABLE A1.1 TYPICAL DATA SET

Student number (STUDNO)	Credits earned (CREDITS)	Credits registered now (CREDSREG)	GPA (GPA)
1	15	15	3.51
2	37	9	2.58
3	18	15	3.60
4	30	16	2.94
5	15	12	3.04
6	0	15	—

There are more complex data structures. One that we will use often has two sets of cases. For example, students 1, 2, and 3 might be from a particular high school, HS1 say, and students 4, 5, and 6 might be from HS2. In this case we have two **groups.** For the purpose of entering the data into the computer, we would still use the one matrix. But we are clearly going to have to indicate to the computer which cases go in which groups: the simplest way to do this is to add another variable, labeled "HS," say, which has the code "1" for cases 1, 2, and 3 and "2" for the cases 4, 5, and 6.

Note that the variables have been given labels. This is so that we will recognize which variable is which in the computer output. The variable "STUDNO" has been included in case we want to tell the computer to ignore a case, form a new group, or alter some incorrectly entered data.

When data are entered into a computer, they are usually entered one case (row) at a time. The computer accepts a string of digits, blanks, and decimals for each case. The information that tells the computer what digits are values of what variables is provided in two steps: the **format** and the **variables list.** The format is sometimes merely specified by indicating to the computer that the numbers are to be read as we normally would except that blanks separate values of different variables. Or we may have a more explicit definition of the format such as "I2, 2X, I3, 2X, I3, 2X, F4.2". This means that the first two digits form the integer which is the first variable, then two blanks, then a three-digit integer variable, then two more blanks, another three-digit integer variable, and two blanks, and finally a decimal number with two digits after the decimal, and one digit before the decimal. (The decimal itself needs a space.) The point is that the computer has to be told exactly where the numbers are in the sequence of codes you enter. The variables list then lists the labels attached to the variables in the same order that the data are entered. In the example it would be: STUDNO, CREDITS, CREDSREG, GPA.

At this point the computer will know what numbers are associated with what variable labels. As you enter each case of data, you keep this same format so as not to confuse the computer.

Data are entered through a **terminal.** A terminal is just a typewriter that is wired into the computer. (Some students may be using microcomputers, in which case the terminal and the computer itself are a single unit.) The electronic organization that lets you, and possibly several others at the same time, enter data into a computer is called an **operating system.** You will have to find out from your instructor how to establish communication with your computer, and this will involve learning how to "sign on" to the computer and how to request the use of a particular statistical computer package. You will also need to know how to save your data sets for future use. It may take some time entering a data set into the computer, but you will not need to do this more than once for each data set. Once it has been stored, you can bring it back to the package with appropriate commands to the operating system.

Output can usually be viewed on the TV-like monitor of the terminal. To get a version that is printed on paper, you have to again use operating system commands.

The steps you should take in getting started are:

Appendix 1 Use of Computers for Statistical Tasks

1. Get approval from your computing center or instructor for using the computer. You will need an account number and a password that lets the computer know that it is really you. (Keep the password to yourself or you will find that your allocated computer time has been used by someone else.)
2. Learn how to use the operating system to store data sets and output files and to print these out on paper.
3. Get the documentation for the statistical package you will be using. Find out how to enter data to a file, store the file, retrieve the file, and print the file.
4. Sometimes you will have a choice between the operating system or the statistical package to enter data files. Your data files will often need editing, such as additions, deletions, or changes. Get advice on whether the operating system or the package itself is best for entering and editing data sets.
5. Try entering a data set such as

    1, 2

    3, 4

    5, 6

    and see if you can find the average of the two variables. This will require you to use a procedure in the statistical package for taking averages. If you get this far without being quite frustrated by the lack of feelings of the computer, take heart. There are very few in this world who have avoided this frustration.
6. Make use of the help resources around you; friends who have already learned are very useful. Most computer centers have a consultant whose job it is to help people use the computer. Your instructor or tutor can help. Very proud people have trouble learning to use computers because they refuse to admit ignorance to anyone. When they realize that knowledge of a particular computer system is not inborn to anyone, they may be more willing to seek help.

You will realize that this small appendix cannot tell you all you will need to know. Every computer system is different, and although there are common aspects to them, having approximately the correct command is not helpful at all. The local system will have to be learned from local sources.

The commands you will be able to make use of for the initial stages of this course are those that accomplish the tasks listed below:

Data entry

Definition of groups of cases

Average and standard deviation

Histogram

Scatter diagram

Regression (one independent variable)

Correlation
Random digit generation
Normal deviate generation

You will have to discover the exact procedures to do these things with the particular package used.

A1.2 STATISTICAL EXPERIMENTATION: MONTE CARLO SIMULATION

The inferential part of statistical methods is concerned with techniques for using sample data to estimate population characteristics. For the purpose of seeing how well (or how poorly) these techniques work, it is useful to experiment with samples from populations whose characteristics are completely known. The way we do this is to have the computer generate samples from a population that we specify, and try to recover the specifications using various techniques. There are two populations that are particularly useful for this purpose. One is a population that consists of equal proportions of 0, 1, 2, . . . , 8, 9. Another is a population that consists of numbers of all sizes, but with relative frequencies that obey a normal curve (with average 0.0 and SD of 1.0).

Most packages readily generate any quantity of sample values selected at random from these two populations. Think of a huge vat with tickets having numbers on them with appropriate relative frequencies, and someone selecting a sequence of these numbers in a way that does not favor any particular tickets. This process would generate a sample, called a random sample, that we use to try to recover the characteristics of the tickets in the vat. In a real-life sampling situation, this corresponds to looking at a part of a population, the sample, to estimate characteristics of the population itself.

For example, we may ask a random sample of 100 people what their voting intentions are, in order to estimate what the outcome of an election will be based on the votes of an entire electorate.

The reason it is handy to have a computer-based method of selecting samples is that we can observe how reliable sample estimates are. In applications we just have one sample to work with—one sample of size 100, say. If we use the average of the sample as an estimate of the average of the population, how wrong might we be? How much error typically occurs in this situation? We have formulas that answer this question in certain situations, but we do not need the formulas if we are willing to do statistical experiments.

Suppose, for example, that a population of 10,000 voters contains 40 percent who intend to vote for Mr. L and 60 percent for Mrs. W. How close to this information could we expect to get from a random sample of 50 voters? We want an artificial mechanism that generates responses to the question "Who do you intend to vote for?"

in proportions just the same as random samples from the population (i.e., 40 percent for L and 60 percent for W, in the long run). One way to do this is to ask the computer to generate a string of digits such as 4, 7, 5, 2, 3, 8, 4, 3, 9, 2, 1, . . . in such a way that each digit has the same relative frequency and such that each digit does not depend on the previous digits. Then we will interpret a 0, 1, 2, or 3 as meaning an intention to vote for Mr. L, and a 4, 5, 6, 7, 8, or 9 as indicating a vote for Mrs. W. This will guarantee, in the long run, that we could get 40 percent for L and 60 percent for W. But our sample is only of size 50, not a very "long run." What might we get in the 50?

The actual outcome of such an experiment is shown below. Five samples of size 50 were produced by the computer and then reinterpreted as voting intentions, according to the scheme proposed.

```
1 6 5 2 3 0 7 5 4 8 8 1 2 8 6 9 8 6 6 5 6 9 3 4 1
L WW L L L WWWWW L L WWWWWWWW L W L
8 3 4 0 6 0 6 4 5 2 1 3 1 9 9 4 7 1 1 1 4 8 0 9 0
W L WLW L WWW L L L L WWWW L L L WW L W L
```
                                              20 L, 30 W

```
7 5 9 6 6 3 9 5 4 9 1 8 9 9 9 6 0 2 7 1 9 4 7 0 5
WWWWW L WWWW L WWWWW L L W L WWW L W
5 6 6 3 7 8 0 2 7 9 8 9 3 7 1 2 8 3 0 6 4 8 1 6 5
WWW L WW L L WWWW L W L L W L L WWW L WW
```
                                              15 L, 35 W

```
1 6 5 0 7 6 0 7 3 1 3 8 4 9 1 0 9 5 4 1 4 3 0 9 3
LWW L WW L W L L L WWW L L WWW L W L L W L
8 2 6 3 8 9 5 3 4 3 8 8 4 6 2 8 8 8 2 2 6 6 1 6 0
W L WLWWW L W L WWWW L WWW L L WW L W L
```
                                              21 L, 29 W

```
6 8 1 1 6 6 3 0 3 5 9 1 1 9 1 3 9 2 1 5 3 5 9 9 4
WW L L WW L L L WW L L W L L W L L W L WWWW
1 1 6 7 1 6 8 5 8 2 4 0 1 7 7 2 4 2 9 7 9 7 3 1 2
L L WW L WWWW L W L L WW L W L WWWW L L L
```
                                              23 L, 27 W

```
5 9 9 4 8 3 1 9 5 0 9 5 8 1 0 4 7 2 0 0 6 8 8 0 8
WWWWW L L WW L WWW L L WW L L L WWW L W
6 3 0 9 3 8 2 0 2 5 4 1 2 0 7 4 2 2 0 6 7 5 9 8 1
W L LWLW L L L WW L L L WW L L L WWWWW L
```
                                              22 L, 28 W

In the five samples, the percentages of L's are 40, 30, 42, 46, and 44. As a rough summary, we might feel that a sample of size 50 would only estimate the population proportion to within about 10 percentage points, or perhaps even a bit worse than that (more samples of size 50 would help to determine this). In other words, if we are concerned about determining who will win the election, a sample of 50 may not be enough. The samples of 50 shown all indicate the correct winner, but judging from the variability of the sample percentage, we would not be surprised to see a sample percentage over 50 percent in a particular sample, even though the population percentage is 40 percent.

By doing the computer experiment above, we have been able to learn something about estimating population percentages from samples. Note that we did not use any formulas for this—we just followed a procedure that mimicked the lottery type of sampling that one might use if one were doing a real opinion poll. Now in this situation it happens that we do have a formula that indicates how much the error is likely to be. When we work it out, we get a typical error of 7 percent. The exact meaning of this is explained in Chapter 5. The point here is that we can do the experiment in situations where we do not know the formula. This technique is called **Monte Carlo simulation** after the famous gambling resort. This racy nomenclature should not detract from the sober purpose we have in mind.

Monte Carlo simulation can solve many statistical problems that are otherwise difficult to solve. But this is an elementary course in statistics from which difficult problems have been carefully excised. Our use of Monte Carlo simulation is to expose statistical phenomena in a way that promotes intuitive absorption. This intuition will help you to avoid mistakes. On the positive side it may give you a taste of the intriguing world of random phenomena. You will find that your intuition needs training in this area.

Let us look at an example of the use of a list of computer-generated numbers from a normal population. Suppose that a population of sacks of potatoes has an average weight of 104 lb with an SD of 3 lb. In a typical sampling situation we would not know this and would be trying to estimate the population average, say, from a sample of 25 carefully weighed bags. How well can we estimate the population average in this situation? Here are five sample averages based on the average of five sets of 25 numbers generated by a computer. The population specified to the computer is that the population distribution be normal and have average 104 and SD 3.

The result of the simulation is the five averages: 103.9, 103.4, 104.5, 104.5, and 103.6. Thus the average seems to be estimated to within about 0.5, or being conservative to within about 1.0 lb. Here again it happens that a formula exists to calculate the typical error of the estimate; the result (see Chapter 5) is that the typical error is 1.3 lb. We were a bit lucky with our samples; they indicated a typical error of less than 1.0. This indicates that such simulations should use more than five samples. If we are using a computer, this is no additional work, and we can do 100 simulated samples in approximately the same time that it takes to do the five samples manually.

One phenomenon that can be discerned from the example which used normal deviates is that the sample average does not vary as much from sample to sample as do the population values. The sacks of potatoes had an SD (a typical deviation from the population average) of 3 lb, whereas the averages had a typical error of about 1 lb. (The individual sample values would vary about the same amount as the population itself, with an SD of 3.0 lb.) This is why averages are thought to be better estimates than individual sample values. Averages of random selections are more stable than the individual selections themselves. To see this happen via Monte Carlo simulation makes the phenomenon more real than to be told that it is so. This sort of demonstration illustrates the main use of Monte Carlo simulation relevant to the material in this text.

The opportunity to do these Monte Carlo studies is provided by the problems grouped under the heading "Computer Projects." The suggestions below provide a gentle introduction to Monte Carlo for those who are attempting to learn the technique before the rest of the material in the text.

EXERCISES FOR FIRST MONTE CARLO EXPERIMENTS

1. Using the table of random digits, Appendix 3, select any 10 groups of five digits. Note the closeness of the averages to 4.5. Calculate the average of all 50 digits by averaging the averages. Note the closeness of this overall average to 4.5. Explain in words the phenomenon you have observed.
2. Go through the calculation steps in Exercise 1 using a computer package. Now repeat the experiment with 10 groups of size 25 instead of size 5. Note the stability of the average of 25 digits as compared to the average of five digits.
3. (a) Round the averages in Exercise 2 to the nearest integer. So you now have 10 averages rounded to an integer in the range 0, 1, . . . , 9. How many of each integer do you have? (That is, 3 of 4's, 4 of 5's, and so on.)
   (b) Get another 10 of these averages of samples of size 25. Now examine the counts of each integer among the 20 averages. State in words what you observe.

If you learn to do Exercise 3 using computer-generated random digits, you will be ready to try some of the computer project problems. But you may have to learn some statistics first.

Appendix 2

List of Applications

Numbers refer to sections or subsections of the text where the application can be found, unless it is preceded by "Ex" for Exercise, or "Pr" for Problem. Exercises are at the end of each section. Problems are at the end of each chapter.

A2.1 SPORTS AND ENTERTAINMENT

Aerobics Ex 3.1.3, Ex 5.3.6
Basketball Ex 2.4.1, Ex 2.4.2, Ex 2.4.3
Carnival Ex 2.7.3, Pr 5.28
Cycling Pr 4.7, 5.5.1, Pr 8.5
Dog races Ex 2.5.3
Entertainment Pr 8.14, Pr 9.21
Firecrackers Pr 9.4
Fitness 5.3.1, Pr 5.11
Gambling and games Ex 7.4.1, Ex 7.4.2, Pr 7.4, Pr 7.24, Pr 7.29, Pr 7.31, 9.1, Pr 9.5
Golf Pr 9.15
Gymnastics Ex 9.6.2
High jumping Pr 7.3

400 Appendix 2   List of Applications

Horse racing 5.3.2
Lotteries 1.1.2, Ex 1.1.2, 7.2.3, 7.4.1, Pr 7.20, Pr 7.30, Ex 9.2.2
Movies 5.2, Ex 5.2.1, 5.4.1, Pr 5.18
Olympic games Pr 6.1, Pr 8.15
Photography 5.4.1, Ex 5.4.1, Pr 5.8
Running 3.1, 3.2, 3.3, 3.4, Ex 3.2.1, Ex 3.2.2, Ex 3.3.1, Ex 3.3.3, Ex 3.4.2, Ex 7.2.4, Ex 7.2.5, 8.2, Pr 8.8
Skiing Pr 9.26
Squash Pr 7.9, Pr 7.27
Student athletes Ex 9.4.5
Television 9.1, Ex 9.1.3, Ex 9.1.4, 9.2.1, 9.2.2
Trivia games Pr 3.7, Pr 3.8
Wine testing Ex 2.3.2, Pr 5.2

A2.2 SOCIAL SCIENCES AND BUSINESS

Advertising and marketing Ex 2.6.3, Ex 4.4.5, Pr 4.5, Pr 5.16, Ex 7.10, Ex 7.4.9, Ex 9.7.1, Pr 9.28
Airline passenger names 1.1.7
Auditing Ex 2.3.5, Ex 8.4.1, Ex 8.4.2, Ex 8.4.3, 9.7
Automobile sales 6.1, Ex 6.1.3, Ex 6.1.4, Ex 6.1.5, 6.2, Ex 6.2.2, Ex 9.6.1, Ex 9.6.2
Banking Pr 8.6
Child behavior 2.1
Demography 1.1.5, 5.5.2, 6.2, Pr 6.6, Pr 7.15
Economics Ex 5.4.3, Pr 6.2, 8.4.2
Education 4.2.3
Employment and unemployment 2.2, 4.2.3, Pr 5.1, Pr 5.5
Family size Pr 4.15, Pr 4.16
Foreign exchange rates Ex 4.5.6, Pr 4.21, Ex 6.2.3
Fund raising Ex 8.2.1, Ex 8.2.2
Hospital use Pr 9.1
Household surveys 7.4.4
I.Q. 2.7, 4.5.2, Ex 4.5.1, Ex 4.5.3, Ex 5.3.2, Ex 7.4.8, Pr 7.19, Pr 8.3
Industrial relations Pr 9.24
Investment 2.3
Language Ex 2.3.1, Pr 2.3, Pr 4.15, Pr 5.25, Pr 7.5, Pr 7.6
Library periodicals Pr 1.4

Marketing surveys Pr 1.1, Pr 1.2, 2.3, 8.6, Ex 8.6.1
Medical insurance Pr 1.5
Mental tests Ex 5.2.5, Pr 5.19, Ex 7.2.6
Operations management Ex 5.1.3
Personnel Ex 4.3.2
Pharmaceuticals 2.3
Political surveys and elections, 1.1.3, 2.3.2, Ex 2.3.6, 2.6, 4.2.1, 7.2.1, 7.3.5, Pr 7.26, 8.1, 8.4.1, Pr 8.1, Pr 9.2, Pr 9.13
Prejudice Ex 9.5.6
Product comparisons Ex 3.1.1, Ex 3.1.2, Pr 3.1, Pr 3.3
Quality control 2.3, Ex 2.3.3, 2.7, 4.3, Ex 4.3.1, Pr 4.4, Ex 7.2.3, 7.3.5, 7.6, Pr 8.10, 9.1, 9.2.2, 9.6.1
Real estate 4.2.2, Ex 4.2.1, 5.3.2, 5.3.3, 5.3.4, 5.3.5, 5.3.6, Ex 5.3.4, Pr 5.23
Salary surveys 2.3.1, Pr 2.1, Pr 2.2, Ex 4.4.3, Pr 4.9, Ex 7.4.7, Pr 7.11, Ex 8.6.3
Sex ratio in families Ex 4.2.2, 7.3.3, 7.3.5, Ex 7.3.6
Socio-economic profiles 5.1.3, Pr 5.4
Teaching methods 2.5, Ex 2.5.2, Ex 9.1.1, Ex 2.5.4, 5.5.1
Textbook sales Ex 1.1.1

A2.3 LIFE SCIENCES AND MEDICINE

Agriculture Pr 2.4, Ex 3.3.4, Pr 3.2, Ex 4.2.3, 5.5.1, 5.5.2, Ex 8.2.4, 9.5.1, Ex 9.5.1, Ex 9.5.3, Pr 9.12, Pr 9.14, Pr 9.23
Animal experiments Ex 7.5.4
Animal lifespans Ex 4.3.6, Pr 4.20, 5.3.1
Body fat 4.2.3, 8.2
Cigarette smoking Ex 2.5
Clinical chemistry 2.4, 2.7, Ex 2.7.1
Conservation Pr 5.6, Pr 8.16
Diet and health 1.1.4, 1.2, Ex 1.2.1, 2.5, 2.6, 5.1.3, 5.3.1, 9.4.2, Pr 9.6
Ecology Pr 9.29
Epidemiology 5.1.3, Ex 5.1.2, Ex 5.1.6, 5.4.3
ESP Ex 7.5.6
Fish size Ex 4.3.7
Forestry 2.3, Pr 4.1, Pr 4.8, Pr 4.18, 7.2.2
Genetics Ex 9.6.3, Pr 9.7, Pr 9.18
Medicine Pr 2.8, Pr 3.5, Pr 5.15, Ex 6.1.2, Ex 9.1.5, 9.4.3, Ex 9.4.2, Ex 9.4.3, Ex 9.4.4, 9.5.2, Ex 9.5.2, 9.7, Pr 9.25

Paleontology 5.4.2
Rat intelligence 4.4, 7.3.8
Reaction times 7.2.1
Sheep weights 4.4.1
Sleep Pr 8.4
Sociology and health Ex 5.1.4

A2.4 NATURAL SCIENCES

Airplane fuel 4.5.1
Geometry 5.26
Mercury pollution Ex 5.3.7, Ex 9.2.4
Oil exploration 9.7
Water resources Pr 5.7
Weather 1.1.2, 2.4, Ex 4.4.2, Pr 4.14, 5.3.5, Pr 6.5, 7.3.3, Pr 9.16

A2.5 GENERAL

Automobiles Pr 5.9, Pr 5.10, Ex 7.3.3, Ex 8.2.5, Ex 9.1.2, Pr 9.19
Aviation Ex 7.3.7
Berry picking 4.4.3
Carpet cleaning Ex 9.2.3
Class size 4.1
Coin collecting Ex 4.3.5
Commuting 4.2.3, 7.5.2, Pr 7.16, 9.1
Computer passwords 7.3.2
Despotism Pr 9.8
Evaluation of lecturers 1.1.6, Ex 1.1.6, Pr 2.5
Food markets 5.1.3, Ex 7.2.7, Ex 7.5.3, Pr 8.13, Ex 9.2.1
Garbage Ex 7.3.9
Gardening Pr 3.4, Pr 4.11, Ex 7.2.2, Ex 9.4.1
Gasoline consumption Ex 3.4.1, 5.1.3, Ex 5.2.3, Ex 5.2.4, Ex 5.3.3, Pr 5.12, Pr 5.22
Height and weight Pr 1.3, Pr 4.12, 4.5.1, Ex 4.5.5, Pr 4.6, 5.2, 5.3.7, Pr 5.3, Pr 5.20
Higher education institutions Pr 8.17
Insurance Ex 7.3.14
Libraries 8.3, 9.3.1, 9.3.2, Ex 9.3.1, Ex 9.7.2, Pr 9.17

Liquor sales Ex 6.1.1, Pr 6.3
Military records Ex 5.1.1
Music and study 9.5.1, Ex 9.5.5, Pr 9.20
Postal and courier service Pr 5.13, Ex 7.2.1
Selection of subcommittees 7.3.1
Sex ratio of mayors Ex 8.3.5
Soft drink production Ex 6.2.1
Student ages Ex 4.4.1
Student expenditures 4.5.3
Student grades Ex 4.3.4, 4.5, Ex 4.5.2, Ex 4.5.4, Pr 4.10, Pr 4.13, Pr 4.17, 5.1.1, Ex 5.2.2, Pr 5.14, Pr 5.21, Pr 5.24, Ex 7.3.12, 7.5.1, Ex 7.5.7, Pr 7.12, Pr 7.17, Pr 7.18, Pr 7.21
Student majors Ex 4.3.3, 5.4.2, Ex 5.4.2, Pr 9.10
Study methods 2.4, Ex 9.5.4
Surveys of students 1.1.1, Pr 2.6, Pr 2.7, Pr 2.10, 7.1, 7.4.4, Ex 7.5.1, Pr 7.8, Pr 7.13, Pr 7.22, Ex 8.3.4, Ex 8.6.2, Pr 8.2, Pr 8.9, 9.4.1, 9.5.2, 9.7
Taxis Pr 2.11, Pr 8.11
Tea party Ex 3.4.3
Textbooks Pr 4.2, Pr 4.3, 5.1.2, Ex 5.1.5, Ex 8.3.1, Ex 8.3.2, Ex 8.3.3
Tossing dice, coins, etc. 7.3.1, 7.3.5, Ex 7.3.1, Ex 7.3.2, Ex 7.3.4, Ex 7.3.5, Ex 7.3.11, Ex 7.3.13, 7.5.2, Pr 7.1, Pr 7.7, Pr 7.10, 9.1, Ex 9.3.2
University research Pr 9.3
Vacation durations Ex 4.4.6

Appendix 3

Table of Random Digits

Appendix 3 Table of Random Digits

40679	92739	78820	17624	57405	61685	72274	45006	62704	49562
00458	69502	30586	52329	86119	54034	51275	84533	26150	86493
09957	63971	01082	06778	49844	24754	87509	13323	65979	02551
04059	15065	84966	20217	16896	74790	55036	53920	25793	20348
57545	39067	94613	37135	98017	01343	52862	11723	18393	97680
59525	65540	90140	05043	81956	57334	90562	72426	01000	50618
07527	33894	77715	15472	33880	52787	78774	02826	28686	76803
86939	32219	81466	09689	79228	51856	39683	78838	28853	73182
60128	60074	53582	57211	92518	08262	84534	94482	35934	79675
09015	46687	71333	27713	44975	83249	14772	21109	85920	21147
75117	06922	76700	85190	03032	54355	11216	71813	70456	71849
65867	21718	81142	22144	69155	77558	59125	86078	17675	86773
57274	47340	24254	21681	09686	33645	14579	43735	81872	88816
04831	31929	68303	75993	39215	77789	35034	33638	04826	01641
53428	15602	16347	96974	14613	86020	56768	00004	80964	52314
23945	16541	89622	05577	69292	76955	16179	73865	36905	78180
39325	51808	88311	39644	64339	88546	33115	13305	82852	85732
96048	56722	68897	70746	09971	28920	38122	78875	72374	11085
66755	27756	89669	84235	07379	42030	33309	84123	60291	71361
59560	33688	79732	23336	71795	44844	03858	30867	32964	77076
41486	51573	22547	12043	86281	25568	44319	38585	10127	21384
34806	11253	12963	11724	00381	50337	64818	65631	67920	41550
58743	12096	40202	84943	34344	20864	09237	60422	68605	70895
51185	43425	76503	36391	26099	12430	89604	68138	50984	80232
09512	23893	92146	16533	18403	32969	51172	12214	66393	25300
48399	50272	38818	28674	55181	46015	18486	03638	32758	76234
24056	48321	40516	99634	71679	97300	31127	84873	94377	41641
99779	30237	84755	14526	14156	73482	46425	48155	35250	61377
05944	53888	49549	61345	39621	95006	65903	85806	65155	79350
22075	20747	00618	68979	59073	22616	19343	72277	47218	83028
71145	73759	74216	02184	53565	73664	56358	03526	66756	32411
35004	70500	20455	46760	35231	70161	01432	95568	22380	81562
81899	81155	23139	83101	51493	38529	60030	09909	97039	71136
08328	21041	42865	31002	01633	61297	68578	66592	04397	87716
49957	26677	24315	86572	51471	45252	67126	08113	60813	28463
29519	93319	82934	50797	19861	49837	49060	12170	76595	27717
35667	42444	89010	25527	66302	10954	32998	89944	00722	32137
00565	79266	56431	14079	36581	49171	62496	30589	33478	03225
84838	30206	35012	30813	44515	98246	71170	05100	71108	43414
39320	45187	68311	27874	53537	35836	54970	64257	35369	40486
71133	33788	86093	89816	51485	07132	69202	90681	29538	04300
90392	24080	71815	32715	57549	03911	92880	39561	27530	55153
18659	89405	80021	34226	47900	82898	26589	95365	61779	08855
83106	90949	00119	02092	75503	52379	16817	79622	11946	33800
90599	40286	24096	56833	42829	38748	75134	95358	12743	66057
30933	06300	66442	80548	31512	87977	95695	03852	45741	89293
88103	23289	39653	96441	46153	40905	69656	69822	31864	19311
40743	66749	21165	36736	53171	36740	74688	42021	37712	24542
46823	85160	23609	33365	40378	61254	71527	04071	33529	02519
84615	50131	64242	76079	73167	92150	46248	78872	99566	52651

Appendix 4

Probability Tables

A4.1 PROBABILITIES FOR THE NORMAL DISTRIBUTION

This table applies to the standard normal distribution, which is the normal distribution with average 0 and SD 1. It applies to any normal distribution for which the variable is expressed in standard units.

$A(z)$ is the area under the normal curve from $-\infty$ to z. $A(z)$ represents a probability. To express $A(z)$ as a percentage (a chance), multiply $A(z)$ by 100.

Note the consequence of the symmetry about $z = 0$. Namely $A(-z) = 1 - A(z)$.

Sec. A4.1 Probabilities for the Normal Distribution

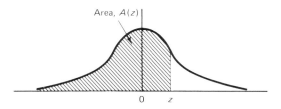

z	$A(z)$	z	$A(z)$	z	$A(z)$
−3.3	0.000				
−3.2	0.001				
−3.1	0.001				
−3.0	0.001				
−2.9	0.002	−0.9	0.184	1.0	0.841
−2.8	0.003	−0.8	0.212	1.1	0.864
−2.7	0.004	−0.7	0.242	1.2	0.885
−2.6	0.005	−0.6	0.274	1.3	0.903
−2.5	0.006	−0.5	0.309	1.4	0.919
−2.4	0.008	−0.4	0.345	1.5	0.933
−2.3	0.011	−0.3	0.382	1.6	0.945
−2.2	0.014	−0.2	0.421	1.7	0.955
−2.1	0.018	−0.1	0.460	1.8	0.964
−2.0	0.023	−0.0	0.500	1.9	0.971
−1.9	0.029	0.0	0.500	2.0	0.977
−1.8	0.036	0.1	0.540	2.1	0.982
−1.7	0.045	0.2	0.579	2.2	0.986
−1.6	0.055	0.3	0.618	2.3	0.989
−1.5	0.067	0.4	0.655	2.4	0.992
−1.4	0.081	0.5	0.691	2.5	0.994
−1.3	0.097	0.6	0.726	2.6	0.995
−1.2	0.115	0.7	0.758	2.7	0.996
−1.1	0.136	0.8	0.788	2.8	0.997
−1.0	0.159	0.9	0.815	2.9	0.998
				3.0	0.999
				3.1	0.999
				3.2	0.999
				3.3	1.000

A4.2 PROBABILITIES FOR THE BINOMIAL DISTRIBUTION

This table presents the probabilities for binomial models with $n = 1, 2, \ldots, 15$ and various values of p. x represents the number of the n outcomes that turn out in a certain way; this way must be the one that has probability p of occurring at each of the n trials.

TABLE A4.2 BINOMIAL PROBABILITIES[1]

n	x	0.05	0.1	0.2	0.3	0.4	0.5	0.6	0.7	0.8	0.9	0.95
2	0	0.902	0.810	0.640	0.490	0.360	0.250	0.160	0.090	0.040	0.010	0.002
	1	0.095	0.180	0.320	0.420	0.480	0.500	0.480	0.420	0.320	0.180	0.095
	2	0.002	0.010	0.040	0.090	0.160	0.250	0.360	0.490	0.640	0.810	0.902
3	0	0.857	0.729	0.512	0.343	0.216	0.125	0.064	0.027	0.008	0.001	
	1	0.135	0.243	0.384	0.441	0.432	0.375	0.288	0.189	0.096	0.027	0.007
	2	0.007	0.027	0.096	0.189	0.288	0.375	0.432	0.441	0.384	0.243	0.135
	3		0.001	0.008	0.027	0.064	0.125	0.216	0.343	0.512	0.729	0.857
4	0	0.815	0.656	0.410	0.240	0.130	0.062	0.026	0.008	0.002		
	1	0.171	0.292	0.410	0.412	0.346	0.250	0.154	0.076	0.026	0.004	
	2	0.014	0.049	0.154	0.265	0.346	0.375	0.346	0.265	0.154	0.049	0.014
	3		0.004	0.026	0.076	0.154	0.250	0.346	0.412	0.410	0.292	0.171
	4			0.002	0.008	0.026	0.062	0.130	0.240	0.410	0.656	0.815
5	0	0.774	0.590	0.328	0.168	0.078	0.031	0.010	0.002			
	1	0.204	0.328	0.410	0.360	0.259	0.156	0.077	0.028	0.006		
	2	0.021	0.073	0.205	0.309	0.346	0.312	0.230	0.132	0.051	0.008	0.001
	3	0.001	0.008	0.051	0.132	0.230	0.312	0.346	0.309	0.205	0.073	0.021
	4			0.006	0.028	0.077	0.156	0.259	0.360	0.410	0.328	0.204
	5				0.002	0.010	0.031	0.078	0.168	0.328	0.590	0.774
6	0	0.735	0.531	0.262	0.118	0.047	0.016	0.004	0.001			
	1	0.232	0.354	0.393	0.303	0.187	0.094	0.037	0.010	0.002		
	2	0.031	0.098	0.246	0.324	0.311	0.234	0.138	0.060	0.015	0.001	
	3	0.002	0.015	0.082	0.185	0.276	0.312	0.276	0.185	0.082	0.015	0.002
	4		0.001	0.015	0.060	0.138	0.234	0.311	0.324	0.246	0.098	0.031
	5			0.002	0.010	0.037	0.094	0.187	0.303	0.393	0.354	0.232
	6				0.001	0.004	0.016	0.047	0.118	0.262	0.531	0.735
7	0	0.698	0.478	0.210	0.082	0.028	0.008	0.002				
	1	0.257	0.372	0.367	0.247	0.131	0.055	0.017	0.004			
	2	0.041	0.124	0.275	0.318	0.261	0.164	0.077	0.025	0.004		
	3	0.004	0.023	0.115	0.227	0.290	0.273	0.194	0.097	0.029	0.003	
	4		0.003	0.029	0.097	0.194	0.273	0.290	0.227	0.115	0.023	0.004

Source: John E. Freund, *Statistics: A First Course*, 3rd ed., © 1981, pp. 418–421. Reprinted by permission of Prentice-Hall, Inc., Englewood Cliffs, N.J.

TABLE A4.2 BINOMIAL PROBABILITIES (*continued*)

n	x	0.05	0.1	0.2	0.3	0.4	0.5	0.6	0.7	0.8	0.9	0.95
	5			0.004	0.025	0.077	0.164	0.261	0.318	0.275	0.124	0.041
	6				0.004	0.017	0.055	0.131	0.247	0.367	0.372	0.257
	7					0.002	0.008	0.028	0.082	0.210	0.478	0.698
8	0	0.663	0.430	0.168	0.058	0.017	0.004	0.001				
	1	0.279	0.383	0.336	0.198	0.090	0.031	0.008	0.001			
	2	0.051	0.149	0.294	0.296	0.209	0.109	0.041	0.010	0.001		
	3	0.005	0.033	0.147	0.254	0.279	0.219	0.124	0.047	0.009		
	4		0.005	0.046	0.136	0.232	0.273	0.232	0.136	0.046	0.005	
	5			0.009	0.047	0.124	0.219	0.279	0.254	0.147	0.033	0.005
	6			0.001	0.010	0.041	0.109	0.209	0.296	0.294	0.149	0.051
	7				0.001	0.008	0.031	0.090	0.198	0.336	0.383	0.279
	8					0.001	0.004	0.017	0.058	0.168	0.430	0.663
9	0	0.630	0.387	0.134	0.040	0.010	0.002					
	1	0.299	0.387	0.302	0.156	0.060	0.018	0.004				
	2	0.063	0.172	0.302	0.267	0.161	0.070	0.021	0.004			
	3	0.008	0.045	0.176	0.267	0.251	0.164	0.074	0.021	0.003		
	4	0.001	0.007	0.066	0.172	0.251	0.246	0.167	0.074	0.017	0.001	
	5		0.001	0.017	0.074	0.167	0.246	0.251	0.172	0.066	0.007	0.001
	6			0.003	0.021	0.074	0.164	0.251	0.267	0.176	0.045	0.008
	7				0.004	0.021	0.070	0.161	0.267	0.302	0.172	0.063
	8					0.004	0.018	0.060	0.156	0.302	0.387	0.299
	9						0.002	0.010	0.040	0.134	0.387	0.630
10	0	0.599	0.349	0.107	0.028	0.006	0.001					
	1	0.315	0.387	0.268	0.121	0.040	0.010	0.002				
	2	0.075	0.194	0.302	0.233	0.121	0.044	0.011	0.001			
	3	0.010	0.057	0.201	0.267	0.215	0.117	0.042	0.009	0.001		
	4	0.001	0.011	0.088	0.200	0.251	0.205	0.111	0.037	0.006		
	5		0.001	0.026	0.103	0.201	0.246	0.201	0.103	0.026	0.001	
	6			0.006	0.037	0.111	0.205	0.251	0.200	0.088	0.011	0.001
	7			0.001	0.009	0.042	0.117	0.215	0.267	0.201	0.057	0.010
	8				0.001	0.011	0.044	0.121	0.233	0.302	0.194	0.075
	9					0.002	0.010	0.040	0.121	0.268	0.387	0.315
	10						0.001	0.006	0.028	0.107	0.349	0.599

TABLE A4.2 BINOMIAL PROBABILITIES (*continued*)

n	x	0.05	0.1	0.2	0.3	0.4	0.5	0.6	0.7	0.8	0.9	0.95
11	0	0.569	0.314	0.086	0.020	0.004						
	1	0.329	0.384	0.236	0.093	0.027	0.005	0.001				
	2	0.087	0.213	0.295	0.200	0.089	0.027	0.005	0.001			
	3	0.014	0.071	0.221	0.257	0.177	0.081	0.023	0.004			
	4	0.001	0.016	0.111	0.220	0.236	0.161	0.070	0.017	0.002		
	5		0.002	0.039	0.132	0.221	0.226	0.147	0.057	0.010		
	6			0.010	0.057	0.147	0.226	0.221	0.132	0.039	0.002	
	7			0.002	0.017	0.070	0.161	0.236	0.220	0.111	0.016	0.001
	8				0.004	0.023	0.081	0.177	0.257	0.221	0.071	0.014
	9				0.001	0.005	0.027	0.089	0.200	0.295	0.213	0.087
	10					0.001	0.005	0.027	0.093	0.236	0.384	0.329
	11							0.004	0.020	0.086	0.314	0.569
12	0	0.540	0.282	0.069	0.014	0.002						
	1	0.341	0.377	0.206	0.071	0.017	0.003					
	2	0.099	0.230	0.283	0.168	0.064	0.016	0.002				
	3	0.017	0.085	0.236	0.240	0.142	0.054	0.012	0.001			
	4	0.002	0.021	0.133	0.231	0.213	0.121	0.042	0.008	0.001		
	5		0.004	0.053	0.158	0.227	0.193	0.101	0.029	0.003		
	6			0.016	0.079	0.177	0.226	0.177	0.079	0.016		
	7			0.003	0.029	0.101	0.193	0.227	0.158	0.053	0.004	
	8			0.001	0.008	0.042	0.121	0.213	0.231	0.133	0.021	0.002
	9				0.001	0.012	0.054	0.142	0.240	0.236	0.085	0.017
	10					0.002	0.016	0.064	0.168	0.283	0.230	0.099
	11						0.003	0.017	0.071	0.206	0 377	0.341
	12							0.002	0.014	0.069	0.282	0.540
13	0	0.513	0.254	0.055	0.010	0.001						
	1	0.351	0.367	0.179	0.054	0.011	0.002					
	2	0.111	0.245	0.268	0.139	0.045	0.010	0.001				
	3	0.021	0.100	0.246	0.218	0.111	0.035	0.006	0.001			
	4	0.003	0.028	0.154	0.234	0.184	0.087	0.024	0.003			
	5		0.006	0.069	0.180	0.221	0.157	0.066	0.014	0.001		
	6		0.001	0.023	0.103	0.197	0.209	0.131	0.044	0.006		
	7			0.006	0.044	0.131	0.209	0.197	0.103	0.023	0.001	

Sec. A4.2 Probabilities for the Binomial Distribution

TABLE A4.2 BINOMIAL PROBABILITIES (*continued*)

n	x	0.05	0.1	0.2	0.3	0.4	0.5	0.6	0.7	0.8	0.9	0.95
	8				0.001	0.014	0.066	0.157	0.221	0.180	0.069	0.006	
	9					0.003	0.024	0.087	0.184	0.234	0.154	0.028	0.003
	10					0.001	0.006	0.035	0.111	0.218	0.246	0.100	0.021
	11						0.001	0.010	0.045	0.139	0.268	0.245	0.111
	12							0.002	0.011	0.054	0.179	0.367	0.351
	13								0.001	0.010	0.055	0.254	0.513
14	0	0.488	0.229	0.044	0.007	0.001						
	1	0.359	0.356	0.154	0.041	0.007	0.001					
	2	0.123	0.257	0.250	0.113	0.032	0.006	0.001				
	3	0.026	0.114	0.250	0.194	0.085	0.022	0.003				
	4	0.004	0.035	0.172	0.229	0.155	0.061	0.014	0.001			
	5		0.008	0.086	0.196	0.207	0.122	0.041	0.007			
	6		0.001	0.032	0.126	0.207	0.183	0.092	0.023	0.002		
	7			0.009	0.062	0.157	0.209	0.157	0.062	0.009		
	8			0.002	0.023	0.092	0.183	0.207	0.126	0.032	0.001	
	9				0.007	0.041	0.122	0.207	0.196	0.086	0.008	
	10				0.001	0.014	0.061	0.155	0.229	0.172	0.035	0.004
	11					0.003	0.022	0.085	0.194	0.250	0.114	0.026
	12					0.001	0.006	0.032	0.113	0.250	0.257	0.123
	13						0.001	0.007	0.041	0.154	0.356	0.359
	14							0.001	0.007	0.044	0.229	0.488
15	0	0.463	0.206	0.035	0.005							
	1	0.366	0.343	0.132	0.031	0.005						
	2	0.135	0.267	0.231	0.092	0.022	0.003					
	3	0.031	0.129	0.250	0.170	0.063	0.014	0.002				
	4	0.005	0.043	0.188	0.219	0.127	0.042	0.007	0.001			
	5	0.001	0.010	0.103	0.206	0.186	0.092	0.024	0.003			
	6		0.002	0.043	0.147	0.207	0.153	0.061	0.012	0.001		
	7			0.014	0.081	0.177	0.196	0.118	0.035	0.003		
	8			0.003	0.035	0.118	0.196	0.177	0.081	0.014		
	9			0.001	0.012	0.061	0.153	0.207	0.147	0.043	0.002	
	10				0.003	0.024	0.092	0.186	0.206	0.103	0.010	0.001
	11				0.001	0.007	0.042	0.127	0.219	0.188	0.043	0.005
	12					0.002	0.014	0.063	0.170	0.250	0.129	0.031
	13						0.003	0.022	0.092	0.231	0.267	0.135
	14							0.005	0.031	0.132	0.343	0.366
	15								0.005	0.035	0.206	0.463

A4.3 PROBABILITIES FOR THE t DISTRIBUTION

The following table provides probabilities for the upper tail of the t distribution, for certain values of the tail cutoff point, t, and for various degrees of freedom, df. Approximate P-values associated with a particular value of a t statistic can be inferred.

The numbers in the column heading are the probabilities. The columns contain the cutoff values.

TABLE A4.3 PROBABILITIES FOR THE t DISTRIBUTION[2]

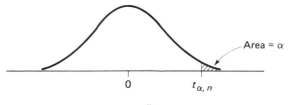

df	0.20	0.15	0.10	0.05	0.025	0.01	0.005
1	1.376	1.963	3.078	6.3138	12.706	31.821	63.657
2	1.061	1.386	1.886	2.9200	4.3027	6.965	9.9248
3	0.978	1.250	1.638	2.3534	3.1825	4.541	5.8409
4	0.941	1.190	1.533	2.1318	2.7764	3.747	4.6041
5	0.920	1.156	1.476	2.0150	2.5706	3.365	4.0321
6	0.906	1.134	1.440	1.9432	2.4469	3.143	3.7074
7	0.896	1.119	1.415	1.8946	2.3646	2.998	3.4995
8	0.889	1.108	1.397	1.8595	2.3060	2.896	3.3554
9	0.883	1.100	1.383	1.8331	2.2622	2.821	3.2498
10	0.879	1.093	1.372	1.8125	2.2281	2.764	3.1693
11	0.876	1.088	1.363	1.7959	2.2010	2.718	3.1058
12	0.873	1.083	1.356	1.7823	2.1788	2.681	3.0545
13	0.870	1.079	1.350	1.7709	2.1604	2.650	3.0123
14	0.868	1.076	1.345	1.7613	2.1448	2.624	2.9768
15	0.866	1.074	1.341	1.7530	2.1315	2.602	2.9467
16	0.865	1.071	1.337	1.7459	2.1199	2.583	2.9208
17	0.863	1.069	1.333	1.7396	2.1098	2.567	2.8982
18	0.862	1.067	1.330	1.7341	2.1009	2.552	2.8784
19	0.861	1.066	1.328	1.7291	2.0930	2.539	2.8609
20	0.860	1.064	1.325	1.7247	2.0860	2.528	2.8453
21	0.859	1.063	1.323	1.7207	2.0796	2.518	2.8314
22	0.858	1.061	1.321	1.7171	2.0739	2.508	2.8188
23	0.858	1.060	1.319	1.7139	2.0687	2.500	2.8073
24	0.857	1.059	1.318	1.7109	2.0639	2.492	2.7969
25	0.856	1.058	1.316	1.7081	2.0595	2.485	2.7874
26	0.856	1.058	1.315	1.7056	2.0555	2.479	2.7787
27	0.855	1.057	1.314	1.7033	2.0518	2.473	2.7707
28	0.855	1.056	1.313	1.7011	2.0484	2.467	2.7633
29	0.854	1.055	1.311	1.6991	2.0452	2.462	2.7564
30	0.854	1.055	1.310	1.6973	2.0423	2.457	2.7500
31	0.8535	1.0541	1.3095	1.6955	2.0395	2.453	2.7441
32	0.8531	1.0536	1.3086	1.6939	2.0370	2.449	2.7385
33	0.8527	1.0531	1.3078	1.6924	2.0345	2.445	2.7333
34	0.8524	1.0526	1.3070	1.6909	2.0323	2.441	2.7284
∞	0.84	1.04	1.28	1.64	1.96	2.33	2.58

TABLE A4.4 PROBABILITIES FOR THE CHI-SQUARE DISTRIBUTION[3]

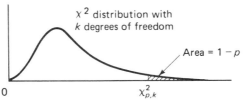

df	0.010	0.025	0.050	0.10	0.90	0.95	0.975	0.99
1	0.000157	0.000982	0.00393	0.0158	2.706	3.841	5.024	6.635
2	0.0201	0.0506	0.103	0.211	4.605	5.991	7.378	9.210
3	0.115	0.216	0.352	0.584	6.251	7.815	9.348	11.345
4	0.297	0.484	0.711	1.064	7.779	9.488	11.143	13.277
5	0.554	0.831	1.145	1.610	9.236	11.070	12.832	15.086
6	0.872	1.237	1.635	2.204	10.645	12.592	14.449	16.812
7	1.239	1.690	2.167	2.833	12.017	14.067	16.013	18.475
8	1.646	2.180	2.733	3.490	13.362	15.507	17.535	20.090
9	2.088	2.700	3.325	4.168	14.684	16.919	19.023	21.666
10	2.558	3.247	3.940	4.865	15.987	18.307	20.483	23.209
11	3.053	3.816	4.575	5.578	17.275	19.675	21.920	24.725
12	3.571	4.404	5.226	6.304	18.549	21.026	23.336	26.217
13	4.107	5.009	5.892	7.042	19.812	22.362	24.736	27.688
14	4.660	5.629	6.571	7.790	21.064	23.685	26.119	29.141
15	5.229	6.262	7.261	8.547	22.307	24.996	27.488	30.578
16	5.812	6.908	7.962	9.312	23.542	26.296	28.845	32.000
17	6.408	7.564	8.672	10.085	24.769	27.587	30.191	33.409
18	7.015	8.231	9.390	10.865	25.989	28.869	31.526	34.805
19	7.633	8.907	10.117	11.651	27.204	30.144	32.852	36.191
20	8.260	9.591	10.851	12.443	28.412	31.410	34.170	37.566
21	8.897	10.283	11.591	13.240	29.615	32.671	35.479	38.932
22	9.542	10.982	12.338	14.041	30.813	33.924	36.781	40.289
23	10.196	11.688	13.091	14.848	32.007	35.172	38.076	41.638
24	10.856	12.401	13.848	15.659	33.196	36.415	39.364	42.980
25	11.524	13.120	14.611	16.473	34.382	37.652	40.646	44.314
26	12.198	13.844	15.379	17.292	35.563	38.885	41.923	45.642
27	12.879	14.573	16.151	18.114	36.741	40.113	43.194	46.963
28	13.565	15.308	16.928	18.939	37.916	41.337	44.461	48.278
29	14.256	16.047	17.708	19.768	39.087	42.557	45.722	49.588
30	14.953	16.791	18.493	20.599	40.256	43.773	46.979	50.892
31	15.655	17.539	19.281	21.434	41.422	44.985	48.232	52.191
32	16.362	18.291	20.072	22.271	42.585	46.194	49.480	53.486
33	17.073	19.047	20.867	23.110	43.745	47.400	50.725	54.776
34	17.789	19.806	21.664	23.952	44.903	48.602	51.966	56.061

TABLE A4.4 PROBABILITIES FOR THE CHI-SQUARE DISTRIBUTION (*continued*)

df	0.010	0.025	0.050	0.10	0.90	0.95	0.975	0.99
35	18.509	20.569	22.465	24.797	46.059	49.802	53.203	57.342
36	19.233	21.336	23.269	25.643	47.212	50.998	54.437	58.619
37	19.960	22.106	24.075	26.492	48.363	52.192	55.668	59.892
38	20.691	22.878	24.884	27.343	49.513	53.384	56.895	61.162
39	21.426	23.654	25.695	28.196	50.660	54.572	58.120	62.428
40	22.164	24.433	26.509	29.051	51.805	55.758	59.342	63.691
41	22.906	25.215	27.326	29.907	52.949	56.942	60.561	64.950
42	23.650	25.999	28.144	30.765	54.090	58.124	61.777	66.206
43	24.398	26.785	28.965	31.625	55.230	59.304	62.990	67.459
44	25.148	27.575	29.787	32.487	56.369	60.481	64.201	68.709
45	25.901	28.366	30.612	33.350	57.505	61.656	65.410	69.957
46	26.657	29.160	31.439	34.215	58.641	62.830	66.617	71.201
47	27.416	29.956	32.268	35.081	59.774	64.001	67.821	72.443
48	28.177	30.755	33.098	35.949	60.907	65.171	69.023	73.683
49	28.941	31.555	33.930	36.818	62.038	66.339	70.222	74.919
50	29.707	32.357	34.764	37.689	63.167	67.505	71.420	76.154

A4.5 PROBABILITIES FOR F DISTRIBUTIONS

The F distributions are labelled by the number of degrees of freedom in the numerator and denominator of the associated F statistic. We present tail probabilities for F distributions for the denominator degrees of freedom ranging from 1 to 60, for each numerator degrees of freedom in the range 1 to 3.

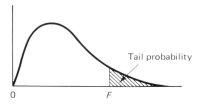

Tail probability	0.25	0.10	0.05	0.025	0.01	0.001
Numerator $df = 1$ Denominator df						
1	5.8	40	160	650	4100	410000
2	2.6	8.5	19	39	99	1000
3	2.0	5.5	10	17	34	167
4	1.8	4.5	7.7	12	21	74
5	1.7	4.1	6.6	10	16	47
6	1.6	3.8	6.0	8.8	14	36
7	1.6	3.6	5.6	8.1	12	29
8	1.5	3.5	5.3	7.6	11	25
9	1.5	3.4	5.1	7.2	11	23
10	1.5	3.3	5.0	6.9	10	21
15	1.4	3.1	4.5	6.2	8.7	17
20	1.4	3.0	4.4	5.9	8.1	15
30	1.4	2.9	4.2	5.6	7.7	13
40	1.4	2.8	4.1	5.4	7.3	13
50	1.4	2.8	4.0	5.4	7.2	12
60	1.3	2.8	4.0	5.3	7.1	12

Sec. A4.5 Probabilities for the F Distributions

Tail probability	0.25	0.10	0.05	0.025	0.01	0.001
Numerator $df = 2$ Denominator df						
1	7.5	50	200	800	5000	500000
2	3.0	9.0	19	39	99	1000
3	2.3	5.5	9.6	16	31	149
4	2.0	4.3	6.9	11	18	61
5	1.9	3.8	5.8	8.4	13	37
6	1.8	3.5	5.1	7.3	11	27
7	1.7	3.3	4.7	6.5	9.6	22
8	1.7	3.1	4.5	6.1	8.7	19
9	1.6	3.0	4.3	5.7	8.0	16
10	1.6	2.9	4.1	5.5	7.6	15
15	1.5	2.7	3.7	4.8	6.4	11
20	1.5	2.6	3.5	4.5	5.9	10
30	1.5	2.5	3.3	4.2	5.4	8.8
40	1.4	2.4	3.2	4.1	5.2	8.3
50	1.4	2.4	3.2	4.0	5.1	8.0
60	1.4	2.4	3.1	3.9	5.0	7.8

Tail probability	0.25	0.10	0.05	0.025	0.01	0.001
Numerator $df = 3$ Denominator df						
1	8.2	54	220	860	5400	540000
2	3.2	9.2	19	39	99	1000
3	2.4	5.4	9.3	15	30	141
4	2.1	4.2	6.6	10	17	56
5	1.9	3.6	5.4	7.8	12	33
6	1.8	3.3	4.8	6.6	9.8	24
7	1.7	3.1	4.4	5.9	8.5	19
8	1.7	2.9	4.1	5.4	7.6	16
9	1.6	2.8	3.9	5.1	7.0	14
10	1.6	2.7	3.7	4.8	6.6	13
15	1.5	2.5	3.3	4.2	5.4	9.3
20	1.5	2.4	3.1	3.9	4.9	8.1
30	1.4	2.3	2.9	3.6	4.5	7.1
40	1.4	2.2	2.8	3.5	4.3	6.6
50	1.4	2.2	2.8	3.4	4.2	6.3
60	1.4	2.2	2.8	3.3	4.1	6.2

Sources and Notes

CHAPTER 1

1. British Columbia Post-Secondary Enrollment Forecasting Committee. *Grade XII Student Survey*. (B.C.F.C. Vancouver, B.C., June, 1981) Questionnaire portions reproduced by permission of B.C.F.C.
2. Western Canada Lottery Foundation. *Lotto West Media Fact Book*. (Winnipeg, Manitoba, June 1984.) This foundation administers both Lotto 6/49 and Lotto West. Lotto 6/49 is a lottery set up by the federal government but administered in western Canada by the Western Canada Lottery Foundation.
3. Clements, Warren. "Nestlings." Cartoon which appeared in the *Toronto Globe and Mail*, 19 September 1984. Reproduced by permission of Warren Clements.
4. Gallup, George H. *The Sophisticated Poll-watchers Guide*. Rev. ed. (Princeton Opinion Press, 1976)
5. Pauling, Linus. *Vitamin C and the Common Cold*. (San Francisco: W. H. Freeman and Company, 1970)
6. Catalogue 84-204. (Ottawa: Statistics Canada)
7. Catalogue 84-001. (Ottawa: Statistics Canada)
8. Ware, J. E., and R. G. Williams. "The Dr. Fox Effect: A Study of Lecturer Effectiveness and Ratings of Instruction." *Journal of Medical Education* 50:149–159. 1975.
9. Kennel, W. B., and Tavia Gordon. "The Framingham Study. An Investigation of Cardiovascular Disease. Section 30: Some Characteristics Related to the Incidence of Cardio-

vascular Disease and Death: Framingham Study, 18-year Follow-Up." (Washington, D.C.: U.S. Government Printing Office, 1974)

10. Galbraith, J. K. *The Age of Uncertainty*. (Boston: Houghton Mifflin Company, 1977)

CHAPTER 2

1. Huxley, T. H. *Collected Essays,* Vol. VIII (London: Macmillan and Company). Volumes I–IX were published during 1894–1908.
2. "Hold the Eggs and Butter." *Time* 26 March 1984, pp. 66–73. Copyright 1984 Time Inc. All rights reserved. Reprinted by permission from *Time*.
3. Gardner, Martin. *aHa! Gotcha: Paradoxes to Puzzle and Delight*. (San Francisco: W. H. Freeman and Company, 1982), p. 26. Copyright © 1982 by W. H. Freeman and Company. All rights reserved. Reproduced by permission.

CHAPTER 4

1. Moore, David S. *Statistics: Concepts and Controversies*. (San Francisco: W. H. Freeman and Company, 1979) Adapted by permission.
2. Deming, W. Edwards. *"Making Things Right."* In *Statistics: A Guide to the Unknown,* 2nd ed. by J. M. Tanur et al. ed. (San Francisco: Holden-Day, 1978) pp. 279–288. Adapted by permission.
3. Huff, Darrell. *How to Lie with Statistics*. (New York: W. W. Norton and Company Inc., 1954) Reproduced by permission.
4. Altman, Philip L., and Dorothy S. Dittmer. *Biology Data Book,* 2nd ed. (Bethesda, Md.: Federation of American Societies for Experimental Biology, 1972). Data reproduced by permission.
5. Thompson, D'Arcy. *On Growth and Form,* Vol II. (New York: Oxford University Press, 1963)
6. Tryon, R. C. *Genetic Differences in Maze-learning Ability in Rats, 39th yearbook, Nat. Soc. Stud. Educ., Part I (1940),* 111–119, as quoted in David Freedman, Robert Pisani, and Roger Purves. *Statistics*. (New York: W. W. Norton & Company, Inc., 1978) Adapted by permission.
7. Gardner, Martin. *aHa! Gotcha: Paradoxes to Puzzle and Delight*. (San Francisco: W. H. Freeman and Company, 1982), p. 114. Copyright © 1982 by W. H. Freeman and Company. All rights reserved. Reproduced by permission.
8. Lindgren, B. W. *Amer. Stat.* 37: 254. 1983.
9. *Toronto Globe and Mail* 17 January 1984.
10. *Toronto Globe and Mail* 26 July 1984, p. B15. The original data has been expressed in terms of U.S. dollars instead of the Canadian dollar equivalents that appeared in the original source. Reproduced by permission.

CHAPTER 5

1. Carroll, Lewis. *Alice's Adventures in Wonderland and Through the Looking Glass.* (New York: Macmillan Company, Fourth Printing, 1966)
2. Mosteller, F., S. E. Fienberg, and R. E. K. Rourke. *Beginning Statistics with Data Analysis,* © 1983 (Reading, Mass.: Addison-Wesley Publishing Company), p. 79, Table 3-3. Reprinted by permission.
3. Daniel, W. W. *Introductory Statistics with Applications.* (Boston: Houghton Mifflin Company, 1977) Data used by permission.
4. Altman, Philip L., and Dorothy S. Dittmer. *Biology Data Book,* 2nd ed. (Bethesda, Md.: Federation of American Societies for Experimental Biology, 1972) Data used by permission.
5. Pearson, E. S., and A. Lee. "On the laws of inheritance in man. I: Inheritance of physical factors." *Biometrika* 2: 357–425. 1903. London. The data are derived from Table XXII and are slightly modified.
6. Matthews, A. D. "Mercury Content of Commercially Important Fish of the Seychelles, and Hair Mercury Levels of a Selected Part of the Population." *Environmental Research* 30: 305–312. 1983. Data used by permission.
7. Chernoff, H. "The Use of Faces to Represent Points in k-dimensional Space Graphically." *J. Amer. Statist. Assn.* 68: 361–368. 1973.
8. Andrews, D. F. "Plots of high-dimensional data." *Biometrics* 28: 125–136. 1972.
9. Fleiss, J. L. *Statistical Methods for Rates and Proportions.* (New York: John Wiley, 1973)
10. Fienberg, Stephen E. *The Analysis of Cross-Classified Categorical Data,* 2nd ed. (Cambridge, Mass: The MIT Press, 1980)
11. Gardner, Martin. *aHa! Gotcha: Paradoxes to Puzzle and Delight.* (San Francisco: W. H. Freeman and Company, 1982), p. 130. Copyright © 1982 by W. H. Freeman and Company. All rights reserved. Reproduced by permission.
12. Tien, T. Y. "China: Demographic Billionaire." *Population Bulletin* vol. 38, no. 2, p. 18. 1983.
13. Ryan, T. A., B. L. Joiner, and B. F. Ryan. *MINITAB Student Handbook.* (Boston: Duxbury Press, 1976) Data used by permission.
14. Freedman, David, Robert Pisani, and Roger Purves. *Statistics.* (New York: W. W. Norton & Company, Inc., 1978) Reproduced by permission.
15. Mendenhall, William. *Introduction to Probability and Statistics,* 6th ed. (Boston: Duxbury Press, 1983) Used by permission.

CHAPTER 6

1. Catalogue 63-007. (Ottawa: Statistics Canada)
2. Catalogue 63-007. (Ottawa: Statistics Canada)
3. Catalogue 63-202. (Ottawa: Statistics Canada)
4. Catalogue 82-541. (Ottawa: Statistics Canada)
5. Catalogue 84-204. (Ottawa: Statistics Canada)

6. Catalogue 32-001. (Ottawa: Statistics Canada)
7. Catalogue 67-001. (Ottawa: Statistics Canada)
8. *The World Almanac & Book of Facts,* 1984 edition, copyright © Newspaper Enterprise Association, Inc., 1983, New York, N. Y. 10166.
9. Catalogue 82-541. (Ottawa: Statistics Canada)
10. Catalogue 82-541. (Ottawa: Statistics Canada)
11. Raddatz, R. L., R. R. Tortorelli, and M. J. Newark. "Manitoba and Saskatchewan Tornado Days, 1960 to 1982." *Environment Canada.*

CHAPTER 7

1. Western Canada Lottery Foundation. *Lotto West Media Fact Book. June 1984.* Winnipeg, Manitoba. (This foundation administers both Lotto 6/49 and Lotto West. Lotto 6/49 is a lottery set up by the federal government but administered in western Canada by the Western Canada Lottery Foundation.) Reproduced by permission.
2. Galton, Francis. *Hereditary Genius.* (London: Macmillan and Co., 1869)

CHAPTER 8

1. *Biometrika Tables for Statisticians, Vol. I.* (New York: Cambridge University Press for Biometrika Trustees: E. S. Pearson and H. O. Hartley, 1962)
2. *The World Almanac & Book of Facts,* 1984 edition, copyright © Newspaper Enterprise Association, Inc., 1983, New York, N. Y. 10166.
3. *The World Almanac & Book of Facts,* 1984 edition, copyright © Newspaper Enterprise Association, Inc., 1983, New York, N. Y. 10166.
4. *The World Almanac & Book of Facts,* 1984 edition, copyright © Newspaper Enterprise Association, Inc., 1983, New York, N. Y. 10166.

CHAPTER 9

1. Gardner, Martin, *aHa! Gotcha: Paradoxes to Puzzle and Delight.* (San Francisco: W. H. Freeman and Company, 1982), p. 117. Copyright © 1982 by W. H. Freeman and Company. All rights reserved. Reproduced by permission.
2. Purdy, D. A., D. S. Eitzen, and R. Hufnagel. "Are Athletes also Students? The Educational Attainment of College Athletes." *Social Problems* 29: 441–446. 1982.
3. Burger, J. "Jet Aircraft Noise and Bird Strikes: Why More Birds Are Being Hit." *Environmental Pollution (Series A)* 30: 143–152. 1983.
4. Wolfe, D. E. "Effects of Music Loudness on Task Performance and Self-report of College-aged Students." *Journal of Research in Music Education* 31: 191–199. 1983.
5. Bowker, Lee K. "Racism and Sexism: Hints Toward a Theory of the Causal Structure of Attitudes towards Women." *International Journal of Women's Studies.* 4: 277–288. 1981.

6. *Marketing* 20 February 1984, p. 2. (Toronto: MacLean-Hunter Publishing Co.)
7. Gardner, Martin. *aHa! Gotcha: Paradoxes to Puzzle and Delight.* (San Francisco: W. H. Freeman and Company, 1982), p. 133. Copyright © 1982 by W. H. Freeman and Company. All rights reserved. Reproduced by permission.

APPENDIX 4

1. Freund, J. E. *Statistics: A First Course,* 3rd ed., © 1981, pp. 418–421. Reprinted by permission of Prentice-Hall, Inc., Englewood Cliffs, N.J.
2. Pearson, E. S., and H. O. Hartley, Biometrika Trustees. *Biometrika Tables for Statisticians, Vol. I.* (New York: Cambridge University Press, 1962)
3. Pearson, E. S., and H. O. Hartley, Biometrika Trustees. *Biometrika Tables for Statisticians, Vol. I.* (New York: Cambridge University Press, 1962)

Annotated Bibliography

1. Fienberg, Stephen E. *The Analysis of Cross-Classified Categorical Data,* 2nd ed. (Cambridge, Mass.: The MIT Press, 1980) Readable and authoritative.
2. Fleiss, J. L. *Statistical Methods for Rates and Proportions.* (New York: John Wiley, 1973) A good first source book for this subject.
3. Freedman, David, Robert Pisani, and Roger Purves. *Statistics.* (New York: W. W. Norton & Company, Inc., 1978) This text provides an excellent introduction to sampling models and avoids algebraic notation in its descriptions of statistical procedures.
4. Gallup, George H. *The Sophisticated Poll-watchers Guide.* (Princeton Opinion Press, 1976) This book includes a comparison of Gallup Poll predictions with actual outcomes concerning the American elections since 1936. In addition, many practical problems of opinion polling are discussed.
5. Gardner, Martin. *aHa! Gotcha: Paradoxes to Puzzle and Delight.* (San Francisco: W. H. Freeman and Company, 1982) The medium is very entertaining, but the messages are often profound. Don't miss this book!
6. Huff, Darrell. *How to Lie with Statistics.* (New York: W. W. Norton & Company Inc., 1954) A classic and entertaining little book on abuses of statistics.
7. Huff, Darrell, and Irving Geis. *How to Take a Chance.* (New York: W. W. Norton & Company Inc., 1959) A popular introduction to some probability problems and games.
8. Moore, David S. *Statistics: Concepts and Controversies.* (San Francisco: W. H. Freeman and Company, 1979) An excellent source of examples and problems. The exposition is at an elementary level.

9. Mosteller, F., S. E. Fienberg, and R. E. K. Rourke. *Beginning Statistics with Data Analysis.* (Reading, Mass.: Addison-Wesley Publishing Company, 1983) Emphasis is on exploring data sets.
10. Ryan, T. A., B. L. Joiner, and B. F. Ryan. *MINITAB Student Handbook.* (Boston: Duxbury Press, 1976) A must for the novice in statistical computing, provided MINITAB is available to them. Explains calculations briefly, and shows how MINITAB can be asked to do the calculations.
11. Tanur, J. M., et al., ed. *Statistics: A Guide to the Unknown,* 2nd ed. (San Francisco: Holden-Day, 1978) Many interesting examples described in a way that does not assume previous familiarity with statistics.
12. *The World Almanac and Book of Facts, 1984.* (Newspaper Enterprise Association, Inc., 1983) Full of lists for sampling. Annual updates provide good time-series data.

Solutions to Selected Problems

Important Note to the Reader

The purpose of the problems in this text is to provide practice in problem solving in statistical contexts. It is not to teach the solutions to the particular problems. The solutions themselves have little information of value to you in applying statistics to new problems (either in real life or on an examination). The only reason for this section of solutions to selected problems is to give you some idea of how well you are doing at preparing yourself for problem solving in statistical contexts. Consequently, for problems that have a numerical answer, the "solution" often consists of the final answer and a hint at the method. The "explain" or "discuss" problems include fuller details to illustrate the intended standard of explanations and discussions. In particular, it will be seen that many problems have more than one correct solution, depending on the point of view taken.

To get the most out of the problems, give them a serious try before peeking at the solution.

Chapter 1

1. **(a)** The first task is to define as carefully as possible which types of game environments you wish to include in your study. For example, you may wish to restrict your attention to those games that are accessible to the general public, age 12 or over, and for which the cost of playing the game is the only access charge. (This would eliminate certain private clubs, bars, and so on.)

The next step would be to devise a way of selecting a group of 10 locations that in some sense is "typical" or "representative" of all the locations that satisfy your definition. Unless the city (i.e., the municipal government) has an information system that could provide a list of all locations satisfying your definition, and will make it available to you, this information may be hard to obtain. Another approach might be to contact companies that distribute or lease out the games. If a few such companies account for most of the games in the city, their lists might provide information on locations. (Better still, their accounts might give the desired information about game popularity. But we will ignore this possibility for now.) If these ideas fail, we may be forced to restrict our attention to the arcades in a certain area of the city, and admit that the results may apply only to this particular market.

These few comments should indicate that the selection problem is both important and difficult. Watch for the selection method called "random sampling" introduced in Chapter 2. **(b)** If the use of the games is seasonal (more young users during the summer months and at Christmas), the date of visit could be an important factor. Age of player may be a factor in the game chosen. Such influences might complicate the gathering of data and the reporting of results. Ideally, the survey should be done during a short interval, such as a week or two, in which the user population is fairly stable. In this case the order would not matter from the point of view of seasonal influences, but of course if several locations were to be visited by one observer on the same day, the time-of-day factor may be important. Our aim should be to avoid ordering the visits according to any particular suspected factor. (The method of randomization could be used here, and is discussed in Chapter 2.) **(c)** To speed the observation process, the efficient coding of information would have to be planned ahead of time. A partial list of the games would be very useful, enabling a simple tabulation of observations. A portable microcomputer would help. One method that would probably not be efficient would be to note down everything that seems relevant in a sort of essay type of record. This kind of information is very hard to summarize. **6.** The Sure Thing lottery gives the holder of each ticket a 1-in-1-million share of the total winnings of \$550,000 (\$100,000 + 9 × \$50,000), which may be deemed to be "worth" a fraction 550,000/1,000,000 of the \$1 cost of the ticket. In other words, the ticket is worth 55 cents. Similarly, a ticket in the Charity Special lottery is "worth" 50 cents. This analysis would suggest that the Sure Thing lottery is slightly better.

However, one could argue that 10 chances in 1 million of winning is inferior to 500 chances in 1 million, and that the amount is not important. The validity of this argument depends on the actual personal value of winning \$1000 in comparison to \$50,000 or \$100,000. If the prize of \$100,000 is considered to be not quite 100 times better than a prize of \$1000, the choice of the Charity Special may be a rational choice.

Of course, if the purchase of a lottery ticket is considered a charitable donation, one should simply pick the lottery for which one has the most sympathy. (Or, perhaps, make a direct donation.) **7.** Some means of guaranteeing confidentiality of responses would help. For example, identification information should not be part of the information requested. The purpose of the study should be clearly explained to the students. If the study seems worthwhile to the students, prank or blank answers are less likely. Proper participation in the study is sometimes encouraged by announcing that the results of the study will be made available to the respondents. Adequate class time must be allowed for the questionnaire to be completed.

Chapter 2

1. No. It is possible that the "random sample" of 100 CEOs includes the CEOs with the 100 highest incomes, which would be unrepresentative in the extreme. Random sampling does not guarantee representativeness, but it tends to produce representativeness. Each possible sample has the same chance of being produced by random sampling, so unusual samples are usually not produced. **6.** A random sample of students should be selected from the registrar's records. If possible, additional information about the students selected should be extracted from the student records such as:

1. The present housing status of the student (i.e., residence, apartment, at home with parents, own home, . . .).
2. Usual method of transportation to off-campus destinations (bus, own car, hitch-hike, . . .).
3. Whatever other information might be available from the student record (sex, age, . . .).

The students sampled could then be contacted by phone to get their opinions. (A mailed questionnaire would not be much cheaper and the nonresponse rate would likely be higher.) The agreement to use the student's file for the purpose of the survey could also be obtained during the telephone call.

The information obtained from the student records would help in ensuring that the sample was representative of the student body, and would help to report which students tended to favor one or other option (i.e., improved transportation or more residences).   **9.** To select a random sample, it is usually necessary to have a list or some such representation of the membership of the population. In Problem 1 of Chapter 1, it is likely that no such list would be available, so that random sampling would be impossible until such a list were compiled.   **13.** The students taking Statistics 101 may be very different from the students taking Statistics 102, so the subsequent success in Statistics 202 is confounded with this difference. Another problem is that there may be different time lags between 101 and 202 and 102 and 202. The basic problem is that the study described is an observational study, not an experiment.

Chapter 3

1. Let the three blends be denoted A, B, and C. One efficient design is described schematically as

$$ABC$$
$$ACB$$
$$BAC$$
$$BCA$$
$$CAB$$
$$CBA$$

which means that subject 1 is presented with a taste of A first, then B, then C, and subject 2 with A first, then C, then B, and so on. Of course, the subject would have to be given no clues as to which was which, and all brews must be equally fresh when tasted. It may be helpful if the server does not know which brand they are serving lest a bias inadvertently affect the taster. The purpose of the design is to balance the effect of order of presentation over the three brands, as this order may possibly influence the comparison. (Each subject would be asked to rank the brands in order of merit.)

The outline of the design task suggests that 10 subjects is the maximum allowed. Usually, there is some cost involved for each subject (at least for the coffee), so there may be an advantage to keeping the number of subjects down to six. In the situation where the use of the extra subjects (up to 10) does not involve much extra cost, there are several schemes that might be used to utilize these. One might be to use the four to compare the two winning brands from the experiment based on the six, that is,

$$AB$$
$$BA$$
$$AB$$
$$BA$$

Another idea, although not quite fair in the context of this problem, is to invite two friends over for coffee and do two replicates of the first design (6 + 6).   **3.** Even if, as in Exercise 1 of Section 3.4, the mileage is determined without error for a particular vehicle, there may well be variation in the mileage achieved for different vehicles of the same model. If the models are to be compared, rather than the vehicles themselves, we must include enough replicates of the model comparison to allow the model difference in consumption to show through. Numerically, it is the average of the model differences, over the replicate pairs, that will indicate the existence of any systematic model difference. (Note the importance of the selection process for choosing the model representatives—we sidestep this issue here.)   **6. (a)** No. Clusters and strata are different in that clusters are ideally just as heterogeneous as the population, while strata are ideally more homogeneous. Moreover, clusters are selected at random from a concrete population of interest, whereas blocks are usually selected from the population of interest in a less formal way, since it is the comparison within blocks that is of most importance. **(b)** More like a stratum, since the aim in blocking is to provide comparisons between treatments of similar experimental units, and the aim in stratification is to choose strata that are homogeneous within. **(c)** Yes, although this is not as crucial to interpretation of the observations as it is with sampling surveys. For example, if we choose a runner as a "block," and observe his performance under various conditions, we would like to think that the runner is typical of runners generally. But this is not the most serious concern in our design—we are more interested in ensuring that the comparison within runners (within blocks) of the various conditions is a fair comparison.

Chapter 4

6. (a) The histogram would have two or three modes, or humps, certainly one for male adults, one for female adults, and possibly one for children. (b) If they were the same species, I would assume that the modes corresponded to trees seeded about the same time, possibly in the aftermath of a blight or weather anomaly. If they were different species, I would conclude that the different species had different growth rates, and the area under each hump would tell me the relative frequency of each species (although only roughly if the separate species distributions overlap).   **7.** The sample is small, but we do get estimates of the average and SD of the 100 cyclists: 27.8 and 1.8, respectively. The normality assumption puts the time 25.7 at $(25.7 - 27.8)/1.8 = -1.17$, which would be the 12th percentile. This suggests that the time of 25.7 would have been about 12th in the race of 100 cyclists. (This is a very rough estimate since the sample values 27.8 and 1.8 are not necessarily very close to the average and SD of the 100 cyclists, and the calculation ignores this. However, there is not much else we can do. We certainly do not want to conclude that 25.7 would be the fastest time among the 100, nor is it likely to be greater than average.)   **10.** (a) Assume normality. 70 is 1.5 SDs below the average of 77.5. Thus the probability of a score below 70 is about 6.7 percent. About 6.7 percent of applicants have been rejected by the test. (b) Normality of the distribution of the scores of the applicants. The assumption that applicants' scores are independent is also necessary, but seems certain to be an acceptable assumption in this case.   **15.** The results obtained by the author were: (a) 64, 69, 68. Average = 67.0, SD = 2.16   (b), (c), and (d)

	(b)	(c)	Difference	(d)
	6	7	+1	7
	8	8	0	8
	5	5	0	5
	7	7	0	7
	6	7	+1	7
	6	6	0	6
	7	8	+1	8
	8	8	0	8
	6	5	−1	6
	9	10	+1	10
Average	6.8		0.30	7.2
SD	1.2		0.64	1.3

(e) $7.2 \times 10 = 72$. This is more than the average of 67 found in part (a) presumably because most errors in (a) were errors of omission. (f) The SD in part (a) on measurement error only. The measurement errors per line are estimated to be about 0.64, the SD of the differences. But the error in the count for the whole paragraph will include errors from each line and this SD should be larger than 0.64. In fact, it is estimated to be 2.16, which supports this idea. [In Chapter 7 this relationship is quantified. We can think of the paragraph total error (a measurement error) as a sum of the 10 independent measurement errors from each line. It will be shown that this SD should be about $\sqrt{10} \times 0.64$, which is 2.02. The value of 2.16 actually observed for this error is amazingly close, considering the small sample on which the estimates are based.]

The SD in part (b), 1.2, is a composite of measurement error and the natural variation of the number of e's per line in English. It should be larger than the pure measurement error, 0.64, which it is, although how much less depends on the relative sizes of the measurement error and the natural variation per line.

The SD in part (d), 1.3, is due to natural variation only and should be less than the SD in part (b), since the latter includes measurement error as well. This is not the case, and this is an anomaly of the relatively small sample size (of 10 lines) that has been used; that is, this unexpected result may be attributed to random variation in the estimation of the SDs.

16. 1 child   *B* or *G*
    2 children   *BB, BG, GB,* or *GG*
    3 children   *BBB, BBG, BGB, GBB, BGG, GBG, GGB, GGG*
    4 children   *BBBB, BBBG, BBGB, BGBB, GBBB, BBGG, BGBG, BGGB, GBBG, GBGB, GGBB, BGGG, GBGG, GGBG, GGGB, GGGG*
(The family denoted *BBG* is one whose two eldest children are boys and the youngest is a girl.)

- For families with one child, 1/1 or 100 percent of sons are eldest sons.
- For families with two children, 3/4 or 75 percent of sons are eldest sons.
- For families with three children, 7/12 or 58 percent of sons are eldest sons.
- For families of four children, 15/32 or 47 percent are eldest sons.

Since the four family sizes are assumed to be equally likely, the population of children contains 10 percent of their numbers in families of one child, 20 percent in families of two children, 30 percent in families of three children, and 40 percent in families of four children. (Think of the population being represented by $\{C, CC, CCC, CCCC\}$ from which the 10 percent, 20 percent, etc. come from 1/10, 2/20, etc.) Thus the proportion of sons that are eldest sons is

$$0.1 \times 1 + 0.2 \times 3/4 + 0.3 \times 7/12 + 0.4 \times 15/32 = 0.613$$

(By symmetry, the same is true of daughters.)
    If we ignore sex, the families can be proportionally represented by $\{C, CC, CCC, CCCC\}$. Hence the proportion of children that are eldest children is $(1 + 1 + 1 + 1)/(1 + 2 + 3 + 4)$ or $4/10 = 0.40$. (We might have also calculated this as $0.1 \times 1/1 + 0.2 \times 1/2 + 0.3 \times 1/3 + 0.4 \times 1/4$.)
    The paradox is explained by noting that there are sons (or daughters) who are not eldest children but who are eldest sons (or daughters).   **23. (a)** The information needed is that the distribution of test scores was a certain shape, such as normal. **(b)** Assuming normality, the 80th percentile is at about 0.85 standard unit (see the normal probability table), that is, at $55 + 0.85 \times 15$, which is about 68. The estimated score for the 80th percentile is 68.

Chapter 5

2. One natural suggestion would be to note the percentage of wines for which the two connoisseurs give identical ratings. This measure is 40 percent for the data given. It has the advantage of being easy to explain, easy to interpret, and easy to calculate. The disadvantage is that it does not reflect the amount of disagreement between the raters when they do disagree. A more subtle difficulty is that the value of the measure depends on the number of categories used, even though this really does not have too much to do with how well the connoisseurs agree. (More categories will tend to reduce the agreement using this measure.) There is room for creativity here. The quality of your suggestion should be judged relative to the purpose you have in mind for the measure.   **5.** For 12 weeks worked, the regression method predicts 14.1 days absent from work. SD_{pred} is 6.7 days. To check the assumptions, draw a scatter diagram with "days absent" on the vertical axis. Check visually that the relationship is reasonably assumed to be linear, that the prediction errors are about the same size over the range of values of X, and that there are no trends to the errors. (The simplest way to calculate the errors of prediction is to draw the regression line on the scatter diagram, and read off the prediction errors.) These assumptions seem to be satisfied in this instance. (The equation of the regression line is $Y = 2.86 + 0.94X$.)   **8.** By direct inspection of the data, or using a star plot, it can be seen that the best film for high-quality pictures depends on the lighting conditions. A simple way to examine these data is to examine a contingency table for each set of lighting conditions.

LIGHTING CONDITIONS = 1

		Film price	
		1	2
Picture quality	0	1	2
	1	2	1

430   Solutions to Selected Problems

LIGHTING CONDITIONS = 2

		Film price 1	Film price 2
Picture quality	0	2	1
	1	1	2

From these tables the dependence on lighting condition can be clearly seen. For lighting conditions = 1, the low-price film is best, but for lighting conditions = 2, the higher-priced film is best.   **10.** Cluster analysis. Multiple regression analysis.   **13. (a)** Use the regression method. Estimate is 59.3 hours. Precision is measured by SD_{pred} = 20 hours. **(b)** Regression prediction corresponding to 1000 km is 59.3 hours. SD of the distribution with this average is SD_{pred} = 20 hours. 48 hours is $(48 - 59.3)/20$ or -0.57 in standard units. Assuming normality, the normal table gives the left-tail probability as 0.28. So 28 percent of the letters traveling this distance take 48 hours or less.   **16.** The slope of the regression line for predicting weekly sales from square feet of display space will provide the answer to the question. The slope depends on the correlation coefficient and the SD of each variable. The slope turns out to be $28.11, which is the estimated return for one additional square foot of display space. (This analysis ignores the sequential nature of the data. The seriousness of this in the present application is a matter of judgment, and would be based on knowledge of experience with weekly sales patterns. For example, if weeks 4, 6, and 9 were special in the overall marketing strategy of the store, the data may be giving a biased picture of the relationship the problem addresses.)   **18.** A graph of the monthly ticket sales (on the vertical axis) and time (month and year, on the horizontal axis) would allow the owner quickly to assess the trend. The order of the months is obviously crucial to any reasonable interpretation, so histograms or averages and SDs are just not appropriate summaries in this case.   **22.** If the car owner fills up her car with gas when the tank gets down to about one-quarter full, she will have traveled about the same number of miles before each fill-up—only slight variation in the "miles driven" variable is allowed. Over this range of values of miles driven, there is only a poor relationship between miles driven and gasoline purchased, and this is confirmed by the correlation of 0.2. If a few fill-ups had been done after only 10 or 20 miles, the correlation would have been much closer to 1, since the amount of gasoline used is indeed well predicted by the miles driven, provided that a reasonable range of values of miles driven is being considered. [Note that SD_{pred} for the regression of gasoline consumption on miles driven would be small, as a percentage of the actual gasoline consumption, even for the case described (where $r = 0.2$).]   **26.** Area is measured by, for example, square feet, while perimeter is measured by feet. A square whose perimeter is 4 ft has an area of 1 ft^2, a square whose perimeter is 8 ft has an area of 4 ft^2, and a square whose perimeter is 12 ft has an area of 9 ft^2. This relationship is not linear, and there is little reason to think the relationship for quadrilaterals would be linear. The more traditional geometric approach to the problem is much more enlightening, even if only approximate rules of thumb are required.   **31. (a)** The regression method gives $160,000 for the predicted value. **(b)** SD_{pred} error is $9000. **(c)** The estimate could be very wrong. The data probably do not extend to 2500 ft^2, and the relationship may not be linear over this range. Also, the errors of prediction could be quite large for the larger houses. Also, any errors in estimating the slope of the regression line would be magnified at values of the independent variable that are so far from the average.

Chapter 6

5. The only feature of the series that appears to be predictable is the serial correlation, which is 0.46 over the 22 pairs of values. As the average of the series is 9.6, and the 1982 value is 10, the 1983 value is forecast to be $10 + 0.46 \times (9.6 - 10) = 9.8$.   **6.** Over the 1960s and 1970s, the birthrate has declined to record low levels, although the decline seems to be at a much slower rate in the 1970s (Figure 1.4). At the same time (1960–1980) the proportion of the population that are women of childbearing age has been increasing (Figure 1.5). However, the latter index fell in 1981–1982 and

Solutions to Selected Problems   431

1982–1983 suggesting that future birthrates may fall more quickly in the 1980s than they did in the 1970s.

Chapter 7

1. There are two natural models; both are correct.

  1. Population: $\{1, 2, 3, 4, 5, 6\}$. Method of sampling: at random, with replacement. The sample size is 2.
  2. Population: $\{2, 3, 3, 4, 4, 4, 5, 5, 5, 5, 6, 6, 6, 6, 6, 7, 7, 7, 7, 7, 7, 8, 8, 8, 8, 8, 9, 9, 9, 9, 10, 10, 10, 11, 11, 12\}$. Method of sampling: at random. The sample size is 1. *Note:* A less tedious way of expressing the population is as follows:

Value	2	3	4	5	6	7	8	9	10	11	12
Frequency	1	2	3	4	5	6	5	4	3	2	1

The expected value of the average outcome for 10 rolls is just the average on one roll (i.e., 7). The average on one roll is 7 because the expected value of the average sum of two rolls of a single die (which has an average of 3.5) is $2 \times 3.5 = 7$.

The chances concerning the sum of two dice having an average sum of 10 or better, in 10 rolls, can be computed either by considering all possible equally likely outcomes (there are about 4 million trillion of them) or by using the normal approximation. Using the latter technique, $(10 - 7)/(1.41 \times 1.71) = 1.20$ standard units, so the chance that the average is 10 or more is approximately 11.5 percent.   **5.** We use labels $1, 2, 3, \ldots, 60$ for the words in the paragraph. A random sample of five numbers from a random number table is (following the procedure of Figure 2.1) 29, 54, 27, 5, and 47. The words in these positions are "in," "the," "and," "is," and "word." The word-length sample is therefore 2, 3, 3, 2, 4. The average is 2.8 and the SD is 0.75, so these are the estimates required. (Note that this sample is quite unrepresentative of its population—this happens occasionally. See also Problem 6.)   **7.** The binomial table (Appendix A4.2) gives the chance of 6 heads in 10 tosses of a fair coin to be 20.5 percent. The chance of 60 heads in 100 tosses is found using the normal approximation. 60 is between 59.5 and 60.5, and converting to standard units, $(59.5 - 50)/5 = 1.90$ and $(60.5 - 50)/5 = 2.10$. The normal table gives 1.09 percent as the chance that a standard normal is between 1.90 and 2.10. Clearly, the 6 out of 10 has the better chance of occurring.   **9.** It can easily be checked that there are 21 possible teams (*ABCDE*, *ABCDF*, *BCDEG*, etc). Each combination will occur a certain number of times during the 21-game season. Although each team is selected an average of once per season, it is very unlikely that each team will be selected exactly once in a particular season. In fact, the number of times a particular team is selected has a binomial distribution with probability $1/21$ and $n = 21$. A look at the binomial table will suggest that several possible teams will never be selected in a particular season. If the "interaction" is a little different for each possible team, the coach may not have a very good basis for his choice.   **13.** Large samples are precise, but they need not be unbiased unless they are selected at random from the population to be described.   **17.** Use a normal approximation to get the chance as 86.3 percent. (This is an easy way to get a C.)   **20. (a)** False. Although the probability that a particular ticket is a winner is, in the absence of any other information, $25/1000$, the chance that any two tickets are both winners is not $(25/1000) \times (25/1000)$ since the sampling must be considered to be without replacement. [This chance would actually be $(25/1000) \times (24/1000)$.] So the probability shown is not correct. Note also that it is impossible for the company's employees to win with each of the 100 tickets, since there are only 25 winning tickets. **(b)** The estimate is 100 times the average winnings per ticket, $2500/1000$, which comes to \$250. The SE of the company's winnings is $\sqrt{100}$ times SD times correction factor for sampling without replacement. The population of prizes consists of 975 0's and 25 100's. The SD of this population is $100 \times [(25/1000) \times (975/1000)]^{1/2}$, or \$15.61. (This uses the formula for the SD of a population of 0's and 1's, then multiplies this times 100 to scale up the answer.) The correction factor is

$[(1000 - 100)/(1000 - 1)]^{1/2}$ or 0.949. Thus the SE of the company's winnings is $10 \times 15.61 \times 0.949 = \$148.16$. This indicates that the company's winnings will be $\$250.00 \pm$ about $\$150$.

25. (a)
1, 2   average is   1.5
1, 5   3.0
1, 6   3.5
2, 5   3.5
2, 6   4.0
5, 6   5.5

Yes, the samples are equally likely, by the definition of random sampling. (b) SD of the above sample averages is 1.19. The SE of the sample average is (SD of $\{1, 2, 5, 6\}/\sqrt{2}$) times the correction factor $[(4 - 2)/(4 - 1)]^{1/2}$ [i.e., $(2.06/1.414) \times 0.82 = 1.19$]. Thus the two numbers are identical. This must be so since both calculations are computing the SD of the sampling distribution of samples of size 2 from the population $\{1, 2, 5, 6\}$. **26.** This is correct since political opinion polls are usually done on large populations using a sample that is a small proportion of the population. Hence the correction factor for the SE will be approximately 1, and the SE estimate will not be changed much by the correction factor. So even though political opinion polls use sampling without replacement (it does not seem reasonable to have a pollster poll the same respondent twice), it may be assumed that the sampling is done with replacement for the purposes of SE calculations.
28. (a) The population is $\{2, 3, 7\}$. Possible samples are as follows:

	(2, 2)	(2, 3)	(2, 7)	(3, 2)	(3, 3)	(3, 7)	(7, 2)	(7, 3)	(7, 7)
Average	2.0	2.5	4.5	2.5	3.0	5.0	4.5	5.0	7.0
SD	0.0	0.5	2.5	0.5	0.0	2.0	2.5	2.0	0.0

(b) The relative frequencies of the nine possible samples, in the long list of samples, should each be about 1/9. That is, each of the nine possible samples appears in the long list with the same relative frequency. (c) The average of the sample averages is 4.0 also. So the sample average is an unbiased estimate of the population average. (d) The average SD is 1.11, which is much less than the population SD of 2.16. So the sample SD is not an unbiased estimate of the population SD. (e) The SD^2 values are 0.0, 0.25, 6.25, 0.25, 0.0, 4.0, 6.25, 4.0, and 0.0. The average of these is 2.33, which is less than 4.67, the population value of V. So SD^2 is not an unbiased estimate of V. However, the average value of $S^2 = 2.33 \times 2 = 4.66$, which except for rounding error, agrees with $V = 4.67$. Thus S^2 is an unbiased estimate of V. (f) The average value of S is the average of 0.0, 0.71, 3.54, 0.71, 0.0, 2.83, 3.54, 2.83, and 0.0, which is 1.57. This is not equal to 2.16, so S is a biased estimate of the population SD. (g) X would tend to be larger than the SD, since the SD uses the deviations of the sample values from the sample's own average, whereas X uses the deviations of sample values from the population average, which is sometimes quite far from the center of the sample. (h) The population average is 4.0. The average squared deviation from 4.0 of the nine possible samples is 4.67. In other words, the average value of X^2 is 4.67. But the average value of S^2 was also 4.67. Thus both X^2 and S^2 are unbiased estimates of V. (However, the estimator X required knowledge of the population average. But when the population average is known, it can be checked that X^2 has much less variability about 4.67 than does S^2. When the population is not known, it can be shown that the SD^2 has much less variability about 4.67 than does S^2.)

The point of all this is to justify using the sample SD as our estimate of the population SD in most of the book while admitting the use of S in the context of analysis of variance (in Chapter 9). **32.** (a) A random sample of size 25 drawn with replacement from the population $\{0, 1, 1\}$ would have the desired characteristics. (b) Yes. Successive flips should be considered independent since there appears to be no likelihood that the coin tosser can avoid making innumerable variations in his throw which will have an unpredictable effect on the outcome of the toss. (The independence has nothing to do with the chance that the coin comes up heads.) So successive throws will not be related in any way in the sense that knowing the outcome of one throw will alter the probability of heads on the next throw. The independence assumption seems reasonable. (c) Using a normal approximation yields a chance of about 5.3%.

Chapter 8

1. B, A, C is more likely in the population of voters than A, C, B, given that the sample reveals the ordering A, B, C, since sample proportions are more variable as their corresponding population parameter values get closer to 0.5. The probability that B, A, C in the population produces A, B, C in the sample is greater than the probability that A, C, B in the population produces A, B, C in the sample.   **5.** Since no value is given for the population SD, and since the sample is very small, we will have to use a t procedure. This requires an assumption of normality; the assumption does not seem unreasonable in this instance. To estimate the average time of the 100 cyclists, a 95 percent confidence interval will be used. It is, since the sample has average 27.8 and SD = 1.8,

$$27.8 \pm t(4 \text{ df}, 95 \text{ percent})1.8/\sqrt{4}$$

which reduces to 27.8 ± 2.50. This is the interval estimate of the average circuit time. (The t value is 2.78.)   **8.** (a) 65.3 to 70.7 (b) About the same. See the formula for the SD. (c) Smaller. The width of the confidence interval (for a sample that is large enough that the normal procedure can be used) depends only on SD/$\sqrt{n}$. Since the SD is about the same for the larger sample, the width of the confidence interval should be reduced by a factor $1/\sqrt{2}$ (since 100 is 2×50).   **11.** The information that an average of 10 taxis per day will not start suggests that the actual number has a binomial distribution with $p = 10/1000 = 0.01$ and $n = 1000$. The normal approximation has average 10 and SD = $\sqrt{1000 \times 0.01 \times 0.99}$. That is, average 10 and SD = 3.15. The number of cars needing service each day would be estimated to be $10 \pm 2 \times 3.15$, that is, 3.7 to 16.3. Since the loss of revenue is $150 per car, the loss is estimated at between $555 and $2455, with a confidence of 95 percent.   **12.** (a) False. There is only one population average, and it is either in the interval or it is not. (b) False. The proportion of heads may be close to its expected value, but the standard error of the number of heads increases with the number of draws. (c) True. Unbiased because the sample is a random sample, and precise because the sample is large and the SE of the sample average is small. (d) True. If the population percentage were known, the sample would usually be unnecessary. However, the variability of the sample percentage does depend on the population percentage. (e) True. One lets 0 and 1 indicate the two items, and the sum of the numbers selected in a sample of size n indicates the number of one of the items. (f) True. The central limit theorem requires the sample to be large to yield normality of the sample average. If the population is normal, the sample average will have a normal distribution even for small samples, but if the population is not normal, the normality of the average of small samples does not follow.   **15.** The question that needs to be considered is whether this is a fantastic coincidence. The information in Problem 1 of Chapter 6 suggests that the improvement from one Olympic meet to the next is a few hundredths of a second. The smoothness of this curve suggests that to prepare for the event, one must take hundredths of a second seriously. It may well be that Carl Lewis runs the 100-m race in times that consistently run close to 9.99 sec. This explanation provides a reasonable alternative to the coincidence theory. The fact that the races resulted the way they did is evidence that Carl Lewis is a consistent runner. (For more on this sort of logic, see Chapter 9.)   **18.** The bias will be zero since we are told that the sample is random and there are no "nonrespondents" which reduce the size of our sample. The "no answer" category is a perfectly good response in this case since the population proportion desired eliminates these cases, too. The precision is estimated by the standard error of a sample proportion, $\sqrt{p(1-p)/n}$. Since p is unknown, the best we can do to assess the precision is consider the worst case, $p = 1/2$. We know that $n = 100$ and since $N = 500$, we can make a correction for sampling without replacement, by a factor 0.90. Thus the SE is no more than $0.90 \times (1/2)/10$ or 0.045, which gives the precision of the estimated proportion.   **20.** (a) The population can be described as a mixture of 1000 subpopulations, where each subpopulation consists of numbers that differ from the actual volume of a particular tree by chance errors of measurement. The subpopulations are all equally represented in the mixed population. (b) Using the t procedure (9 df), a 95 percent confidence interval for the average tree volume is

$$\frac{10.0 \pm (2.26)(2.00)}{\sqrt{9}} = 10 \pm 1.5$$

So the 95 percent confidence interval for the woodlot wood volume is $10,000 \pm 1500$ m². (c) The sampling distribution of the sample average is known to be approximately normal with average equal to the population average and SD equal to the population SD divided by $\sqrt{10}$. However, the first

Chapter 9

2. Let us begin by giving Mr. Nicely the benefit of the doubt and suppose that the proportion supporting him is just over 0.50. How surprising would it be to get only 45 out of 100 supporting Mr. Nicely in a random sample of 100? If this is a very small probability, the evidence will have refuted Mr. Nicely's claim. Otherwise, we would have to accept the claim.

To compute the probability in question, use the normal approximation. 45 out of 100 is $(45 - 50)/5$ or 1 SD below the population average, assuming the population value of 50 percent. The P value associated with the 50 percent is 0.16, so Mr. Nicely's claim is credible. The answer to the question posed is "no," we cannot refute conclusively Mr. Nicely's claim.   **4.** No. The zeros in the nonnegative data indicate that the distribution of the data is not normal. So the t procedure is invalid.   **5. (a)** 5.26 percent. **(b)** Use binomial with $p = 0.474$, $n = 6$. The probability is 0.209. **(c)** The expected values of the frequencies are 200, 1800, and 1800. The chi-square statistic is

$$\frac{(225 - 200)^2}{200} + \frac{(1732 - 1800)^2}{1800} + \frac{(1843 - 1800)^2}{1800} = 6.72$$

Comparing this value with the chi-square distribution on 2 df shows that the P value for the test is less than 0.05. Conclude that the wheel is not properly balanced.   **10.** The expected values are

$$\begin{array}{cc} 17.1 & 22.9 \\ 12.9 & 17.1 \end{array}$$

so the chi-square statistic is

$$\frac{(15 - 17.1)^2}{17.1} + \frac{(25 - 22.9)^2}{22.9} + \frac{(15 - 12.9)^2}{12.9} + \frac{(15 - 17.1)^2}{17.1} = 1.05$$

Comparing the chi-square distribution with 1 df, P is about 0.30, and one concludes that section and major are independent.   **13. (a)** No. Eleven probabilities are calculated from 11 different distributions. Note that the probabilities do not add to 1.0 **(b)** The range 0.3 to 0.5 (or possibly 0.25 to 0.55) seems far more likely than other values to produce a sample like the one observed. **(c)** We calculate the probability that a proportion in the population of 0.6 would yield a sample of 25 that had 10 or fewer in favor of a tax increase.

$$z = \frac{0.4 - 0.6}{\sqrt{0.6 \times 0.4/25}} = 2.04$$

$P < 0.05$, so we reject the hypothesis of 0.6. We conclude that the evidence is against a population proportion greater than 0.6.

This concurs with the impression gained in part (b).   **16.** The test to be performed is whether the differences have a zero expectation. A one-sample t test does this. The average difference is 0.46, and the SD of the difference is 1.08. Thus the t statistic is $(0.46 - 0)/(1.08/\sqrt{6}) = 1.04$. Compare this with the t distribution on 6 df, to get a P value (two-tailed test) of over 60 percent. Accept the unbiasedness of differences. Conclude that John's guesses are unbiased.
19. (a) Assuming normality of the lifetimes, the SD must be such that $15 - 10$ years corresponds to a normal standard units value of 1.3 (from normal table, using the 10 percent figure in the problem). So the SD of lifetimes is $5/1.3 = 3.85$ years. Thus $(20 - 10)/3.85 = 2.60$ is exceeded with probability 0.0046. So about 0.0046 of the cars last more than 20 years.

The model must be very approximate in the tails of the distribution since a proportion 0.0046 of the cars are predicted to last a negative number of years. No, the assumption of normality is not reasonable, especially for calculating these tail probabilities. **(b)** We test whether the average lifetime might reasonably be believed to be 10 years. The appropriate normal approximation test uses $z = (8.5 - 10.0)/(4/\sqrt{30}) = 2.05$. The P value for the test is less than 0.05, so the manufacturer's claim is not believable. **(c)** $0.33 \pm 2 \times \sqrt{0.33 \times 0.67/30}$, which is 0.16 to 0.50. This is a 95 percent confidence interval for the population proportion of cars that last 10 years or more.   **23.** A chi-square test of independence should be used here.   **30.** We perform the hypothesis test of the equality of the two runners' times. This is the same as the test that the average time from a very large number of timers is the same for both runners, or, in other words, that the two times for each runner are random samples of size 2 from a single population.

H: actual time for runner 1 = actual time for runner 2

A: actual time for runner 1 is less than actual time for runner 2

Average and SD for runner 1:   10.06, 0.040

Average and SD for runner 2:   10.21, 0.030

$$\text{SE of difference} = \sqrt{\frac{(0.040)^2}{1} + \frac{(0.030)^2}{1}} = 0.05$$

$$t = \frac{10.21 - 10.06}{0.05} = 3.0$$

Compare with t on $(2 - 1) + (2 - 1) = 2$ df. $P_{approx} = 0.20$. "Not statistically significant." Conclude that runner 2's claim is credible and has merit.

32. (a)

$$z = \frac{0.311 - 0.296}{\sqrt{(0.31)(0.69)/10{,}000 + (0.30)(0.70)/10{,}000}}$$

$$= \frac{0.015}{0.006} = 2.5$$

So P for a two-tailed test is 0.025, and the proportions in the population are shown to be different. "Statistically significant," since 0.025 is less than 0.05. **(b)** This result is probably of no importance whatsoever to the wholesaler. He may be interested in the fact that both proportions are about 0.3, but the difference between the population proportions cannot be of any importance since it is so small.

Index

Note: To locate exercises and problems by their application area, see Appendix 2, List of Applications.

A

Acceptance error of hypothesis test, 331
Accuracy, 76–78
Addition rule for combining probabilities, 235
Adjustment of averages
    (*see* Analysis of covariance)
Agreement, 140
    association and correlation, 145
Alternative, 328
Analysis of covariance, 173–74
    blood-pressure example, 174
Analysis of variance, 177, 352–58
Association, 138
    agreement and correlation, 145
    positive and negative, 140
Augmented scatter diagram, 169
Autoregressive model (*see* Time series)
Average absolute deviation (*see* Mean absolute deviation)
Average absolute error (*see* Mean absolute deviation)
Average and averages:
    compared to median, 94–95
    compared to median and mode, 97
    expected value of, 248
    normal approximation to distribution of, 259–67
    for prediction, 151–55
    to reduce confusion, 57–59
    to summarize a distribution, 91–96
    physical meaning of, 96
    probability distribution of a sample average:
        average and standard deviation, 263
        for a normal population, 264
    sampling distribution of a sample, 251
    sensitivity to outliers, 100
    standard error of sample, 248
    use in regression, 150–55
        horse-racing example, 151–52

B

Baby boom, 6
Balanced block designs:
    complete, 61
    incomplete, 61

Bias, 76–78
Binomial experiment, 239
Binomial model for probabilities, 240–41, 278
   histogram, 241
   normal approximation, 269
   table, 408–11
Birthrate, 6–8
Blocking, 59–62

C

Causation, 37, 54
   and association, 138
   eye-infection example, 138
Center, measures of, 93–96
   average, 94
   median, 94
Central limit theorem, 260
Chernoff faces, 170–71
Chi-square distribution, 347
Chi-square tests:
   of independence, 358–61
   of proportions, 344–48
Cholesterol study, 31, 37
Cluster, 25–26
Cluster analysis, 172–73
Combinations, 230–32
Comparisons, 29–30
Compatibility of events (see Incompatibility of events)
Complementarity rule for probabilities, 235
Computers in statistics, 8–9
Confidence interval:
   for an average, 293–97
      using normal distribution, 294
      using t distribution, 309
   choice of confidence level for, 295
   coverage property, 294
   interpretation, 295
   as an interval estimate, 295
   for a proportion, 300
Confidence level (see Confidence interval)
Confounding factors, 34–37
Contingency tables, 175–78
   bike example, 175–76
   dependence in, 178

fruit farm example, 177
Continuity correction, in normal approximation to binomial, 268–69
Correlation, 140–48
   agreement and association, 145
   of averages, 148
   coefficient, 140–42
      calculation procedure, 142
      definition, 142
      interpretation, 144
      linearity, 146
      scatter diagram summary, 144, 146
Critical P value, 329
Crossover design, 60

D

Data set terminology, 133–34, 392–93
Decile, 111
Degrees of freedom:
   analysis of variance, 356–57
   t distribution, 309–10
Demographic trends, 6
Density function, 123
Dependence and independence, 178
Dependent samples (see Paired data)
Dependent variable, 55, 137
Descriptive methods:
   for one variable, 73–112
   for time series, 199–216
   for two variables, 132–80
Descriptive statistics (see Statistics, the discipline)
Differencing (see Time series)
Discriminant analysis, 173

E

Echo boom, 6
Empirical distributions:
   description, 111–12
Error, 76
Estimation of parameters, 243–59, 290–312
   averages, 291–97, 307–10
   proportions, 299–302

Estimation of probability distributions, 310–12
Expected value, 245–48
Experimental units, 55
Experiment with experiments, 33
　contrasted with surveys, 31–37
　design of, 54–65
　of teaching methods, 32

F

Factor, causal, 55
Factor analysis, 177
Factorial design, 63–64
　levels of a factor, 63–64
F distribution, 355
Film critics' example, 140–46
Fisher, Sir Ronald, 13
Five-number summary, 146
Forecasting (*see* Time series)
Framingham study, 12
Frequency distribution, 80–81

G

Galton, Sir Francis, 13
Graphs (*see* Augmented scatter diagram; Chernoff faces; Histogram; Oval diagram; Residual plot; Scatter diagram, scattergram; Star plots)

H

Hempel's Ravens, 371
Heteroscedasticity (*see* Regression)
Histogram, 80–88
　axis break, 86
　comparison, 93
　complications in construction, 83–87
　as a descriptive technique, 91
　features, 82–83
　grouped data, 83–84
　for mixed-data types, 178, 180
　open-ended intervals, 87
　ordered data, 87–88
　for qualitative data, 178, 180
　relative frequency, 85
　smoothed, 92
　summary measures, 91–101
　use of area for frequency, 85
Homoscedasticity (*see* Regression)
Huxley, T. H., 20
Hypothesis testing, 328
　alternative in, 328
　for an average, 333–39
　　normal tests, 333–35
　　t test, 335–37
　critical P value, 329
　　choice of, 367–68
　decision in, 329
　errors in decisions from, 331
　hypothesis in, 328
　logic of, 324–33, 366–70
　outcomes, 331
　for a percentage, 339–40
　　normal test, 338–39
　　binomial test, 339–40
　prior knowledge in, 368–69
　P value, 328
　sampling distribution for, 328
　sample size, effect, 369–70
　statistic in, 328
　2 by 2 contingency table tests, 344–48
　two-sample tests, 340–51
　　normal and t tests, 340–44
　when required, 366–67

I

Imprecision, 78 (*see also* Precision)
Incompatibility of events, 233 (*see also* Addition rule for combining probabilities)
Independence, 178, 234–35 (*see also* Multiplication rule)
Independent variable, 54, 137
Inferential statistics (*see* Statistics, the discipline)
Instrument, 74
Interaction, 64

Index **439**

Interquartile range, 111
Interval estimate (*see* Confidence interval)
Invariance, 100–101
I.Q.:
    example of normal distribution, 109–10
    variation-in-measurements example, 42

L

Large sample theory, 293
Lecturer effectiveness and ratings of instruction, 8
Linearity:
    measured by correlation, 144, 146
Logic:
    of hypothesis testing, 324–33, 366–70
    in study design, 18–21
Long-run relative frequency, 223, 229
Lotteries, 3–5, 10, 245–47

M

Management:
    of airlines, 9
    by exception, 19
Many-sample tests, 352–61
Mathematical notation, 121–22
Matrix, 134
Mean absolute deviation (MAD), 99–100
    compared with standard deviation, 99
    possible use in regression, 153–54
Measurement, measurements, 74
    qualitative, 75
    quantitative, 75
    scales of, 75–76
Median, 94
    compared to average, 94
    compared to average and mode, 97
    insensitivity to outliers, 100
    physical interpretation, 96
Median test, 363
mn rule, 232
Mode, 82
    compared to average and median, 97

Models and modeling, 222–23
    sampling, 227, 230
    probability, 237–39
Monte Carlo simulation, 395–98
Multiple comparison tests, 356
Multiplication rule for probabilities, 235

N

Nonparametric probabilities, 270–71
Nonparametric tests, 362–65
    one-sample median test, 362–63
    sign test, 363–65
    two-sample median test, 365
Nonresponse in surveys:
    bias, 40
Normal curve, normal distribution, 103
    approximation, 103–7, 266
    formula, 104
    shape, 104
    standard, 122
    table, 406–7
        use of table, 107–10
    use of standard units, 105–7
Normal probability distribution:
    as approximation:
        to binomial distribution, 268–69
        to known distributions, 268
        to unknown distributions, 266
    and central limit theorem, 260
    as a model for the distribution of averages, 259–67
    table, 406–7
Normal values, 30

O

Observational study (*see* Surveys)
One-sided test, 342
Outlier:
    definition, 101
    effects on summary measures:
        averages and medians, 97
        correlation coefficient, 146

Outlier (*cont.*)
   standard deviation, 101
Oval diagram, 145

P

Paired data:
   confidence intervals using, 296–97
   in hypothesis testing, 349–51
   regression and correlation, 136
Parameter (*see* Population)
Pearson, Karl, 13
Percentile, 110
   of the normal distribution, 110
   percents and percentiles, 110
   related to area under a curve, 108
Placebo, 6, 40, 55
Point of averages, 155
Polls (*See also* Surveys)
   Gallup, 5–6
   Harris, 5
   Literary Digest, 5–6
Population, 22
   concrete, 222
   hypothetical, 223
   parameter, 224, 244
      average and SD, 244
   subpopulations of a, 25–26
Precision, 76–78
Predicted variable, 137
Prediction, 137
   error:
      calculation, 157–59
      definition, 158
      in regression (SD_{pred}), 157–59
      using average, 154
   regression as, 150
Predictor variable, 137
Principal components, 173
Probability, 228–42
   combination of probabilities, 235
   distribution, 230
   and long-run relative frequency, 229–30
   and unexplained variation, 228–29
Public lotteries (*see* Lotteries)

P value, 328
   and critical P value, 329

Q

Quality control examples, 80–88
Quartile, 111
Questionnaire:
   of higher education intentions, 1–3

R

r, 185 (*see also* Correlation, coefficient)
Random digits and random numbers, 23
   table of, 404–5
   use in random sampling, 24
Randomization, 56
Random sample (*see* Sample and sampling)
Random selection and random allocation, 32–33
Range, 96–98
Regression, 150–66
   assumptions, 159–62
      absence of outliers, 160
      histogram of errors, 160
      homoscedasticity of errors, 159–60, 163
      independence of errors, 160
      linearity of fit, 159
   causality in, 162–64
      house appraisal example, 162–63
   and correlation of averages, 166
   effect, 164
   fallacy, 164
   line, 155–57
      equation, 157
      intercept, 157
      and SD line, 155–56
      slope, 155
      through point of averages, 155
   method, 156
   multiple, 173
   as prediction, 150
   prediction error in, 157–59

extrapolation, 162
rental space example, 152–53
toward the mean, 166
use of averages in, 151–55
Rejection error in hypothesis tests, 331
Remedial courses:
example of Simpson's paradox, 34–36
Replication, 55
Residual plot:
for regression, 161
for time series, 209
Residuals:
for regression, 161
for time series, 209–13

S

Sample and sampling, 21
cluster, 25–26, 258
compared to stratified, 26, 256–259
models, 227
random, 25, 225–26
simple random (SRS), 22–25, 225–26
from infinite populations, 229
statistic, 244
expected value of a, 245–48
in hypothesis testing, 328
standard error of a, 248–53
value of, 244
stratified, 25–26, 256
compared to cluster, 26, 256–59
systematic, 22
variation, 225–26
with and without replacement, 23, 28
Sample size, 303–6
for estimating averages, 305–6
for estimating percentages, 303–5
Sampling distribution of a statistic, 251
in hypothesis testing, 328
of a sample average:
large sample theory, 252, 293
t-distribution, 307–10
of a sample proportion, 252, 302
Scatter diagram, scattergram:
one-dimensional, 94

two-dimensional, 135
SD (*see* Standard deviation)
SD line, 155–56
SD_{pred}, 157–59
calculation, 159
Seasonal effects in time series (*see* Time series)
Serial correlation (*see* Time series)
Sign test, 364
Simple random sampling (*see* Sample and sampling)
Simpson's paradox, 36, 178–79
Skewness, 82–83
Smoothing of time series (*see* Time series)
Spread of a distribution, 93
interquartile range, 111
measures of, 96–100
SRS (*see* Sample and sampling, simple random)
Standard deviation (SD), 98–100
compared with MAD, 99
invariance property, 101
n versus $n-1$ definition, 98–99
population, 249
estimation of, 250
relationship to variance, 249
sample, 250
of a sum of independent variables, 249
use in empirical rule, 99
Standard error:
of a sample average, 248
of a sample proportion, 248
in sampling without replacement, 254–56
Standard units, 105–6
compared to ordinary units, 105
conversion to standard units, 106
Star plots, 171
Statistic (*see* Sample and sampling)
Statistical significance, 331–32, 370
Statistics, the discipline:
definitions, 13–14
descriptive, 222
history of, 12–13
inferential, 222
learning the principles of, 12
problem solving in, 10–11
training in, 10–12
Stratum, 25–26

Student grades:
  effect of averaging, 266
Studies:
  experiments and surveys, 31–37
  with human subjects, 38–41
Surveys (*see also* Polls)
  of CEOs, 22–24
  contrasted with experiment, 31–37
  of higher education intentions, 1–3
  of households, 26
  of opinions about tax increases, 39–40
  of public opinion, 5–6
  of textbooks, 10

T

Tail of a distribution:
  abrupt, 83
  in skewed histograms, 82–83
t distribution, 307–10
  statistic, 308
Time series, 8, 199–213
  assessment of fit, 209–13
  autoregressive model, 211
  differencing, 207
  examples:
    birthrates, 212–13
    car production, 200–212
  eyeball fit, 207
  fit and residual, 205
  forecasting, 206
  moving average, 203–6
    order of, 206
  patterns of residuals, 209–13
  predictability of residuals, 209–13
  seasonal adjustment, 200–207
  serial correlation, 210
  smoothing, 205
  trends, 200–207
    linear, 204
    quadratic, 207
Treatment, 55
Trends (*see* Time series)
Two-sided test, 342

U

Unbiased measurement, 78
Unexpected tiger, 44
Unexplained variation, 76
  and probability, 228
Unimodal, 82

V

Validity, 79
Variance, 249
  between-group, 354–55
  relationship to SD, 249
  of a sum, 249
  within-group, 354–55
Variation in measurements:
  sources of, 41–43
Venn diagram, 235
Vitamin C therapy for colds, 6

W

Weather forecasting, 3